THE ILLUSTRATED ENCYCLOPEDIA OF
DINOSAURS

THE ILLUSTRATED ENCYCLOPEDIA OF
DINOSAURS

The ultimate reference to 355 dinosaurs from the Triassic, Jurassic and Cretaceous
periods, including more than 900 illustrations, maps, timelines and photographs

DOUGAL DIXON

HERMES
HOUSE

This edition is published by Hermes House, an imprint of Anness Publishing Ltd,
108 Great Russell Street, London WC1B 3NA; info@anness.com

www.hermeshouse.com; www.annesspublishing.com; twitter: @Anness_Books

If you like the images in this book and would like to investigate using
them for publishing, promotions or advertising, please visit our website
www.practicalpictures.com for more information.

© Anness Publishing Ltd 2015

A CIP catalogue record for this book is available from the British Library.

Publisher: Joanna Lorenz
Editorial Director: Helen Sudell
Editor: Simona Hill
Designer: Nigel Partridge
Illustrators: Andrey Atuchin, Peter Barrett, Alain Beneteau,
    Stuart Carter, Julius Csotonyi, Anthony Duke
Production Controller: Rosanna Anness

CONTENTS

THE WORLD OF DINOSAURS 66

INTRODUCTION

The science of paleontology (the study of dinosaurs) is developing at an exciting speed. New discoveries are being made so quickly, that before this book is on the bookshelves there will have been an overwhelming number of new finds and developments in the understanding of the subject. Each week there is something new to report, whether it be a skeleton that constitutes an entirely new branch of the dinosaur evolutionary tree, or some indication of the dinosaurs' life gleaned through new finds of footprints or feeding traces. Microscopic analysis of fossilized dinosaur dung has recently provided new information about dinosaur diet and the plants of the contemporary landscape.

In total about 500 dinosaur species have been identified. The majority of these are based on only a scrap of bone or a tooth or some such small piece of evidence. This figure represents those species that have been found, excavated, studied and described scientifically. Many more are likely to be found by the next generation of palaeontologists.

Our knowledge of dinosaurs is weighted towards those that we know existed close to rivers, or in deserts, or on the banks of lakes or lagoons – places where dead bodies are likely to have been buried

quickly and fossilized. We do not have direct evidence yet of the dinosaurs that lived on mountains or highland forests, or other places that were far from quick burial sites.

Estimates have been made about the number of dinosaurs that actually existed. Mathematical formulae have been applied using such factors as the speed at which new discoveries are being made today compared with the speed at which they were made by earlier palaeontologists, the larger areas that are now being explored worldwide compared with the early history of the science, the different habitats and land areas that existed during dinosaur times, and so on. One result estimated that 1,500 dinosaur species existed.

To give a rounded view of what the Age of Dinosaurs was really like, we have included the contemporaries of the dinosaurs in this encyclopedia. The Age of Dinosaurs was the age of reptiles, and besides the dinosaurs, other major reptile groups were present. In the sky were the pterosaurs – the masters of the air until birds evolved from the

Below: An armoured Gastonia *lashes out at a pair of attacking* Deinonychus.

dinosaurs about halfway through the dinosaur age. In the sea there were unrelated groups of swimming reptiles – long-necked plesiosaurs, fish-shaped ichthyosaurs, giant swimming lizards called mosasaurs, and turtle- and walrus-like shellfish-eaters called placodonts. These were important groups of animals. Although they were not dinosaurs, and most of them were in no way related, they are often referred to as "honorary dinosaurs". Their presence helps to show the diversity and richness of the animal life in dinosaur times.

The general introduction provides an overview of the Age of Dinosaurs, introducing the key areas of research that have helped paleontologists to paint a picture of what the world was like in dinosaur times. From fossil evidence, we can say with certainty much about how the dinosaurs lived, what food they ate, whether they lived in groups, had family networks, and what the landscape looked like.

The second section of the book is an encyclopedia of all the dinosaurs that have been identified. The 355 entries are arranged chronologically. The Age of Dinosaurs, known as the Mesozoic Era (a reference to the rock formations in which the dinosaurs have been found) is divided into three periods – Triassic (representing the earliest period of dinosaur history), Jurassic and Cretaceous (the latest period of dinosaur history). These periods of time are subdivided further.

Below: A herd of Monoclonius *flee a forest fire. Little dramas like these can be deduced from the fossil record.*

Above: A mother Shanxia *guards her offspring against a desert sandstorm.*

Fascinating information about each dinosaur entry is provided and is accompanied by a concise description of the features that make the animal distinctive. A fact box lists some of the technical data, such as the period of history when the animal lived, its dimensions and its discoverer. Each is illustrated with a beautiful watercolour that shows what the animal looked like. The appearance is based on the evidence available, sometimes gained from only a single bone, utilizing studies of related animals to make the best attempt possible of a restoration of the living beast.

In science an animal is known by its scientific name, or its "binomial". For example, humans are scientifically known as *Homo sapiens*. *Homo* is the genus name and *sapiens* is the species name. The names, usually derived from Latin or classical Greek, are always italicized with only the genus name capitalized. For dinosaurs it is customary to use only the genus name in popular literature. *Tyrannosaurus rex*, however, is so evocative that often both are used. Once the genus name has been introduced, it can then be referred to by its genus initial along with its species name. Hence *T. rex*. In many instances a particular dinosaur genus has several species. This should help to explain the usage of names in this encyclopedia.

Here you will find animals that have never been portrayed before, with facts and illustrations based on the analysis of the most up-to-date scientific papers.

1 *Oviraptor.*
2 Pterosaur.
3 *Velociraptor.*
4 Baby *Oviraptor.*
5 *Ornithoides.*

THE AGE OF DINOSAURS

The dinosaurs lived between 220 and 65 million years ago. Science can give us a good idea of what the Age of Dinosaurs was like. For the past two centuries, dinosaur bones and traces have been dug up and studied, and every discovery adds some new information to what we already know.

The physical bones, mineralized and preserved as fossils, provide the main clues to what the dinosaurs looked like. However, the scientific interpretation of these fossils has changed over the decades – what we would construct now from the bones would not be recognizable to the Victorian scientists who first unearthed them.

The soft anatomy of skin, muscles and tissues is rarely fossilized, but when it is, it provides a valuable insight to the anatomical make up of the dinosaur. When we compare this with the anatomy of living animals, it is possible to understand how that dinosaur functioned.

Other lines of evidence show us how dinosaurs lived. Fossil dung can be analyzed to establish what they ate. We can also examine tooth marks on the bones of the animals killed and eaten by the dinosaurs.

Fossil footprints are the remains of dinosaur movements. A whole branch of the science of palaeontology is dedicated to the study of these trackways and what they can tell us about how dinosaurs moved. There is also evidence of their family groups and the social structure within which dinosaurs lived. Over the years eggs and nests have been found in various places around the world, and these help paleontologists build up a picture of family life and colonies.

Of course, dinosaurs lived in more than one place and collectively lived over millions of years. Different habitats were home to different dinosaur types, and these environments produced different modes of fossilization of their remains. Nonetheless all these lines of evidence give us an image of what conditions were like for the dinosaurs, and for the other animals that lived alongside them. Consequently, we can produce a landscape of dinosaur life within known habitats, and be confident in its accuracy.

Welcome to the Age of Dinosaurs.

Left: The Gobi Desert, 100 million years ago, late Cretaceous period. Dinosaurs hunt for food, fight over prey and look after their young.

THE GEOLOGICAL TIMESCALE

Geological time is an unbelievably massive concept to grasp. Millions of years, tens of millions of years, hundreds of millions of years. These unfathomable stretches of time are often referred to as "deep time". This is the scale that palaeontologists and anyone interested in dinosaurs must use.

When did dinosaurs appear? About 220,000,000 years ago. And when did they die out? About 65,000,000 years ago. It is easier when we say 220 million and 65 million, but we could use a better system.

To make the concept easier, geologists and palaeontologists have always split geological time into named periods. It is the same when we talk about human history. We can say 150 years ago, or 200 years ago, or 600 years ago, but it gives a clearer idea of the time if we say Victorian London, or Napoleonic Europe or Pre-Columbian North America – then we can put events into their chronological context.

Geological time periods are named after the rock sequences that were formed at that time, and the names

Below: The age of the Earth is so immense that it can only be shown diagrammatically in some kind of distorted image. The Earth's origin can be placed about 4,600 million years ago, but the part that really interests palaeontologists begins about 590 million years ago. Since that time the geography of the Earth has changed, with the continents constantly moving to new positions.

were given by the scientists (mostly Victorian, about 150 years ago) who first studied them in the regions in which they typically outcrop.

The periods that concern anyone interested in dinosaurs are the Triassic (so named because the rocks of that time were identified as three separate sequences in Germany), the Jurassic (named after the Jura Mountains, on the border of France and Switzerland, where they were first studied), and the Cretaceous (named after the Latin for chalk, the most prominent rock type formed at that time in southern England). Collectively the three periods are known as the Mesozoic era. Each period encompasses tens of millions of years, so is further subdivided into stages for ease of reference. The stages are given at the bottom of the facing page, along with the actual number of years that they entailed, so that they can be referred to when necessary. The stages are then divided into zones, but these time divisions are too small to be of any interest to us here.

Above: A fossil forms when an animal dies and its body falls into sediment that is accumulating at the time. The body is buried and the organic matter of the hard parts is transformed into mineral at the same time as the sediment is transformed into sedimentary rock. The pterosaur shown must have died while it was flying over a shallow lagoon in late Jurassic times. It sank to the bottom of the lagoon, where it was buried by contemporary sediment.

There are two ways in which geological events are dated. The first is "relative dating" – placing events on the geological time scale in relation to each other. This is the principle involved in most studies of the past. Fossil A is found in rocks that lie

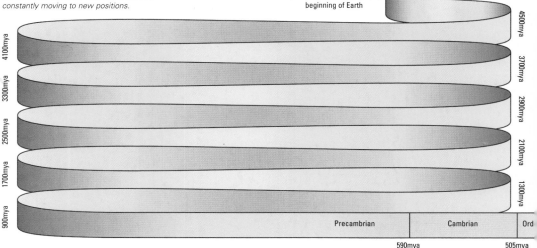

4,600 million years ago (mya), beginning of Earth

4100mya	4500mya
3300mya	3700mya
2500mya	2900mya
1700mya	2100mya
900mya	1300mya

Precambrian	Cambrian	Ord
	590mya	505mya

above those in which fossil B is found. That means that fossil B is older than fossil A; in an undisturbed sequence of sedimentary rocks the oldest is always on the bottom. If a fossil is found in a rock on another continent from which that fossil is usually found, then the two rocks will be of the same age, even if there are no other clues to the age of the rocks. The fossil dates the rock.

The second type of geological dating is absolute dating. This is much more tricky, and involves studying the decay of radioactive minerals in a particular rock. A radioactive mineral breaks down at a particular known rate. If we can measure the amount of that mineral remaining, and compare it with the amount of what is called the "decay residue", we can tell how long

Below: The three periods of the Mesozoic era, the Triassic, the Jurassic and the Cretaceous, are the periods in which the dinosaurs lived. They evolved in the latter part of the Triassic and died out at the end of the Cretaceous. These periods are further divided into stages.

When geologists refer to different parts of a period, they talk about "upper" Cretaceous or "lower" Jurassic. This is a reference to the rock sequence in which the rocks formed. When we talk about the events that took place at these times, we use the terms "late" Cretaceous or "early" Jurassic instead.

Above: Geological periods are defined by the fossils found in the sedimentary rocks formed at that particular time. Sedimentary rocks are those that are built up from layers of mud and sand, and have been compressed and solidified over time. Those shown here represent an angular unconformity between two rock formations: Triassic rocks are the horizontal ones lying above Devonian rocks, which are inclined at 40 degrees. These were laid down horizontally, but have been tilted by movements under the surface.

it has been decaying and how long ago it formed. Several radioactive elements are used in this method.

One disconcerting aspect about geological time, however, is that the absolute dates keep changing. This is because the science used to determine them becomes increasingly sophisticated and precise with developments in technology and understanding. A century ago we were talking in terms of tens of millions of years, whereas nowadays the same periods are talked of in hundreds of millions of years. This is why dates may differ in various dinosaur books.

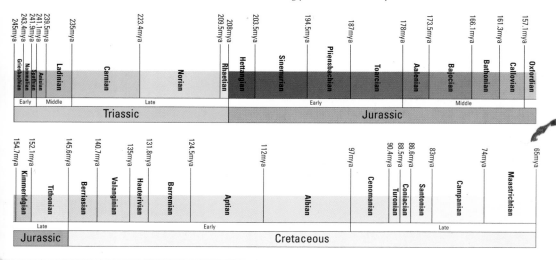

Timescale (Triassic and Jurassic):
245mya / 243.4mya / 241.9mya / 241.1mya / 239.5mya — 235mya — 223.4mya — 209.5mya / 208mya — 203.5mya — 194.5mya — 187mya — 178mya / 173.5mya — 166.1mya — 161.3mya — 157.1mya

Stages: Griesbachian, Nammalian, Spathian, Anisian, Ladinian, Carnian, Norian, Rhaetian, Hettangian, Sinemurian, Pliensbachian, Toarcian, Aalenian, Bajocian, Bathonian, Callovian, Oxfordian

Divisions: Early | Middle | Late — **Triassic** ; Early | Middle — **Jurassic**

Timescale (Jurassic and Cretaceous):
154.7mya / 152.1mya — 145.6mya — 140.7mya — 135mya — 131.8mya — 124.5mya — 112mya — 97mya — 90.4mya / 88.5mya / 86.6mya — 83mya — 74mya — 65mya

Stages: Kimmeridgian, Tithonian, Berriasian, Valanginian, Hauterivian, Barremian, Aptian, Albian, Cenomanian, Turonian, Coniacian, Santonian, Campanian, Maastrichtian

Divisions: Late — **Jurassic** ; Early | Late — **Cretaceous**

Timescale (overall):
...ian | Devonian | Carboniferous | Permian | Triassic | Jurassic | Cretaceous | Tertiary | Present
408mya | 360mya | 286mya | 248mya | 213mya | 144mya | 65mya

EARLY EVOLUTION

Where did life come from? We are not quite sure, but it seems that living things of one kind or another have been around since the Earth was cool enough to have liquid water on its surface. The process of evolution has meant that there has been an uninterrupted stream of living creatures ever since then.

What is life? There are several definitions, but each agrees that a living thing absorbs materials and energy, grows and reproduces. The tiniest bacteria and single cells conform to this definition, and these are the living things that existed way back when the Earth had just begun to cool.

By far the greatest part of Earth's history is encompassed by Precambrian time, but there is little direct evidence about what living things were like then. Bacteria and single-celled organisms do not leave much in the way of fossils. However, we have indirect evidence that things lived then, and gradually evolved into soft-bodied, multi-cellular creatures during that period. This vast span of time is called the Cryptozoic, meaning "the time of hidden life". The end of the Precambrian period (590 million years ago) and the beginning of the fossil record proper is usually marked by an event called the "Cambrian explosion".

At this time, the beginning of the Cambrian period, evolution perfected the hard shell. Organisms had the ability to absorb the mineral calcite from the seawater and lay it down as a living shell, or from organic

compounds they built up a kind of natural plastic called chitin – the material from which our fingernails are made. This had two results. First it meant that there was suddenly a kind of evolutionary arms race. Animals had always been hunting and eating one another. Now some animals could defend themselves, and consequently the hunters evolved new structures and techniques to get the upper hand.

Evolution in the ocean
Suddenly the oceans (for all life was in the oceans at this time) were full of all kinds of creatures that had not existed before. And what strange beasts they were! There were animals with many legs or none, with shelled heads, with shelled tails, with spikes, and with burrowing tools – it was as if nature was trying out anything just to see what worked. By the end of the Cambrian period this vast array of strange beings had whittled itself down to a dozen or so well-tried evolutionary lines that have

Below: Evidence of life in the Precambrian period (more than 590 million years ago) is vague. However, what we do know is that all the major evolutionary lines of living things had evolved by the dawn of the Cambrian

continued until the present day. The second result of evolution producing hard shells was based on the fact that hard-shelled animals leave good fossils. The history of life from that point forward is well documented, which is why the time span from the Cambrian to the present day is called the Phanerozoic (meaning "obvious life").

One of the surviving strands of life consisted of worm-like animals with a nervous system running down their length, supported by a jointed framework. The brain was at the front, protected by a box. The mouth and sensory organs were also at the front. From simple animals like this evolved the first vertebrates, the first animals with backbones.

The first vertebrates
Fish were the first vertebrates that we would recognize, and they came to prominence in Ordovician and Silurian times (505–408 million years ago).

The first fish are known as the "jawless fish". Rather like the modern

period (590 million years ago) and were leaving their imprint on the Earth, as well-preserved fossils.

The vast majority of animals at this time, both living and fossil, are invertebrates.

Soft-bodied

Chordates
Brachiopods
Echinoderms
Hemichordates
Sponges
Jellyfish
Corals
Flatworms
Molluscs
Annelids
Arthropods

PRECAMBRIAN, 590 MILLION YEARS AGO   CAMBRIAN, 590–505 MILLION YEARS AGO

lamprey, they had a sucker instead of jaws, and they probably lived by sucking up nutritious debris from the seafloor as they swam along. A fin along the underside of the tail ensured that they swam head down. Proper jaws and a more organized skeleton then developed. The first skeletons were not made of bone but of cartilage. The cartilaginous fish are represented today by the sharks and rays. They appeared in Silurian times.

The next stage was the evolution of bone around the cartilage framework. Bone formed the skeleton and also the armour plates for protection. Then came the kinds of scales we would recognize from the fish we see today. These bony, armoured fish, and the scaly fish, appeared in the Devonian period. By this time there were so many different fish that the Devonian period (408–360 million years ago) is often termed the "Age of Fishes".

A changing environment

Meanwhile, changes were taking place out of the water. In early Precambrian times the atmosphere was a bitter mix of noxious gases, which is why all early life evolved in the sea. Gradually the by-products of early living systems were seeping into the atmosphere and changing it. Oxygen, a product of the photosynthesis process, which keeps plants alive,

started to build up in the atmosphere and make land habitable. The first green patches of land appeared between tides probably during Silurian times. When plants pioneered life on land, animals followed behind. One kind of fish developed lungs to enable it to breathe the oxygen of the atmosphere. It also developed paired muscular fins that would allow it to crawl on a solid surface as well as swim in the water. The vertebrates were poised to take a step on to the land. As soon as the continents became habitable, life spread there from the oceans, and a vast array of animals have been present ever since.

Life on land

Creatures had been venturing out on to land for hundreds of millions of years. Tracks of arthropods (the first land-living animals) are known from beach sediments formed in Ordovician times. There are mysterious marks from dry land deposits that look like motor-cycle tracks back in the Cambrian period. They seem to have been tentative explorations, but it does not appear that animal life out of the water was permanent until plants had gained a foothold. Insects and spiders infested the primitive early plants that clothed the sides of streams in the Silurian period. The first fish ventured out in the subsequent Devonian time.

It is not clear why fish first appeared on land. Some scientists say that the newly evolved arthropod fauna that had established itself among the plants was too tempting a food source to be ignored. Others suggest that land-living was an emergency measure – if a fish became trapped in a drying puddle of water it would need to be able to survive and travel over land to find more water in which it could live. There is also a theory that the waters became too dangerous due to predatory animals; there were clawed arthropods as big as alligators at the time, and some fish found it safer to take up a land-living existence.

Eusthenopteron was typical of the kind of fish that was able to spend time on land. The major adaptation was the lung. Fish normally breathe through gills – feathery structures that can filter dissolved oxygen from the water. Now lungs enabled oxygen to be extracted directly from the air. Then there was the manner of

Below: By Ordovician (505–438 million years ago) and Silurian (438–408 million years ago) times the backboned animals had evolved, in the form of the most primitive fish. The backbone supported the whole body, the limbs were arranged in pairs at the side, and the brain was encased in a box of bone. The next stage came when these swimming animals evolved to be able to breathe air.

Jawless fish

Cartilagenous fish

ORDOVICIAN, 505–438 MILLION YEARS AGO

SILURIAN, 438–408 MILLION YEARS AGO

locomotion. A typical fish's fin consists of a ray of supports with a web between, spreading from a muscular stump. In *Eusthenopteron* and its relatives the fins consisted of muscular lobes, supported by a network of bones, with the fin material forming mere fringes along the edge. Two pairs were arranged on the underside of the body, and they could be used both for swimming and for heaving the animal across open land.

The first amphibians

By the end of the Devonian period, the next stage in the evolution of land vertebrates had been accomplished, and the first amphibian-like animals appeared. These animals were much more complex in their variety and relationships than the single term "amphibian" implies. *Ichthyostega* was one of the earliest of these animals. The difference between *Ichthyostega* and the lobe-finned fish was in the limbs. Now the fins were clearly jointed, with leg and toe bones. It seems likely that they evolved for pulling the animal along through weeds in shallow water, but they were ideal for clambering on land as well. The *Ichthyostega* foot was odd by

Below: .Many of the evolutionary lines that existed at the start of the Precambrian continued to evolve. Some evolutionary lines, such as the armoured fish, ceased to exist. Other lines split with new creatures evolving and beginning new evolutionary lines.

modern standards because there were eight toes. The standard arrangement of a maximum of five toes for a land-living vertebrate had yet to be established. For all its land-living abilities, *Ichthyostega* and its relatives still had the head and tail of a fish, and had to return to the water to breed.

The next great advance in evolution was the ability of animals to breed on land. This was achieved by the first of the amniotes, named after the amnion – the membrane that contained the developing young within the egg. A hard-shelled egg evolved, that nourished the young in what was essentially a self-contained watery pond, that could be laid away from the water. At last the link with the ancestral seas was severed. Early examples include *Westlothiana* from Scotland and *Hylonomus* from Nova Scotia, both dating from the early Carboniferous period.

The first reptiles

The true reptiles then established themselves along a number of evolutionary branches. In the simplest form of classification they can be classified by the number and arrangement of holes in the skull behind the eye socket. The anapsids had no such holes because the skull was a solid roof of bone behind the

eye. The anapsids were prominent in the Permian period in a group called the pareiasaurs. Modern relatives of pareiasaurs include tortoises and turtles.

The synapsids, however, had a single hole in the skull at each side. They became the mammal-like reptiles, the major group of the Permian period. When they died away in Triassic times they lived on as the humble mammals, and did not really come to prominence again until the Tertiary period.

The diapsids were different because they had two holes behind the eye. Modern diapsids include snakes, crocodiles and lizards. However, a group of Mesozoic (the combined Triassic, Jurassic and Cretaceous periods) diapsids though, was much more important. They were the dinosaurs.

Dinosaur evolution

The dinosaurs evolved from the diapsid line that we call the archosaurs, meaning the ruling reptiles. Other archosaurs were the pterosaurs, and the crocodiles and alligators that we have today. A typical Triassic archosaur was a swift, two-footed, running meat-eater, usually no bigger than a wolf and usually much smaller. In fact an advanced archosaur would have looked very much like a typical,

Jawless fish

Cartilagenous fish

Sharks

Armoured fish

Chordates
Brachiopods

Bony fish

Echinoderms

Hemichordates

Sponges

Jellyfish

Corals

Flatworms

Molluscs

Annelids

Arthropods

DEVONIAN, 408–360 MILLION YEARS AGO

CARBONIFEROUS, 360–286 MILLION YEARS AGO

small, meat-eating dinosaur. What made the dinosaur different from its archosaurian ancestor lay mostly in the structure of the leg and hips.

Most reptiles have legs that stick out at the side, with the weight of the animal slung between them. This gives the animal a sprawling gait. To enable it to run quickly it must throw its body into S-shaped curves to give the sideways-pointing limbs the reach that is needed. In contrast, a dinosaur's leg was straight and vertical. It was plugged into the side of the hip where it was held in place by a shelf of bone.

This meant that a dinosaur's weight was at the top of the leg, and transmitted straight downwards. Vertical limbs can support a greater weight than sprawling limbs. This is the arrangement that we see in a typical mammal, and this upright stance is seen in all dinosaurs, whether two- or four-legged.

Saurischians and ornithischians

So, the first dinosaur was probably like an archosaur, and good at running. From there dinosaur evolution

diverged into two main lines – the saurischia and the ornithischia. The difference between the two lines lies in the structure of the hips.

The saurischia had hip bones arranged like those of a lizard, a structure that radiated from the leg bone socket, with a pubis bone that pointed down and forward. This line is further divided into two groups; the first group developed along the evolutionary line pioneered by the earlier archosaurs, the two-footed hunters. They were termed theropods

Below: As the fish developed into more complex forms, some became land dwellers, with jointed limbs and lungs able to breathe the air. These became the first amphibians. From them, evolved animals able to live on land all the time, without resorting to water at any stage in their growth. The reptiles, with their hard-shelled eggs, represented this stage, and they diversified into all kinds of land-living types.

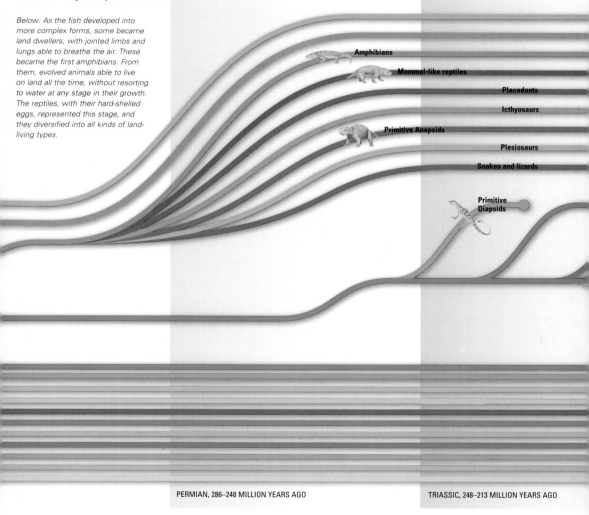

Amphibians

Mammal-like reptiles

Placodonts

Icthyosaurs

Primitive Anapsids

Plesiosaurs

Snakes and lizards

Primitive Diapsids

PERMIAN, 286–248 MILLION YEARS AGO

TRIASSIC, 248–213 MILLION YEARS AGO

or "beast-footed", by the Victorian scientists who detected a similarity between their foot bones and the bones of mammals. All the meat-eating dinosaurs were theropods, from small chicken-size scampering insect-eaters, to massive 15m- (50ft-) long beasts.

The other saurischian group are the sauropods, meaning "lizard-footed", and so called because of the similarity in the structure of the foot to that of a modern lizard. They were the huge, long-necked plant-eaters of the dinosaur world, and their body shape evolved in response to a changing vegetarian diet. The shape of the saurischian hip, with its forward-pointing pubis bone, meant that the big digestive system of a plant-eater had to be carried in front of the hind legs. The result is an animal that would be unable to walk solely on its hind legs, and in response the smaller front legs became stronger to take the weight. This development reduced the mobility of the animal, and so a long neck developed to enable it to reach enough food. As a group of dinosaurs sauropods encompass the biggest land animals that ever existed.

Below: The dinosaurs, once they evolved, soon developed into a number of distinctive groups. Some were meat-eaters, others were plant-eaters. Some moved on four legs and others on two. They were the most significant land animals of the time – between the late Triassic and the end of the Cretaceous. However, at the end of the Cretaceous they, and many other animal groups, became extinct.

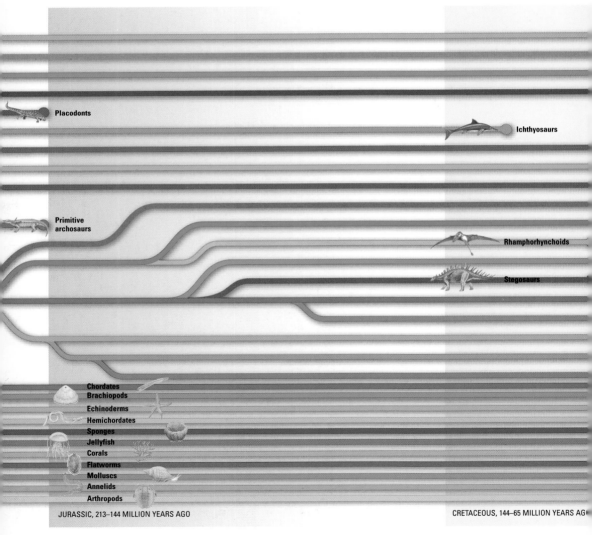

Placodonts

Ichthyosaurs

Primitive archosaurs

Rhamphorhynchoids

Stegosaurs

Chordates
Brachiopods
Echinoderms
Hemichordates
Sponges
Jellyfish
Corals
Flatworms
Molluscs
Annelids
Arthropods

JURASSIC, 213–144 MILLION YEARS AGO

CRETACEOUS, 144–65 MILLION YEARS AGO

The second line of dinosaurs was the ornithischians. They were plant-eaters but had a different arrangement of hip bones. The pubis bone was swept back and lay along the backward-facing ischium bone. This meant the typical ornithischian could carry the weight of its body beneath the hips, and so it could still walk on its hind legs balanced by the heavy tail. A typical two-footed ornithischian was the ornithopod, the bird-footed dinosaur with three splayed toes.

The sauropods and ornithopods also had different eating methods. The sauropods could not chew their food they had to eat so much that they would not have had time to. Their teeth showed that they raked leaves and twigs from the branches and swallowed what they took without processing it. In contrast, ornithopods had teeth that could chew food thoroughly before swallowing it.

Other developments from the basic ornithopod involved the development

of armour. The stegosaurs had plates, the ankylosaurs had armour, and the ceratopsians had horns.

Mammals

After 160 million years, the dinosaurs became extinct, but not before the theropods gave rise to the birds. The way was open for the mammals. Since the end of the Cretaceous, the mammals have expanded and occupied all ecological niches once occupied by the dinosaurs.

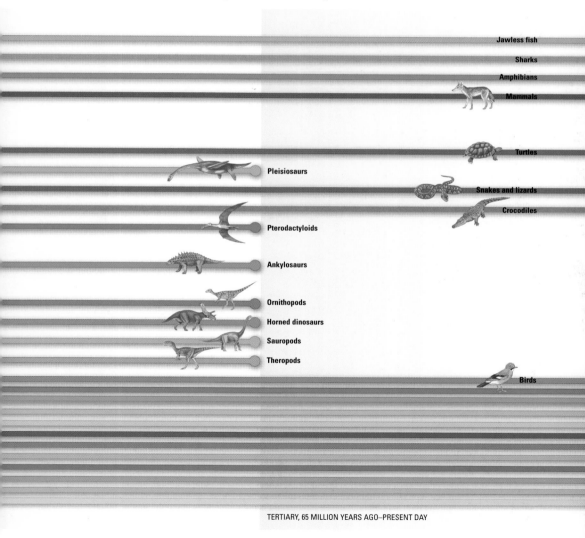

Jawless fish

Sharks

Amphibians

Mammals

Turtles

Pleisiosaurs

Snakes and lizards

Crocodiles

Pterodactyloids

Ankylosaurs

Ornithopods

Horned dinosaurs

Sauropods

Theropods

Birds

TERTIARY, 65 MILLION YEARS AGO–PRESENT DAY

DINOSAUR CLASSIFICATION

The various dinosaurs evolved from common ancestors – in technical terms they were "monophyletic".
Early in their evolution they split into two major evolutionary lines, and these in turn split into a number
of different families, each with its own character and specialization.

The dinosaurs fall into two major groups –
the saurischians and the ornithischians. The
saurischians are divided into the plant-eating
sauropodomorphs and the meat-eating
theropods, while the latter are divided into a
number of different plant-eating types. Note
that the formal classifications (e.g. Theropoda)
are used interchangeably with the less formal
(e.g. theropods) throughout the book. This is
customary in palaeontology.

Saurischia
Lizard-hipped.

Dinosauria
The ruling reptiles are characterized by:
• The number of bones in the skull.
• The presence of a flange on the upper arm
bone that held powerful muscles.
• Three or fewer finger bones in the
fourth finger.
• Three or more vertebrae fixed to the
hip bones.
• A hole rather than a socket in the hip for the
leg bone.
• A ball-like head on the thigh bone.
• A strong joint between the foot bones
and the bones of the hind leg.

Ornithischia
Bird-hipped,
plant-eaters.

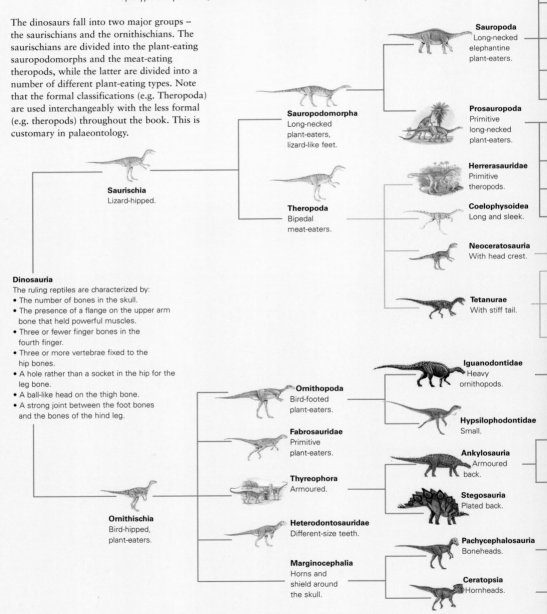

Sauropodomorpha
Long-necked
plant-eaters,
lizard-like feet.

Theropoda
Bipedal
meat-eaters.

Sauropoda
Long-necked
elephantine
plant-eaters.

Prosauropoda
Primitive
long-necked
plant-eaters.

Herrerasauridae
Primitive
theropods.

Coelophysoidea
Long and sleek.

Neoceratosauria
With head crest.

Tetanurae
With stiff tail.

Ornithopoda
Bird-footed
plant-eaters.

Fabrosauridae
Primitive
plant-eaters.

Thyreophora
Armoured.

Heterodontosauridae
Different-size teeth.

Marginocephalia
Horns and
shield around
the skull.

Iguanodontidae
Heavy
ornithopods.

Hypsilophodontidae
Small.

Ankylosauria
Armoured
back.

Stegosauria
Plated back.

Pachycephalosauria
Boneheads.

Ceratopsia
Hornheads.

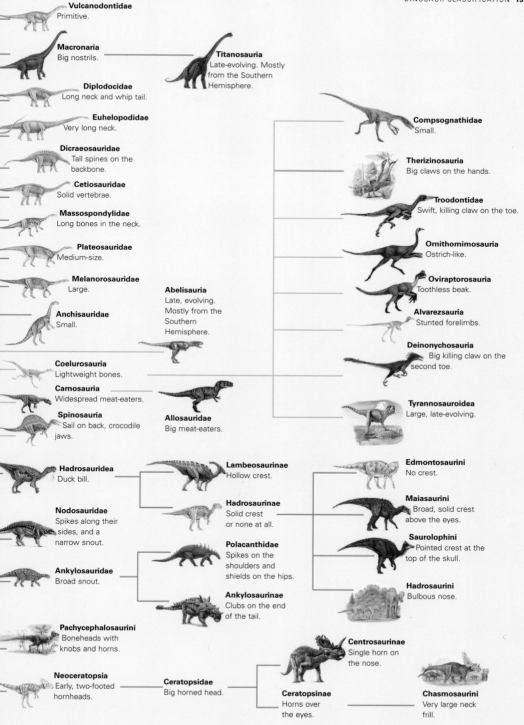

Vulcanodontidae
Primitive.

Macronaria
Big nostrils.

Titanosauria
Late-evolving. Mostly
from the Southern
Hemisphere.

Diplodocidae
Long neck and whip tail.

Euhelopodidae
Very long neck.

Dicraeosauridae
Tall spines on the
backbone.

Cetiosauridae
Solid vertebrae.

Massospondylidae
Long bones in the neck.

Plateosauridae
Medium-size.

Melanorosauridae
Large.

Anchisauridae
Small.

Abelisauria
Late, evolving.
Mostly from the
Southern
Hemisphere.

Coelurosauria
Lightweight bones.

Carnosauria
Widespread meat-eaters.

Spinosauria
Sail on back, crocodile
jaws.

Allosauridae
Big meat-eaters.

Compsognathidae
Small.

Therizinosauria
Big claws on the hands.

Troodontidae
Swift, killing claw on the toe.

Ornithomimosauria
Ostrich-like.

Oviraptorosauria
Toothless beak.

Alvarezsauria
Stunted forelimbs.

Deinonychosauria
Big killing claw on the
second toe.

Tyrannosauroidea
Large, late-evolving.

Hadrosauridea
Duck bill.

Lambeosaurinae
Hollow crest.

Edmontosaurini
No crest.

Hadrosaurinae
Solid crest
or none at all.

Maiasaurini
Broad, solid crest
above the eyes.

Nodosauridae
Spikes along their
sides, and a
narrow snout.

Polacanthidae
Spikes on the
shoulders and
shields on the hips.

Saurolophini
Pointed crest at the
top of the skull.

Ankylosauridae
Broad snout.

Ankylosaurinae
Clubs on the end
of the tail.

Hadrosaurini
Bulbous nose.

Pachycephalosaurini
Boneheads with
knobs and horns.

Centrosaurinae
Single horn on
the nose.

Neoceratopsia
Early, two-footed
hornheads.

Ceratopsidae
Big horned head.

Ceratopsinae
Horns over
the eyes.

Chasmosaurini
Very large neck
frill.

THEROPODS

All the meat-eating dinosaurs belonged to the theropod group. They all conformed to a similar body plan.
They had a small body and walked on two legs. The head was held out at the front and was balanced by
a heavy tail. The arms were small and used for grasping or killing. The hand usually had three fingers.

The theropods appeared at the beginning of the Age of Dinosaurs, in the latter part of the Triassic period. In appearance the early theropods would have resembled their thecodont ancestors. The thecodonts are the reptile group that had teeth in sockets, rather than in grooves as lizards had. The main differences would have been in the stance of the legs and the structure of the skull. The thecodonts had been active hunters, and the early theropods carried on this tradition.

A theropod is an ideal shape for a hunter. The head, jaws and teeth are carried well forward, and are the first part of the animal to make contact with its prey. The arms and the claws are also well forward. The body is quite small, as befits a meat-eating animal. The legs are powerful, with strong muscles working against the bones of the lizard-like hips. The tail is big and heavy, used for balancing and keeping the upper body well forward.

We know that at least some of the theropods were warm-blooded. The term warm-blooded does not

necessarily refer to the temperature of the blood, but instead implies that a mechanism exists that keeps the animal's body at the same temperature regardless of the temperature of its surroundings. Nowadays warm-bloodedness is found in mammals and birds, the direct descendants of the theropods.

Warm- or cold-blooded?

Whether or not a dinosaur was warm-blooded is of great significance to understanding its lifestyle. A warm-blooded animal needs ten times as

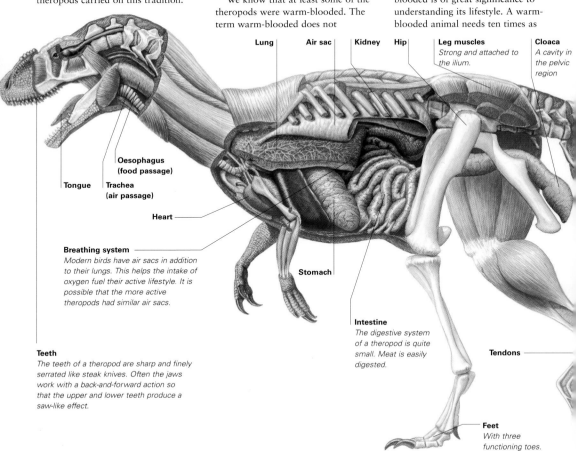

Lung   **Air sac**   **Kidney**   **Hip**

Leg muscles
Strong and attached to the ilium.

Cloaca
A cavity in the pelvic region

Oesophagus (food passage)

Tongue   **Trachea (air passage)**

Heart

Breathing system
Modern birds have air sacs in addition to their lungs. This helps the intake of oxygen fuel their active lifestyle. It is possible that the more active theropods had similar air sacs.

Stomach

Intestine
The digestive system of a theropod is quite small. Meat is easily digested.

Tendons

Teeth
The teeth of a theropod are sharp and finely serrated like steak knives. Often the jaws work with a back-and-forward action so that the upper and lower teeth produce a saw-like effect.

Feet
With three functioning toes.

much food to fuel its system than does a cold-blooded animal of the same size. So, a cold-blooded, meat-eating dinosaur could make one kill, eat as much as it wanted, and then rest for weeks, like a modern python. However, a warm-blooded, meat-eating dinosaur would have to hunt nearly all the time. Warm-blooded animals also need insulation to regulate their body temperature. In mammals this is formed from hair, while in birds it is formed from feathers. We have evidence that some of the later meat-eating dinosaurs were covered in hair or feathers. Lake deposits at Liaoning have fossils that show this, and skeletons of some of the ostrich-mimic (ornithomimid) dinosaurs have pores along the arm bones that indicate where feathers were attached. They are all small, late dinosaurs, from the Cretaceous period, but we do not know how far back they had these features. Big dinosaurs, even if they were warm blooded, would not need as much insulation – a modern elephant, for example, has very little hair.

Therizinosaurs

The therizinosaurs are a side-group of the theropod family. They had heads that looked more like the heads of plant-eaters, and the hip bone was similar in shape to a bird's. The hands were armed with powerful claws, probably used for food gathering in some way.

Left: Alxasaurus is a typical example of a therizinosaur.

Alvarezsaurids

The alvarezsaurids, such as *Mononykus*, were a group of small, bird-like theropods. They had remarkable forearms, which were stunted and carried a single big claw. The arms looked like atrophied wings but their purpose is still a mystery.

Left: Shuvuuia stood at less than 1m (3ft) tall.

Skin covering *Usually scaly, but sometimes feathered in small types of dinosaur.*

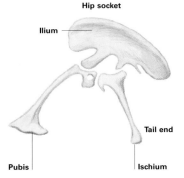

Hip socket

Ilium

Hip *(left) A theropod meat-eater had the saurischian hip – with the arrangement of bones like those of a lizard. At each side there was a top bone called the ilium, which held the leg muscles. Pointing forward and down from the hip socket was the pubis, which took the weight of the animal as it lies down. Sweeping down and back was the ischium, to which the tail muscles and the inside leg muscles were attached.*

Tail end

Skull *(above) A dinosaur's skull is a lightly-built framework of struts of bone. The teeth constantly grow and drop out, each one being replaced by another that is growing beneath it.*

Pubis

Ischium

SAUROPODS AND PROSAUROPODS

The sauropods were the big plant-eaters of dinosaur times, with heavy elephantine bodies, long necks and whip-like tails. They evolved from the prosauropods in the late Triassic period and became the main plant-eaters of Jurassic times. They existed until the end of the Cretaceous period.

The only obvious resemblance between a sauropod and its theropod relative was in the arrangement of the hip bones. They both had hips like a lizard's, with the pubis bone pointing down and forward. As with meat-eaters, the innards had to be carried forward of this pubis bone, and since the sauropods had huge, plant-digesting intestines, these innards tended to be very heavy. Since the heavy body in front of the hips made it very difficult for the animal to balance on its hind legs, sauropods were nearly always four-footed. The front legs were usually shorter than the hind legs, suggesting that their ancestors were two-footed. The neck was particularly long and carried a small head. The tail was very long and often ended in a fine whip point that may have been used as a weapon.

There is a great deal of controversy about whether the sauropods were warm- or cold-blooded. The argument for warm-bloodedness is their close-ness to the theropod line, some of which were definitely warm-blooded. On the other hand the sheer size of the animal – a beast weighing tens of tonnes would be able to conserve its heat more easily than a small animal, and would not need a warm-blooded metabolism. Then there was the problem of being able to eat enough food through the little jaws to fuel a warm-blooded lifestyle. A compromise theory suggests that what we regard as warm-blooded and cold-blooded are really just the extreme ends of a graded sequence. It may be that the theropods lay close to the warm-blooded end, and the sauropods close to the cold-blooded end. In any case it is unlikely that the sauropods would have been covered in feathers or hair.

Above: The skin impression left by a sauropod shows a relatively smooth skin covered by scales.

Liver

Large intestine

Ligament
Supporting the weight of the neck.

Circulation system
The heart of a sauropod had to be particularly powerful to pump blood from one end to the other, and to send it up to the heights reached by the neck. It has been suggested that there may have been several hearts along the length of the neck to help the circulation.

Gizzard
The digestive system is much bigger than that of a meat-eater's. The lack of chewing ability meant that sauropods swallowed stones frequently, and these stones gathered in a gizzard to grind up the food as it entered the digestive system. Modern plant-eating birds, such as turkeys or pheasants, also do this since they cannot chew with their beaks.

Small intestine

Caeca
Sac-shaped intestine extensions.

There is also some dispute as to what did cover the sauropods. Conventional restorations show a wrinkled, elephant-like skin. Isolated remains, however, indicate the possibility of a scaly skin like a lizard's, and even horny spikes down the back. These issues are still disputed.

Prosauropod teeth

The teeth of a prosauropod were leaf-shaped, overlapping and coarsely serrated, like vegetable shredders. They had evolved to rip up coarse plant material. In comparison those of the later sauropods were quite simple.

The teeth of a sauropod were arranged for raking and combing rather than for chewing. Some dinosaurs, such as Diplodocus, *had peg-like teeth, while others like* Camarasaurus, *had teeth that were spoon-shaped. A feature of the jaw bone is the low position of the joint. As in most plant-eating dinosaurs, this allowed for good leverage to break off tough food.*

Teeth

Skin covering

Ligament

Spines
May have existed in some species.

Cloaca

Gristle
Wedge to take the animal's weight.

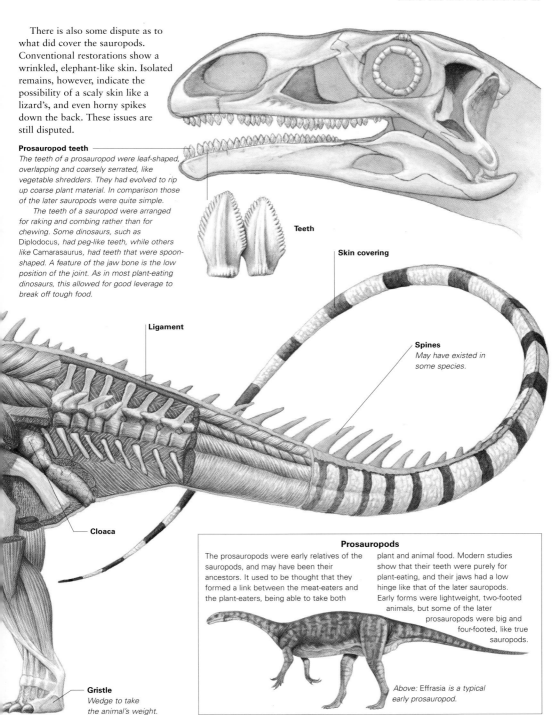

Prosauropods

The prosauropods were early relatives of the sauropods, and may have been their ancestors. It used to be thought that they formed a link between the meat-eaters and the plant-eaters, being able to take both plant and animal food. Modern studies show that their teeth were purely for plant-eating, and their jaws had a low hinge like that of the later sauropods. Early forms were lightweight, two-footed animals, but some of the later prosauropods were big and four-footed, like true sauropods.

Above: Effrasia *is a typical early prosauropod.*

ORNITHOPODS

The second major group of dinosaurs were those with hip bones arranged like those of a bird, with the pubis bone swept back. They evolved into a number of different lines, including the armoured and horned types, but the basic anatomy is seen in the two-footed plant-eating types – the ornithopods.

Ornithopods were all plant-eaters. The basic "bird-hipped" dinosaur type was the two-footed plant-eater. Examples ranged from small, chicken-sized animals up to monsters 15m (50ft) long or more.

An ornithopod, being a plant eater, had a large, complex, digestive system. This was carried well back, beneath the hips – the new arrangement of hip bones allowed for this. That meant that this animal, even though it was a plant-eater with heavy intestines, could still walk on its hind legs, balanced by a heavy tail, just like the theropods.

From a distance, a lightweight ornithopod may have looked like a theropod, but there were certain differences. For one thing the body of the plant-eater was much bigger, giving it a pot-bellied appearance. Then there were the hands. A theropod had no more than three fingers on the hand, whereas the ornithopod had the full complement of five. The head was also quite different. The jaw had a chewing system that was different from that of any other dinosaur seen so far.

The ornithopods evolved early in the age of dinosaurs, but for the Triassic and Jurassic periods were mostly quite small animals. The long-necked sauropods were the big plant-eaters of this time. Then, in the

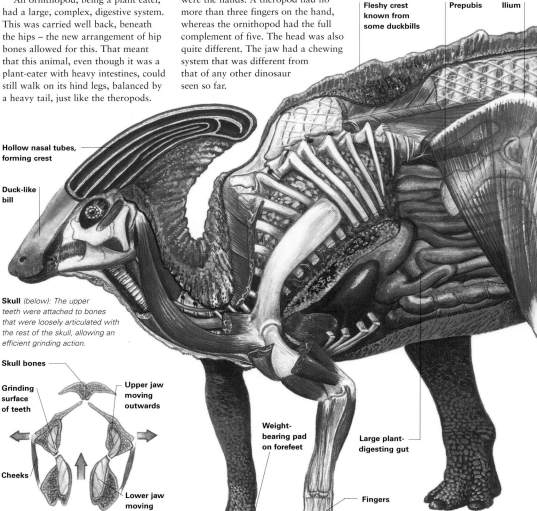

Fleshy crest known from some duckbills

Prepubis

Ilium

Hollow nasal tubes, forming crest

Duck-like bill

Skull (below): The upper teeth were attached to bones that were loosely articulated with the rest of the skull, allowing an efficient grinding action.

Skull bones

Grinding surface of teeth

Upper jaw moving outwards

Cheeks

Lower jaw moving upwards

Weight-bearing pad on forefeet

Large plant-digesting gut

Fingers

Cretaceous, they blossomed. They spread everywhere, at the expense of the sauropods, and some attained great size. The biggest ones became too heavy to spend much time on their hind legs and so they moved about on all fours (as youngsters they still scampered around as two-footed animals).

The ornithopod head was totally different from that of the other dinosaurs. The jaws had batteries of grinding teeth, usually with wrinkled enamel, and they were constantly replaced to ensure that there was always a grinding surface. In the advanced forms the upper jaws

were fitted loosely into the skull. This meant that as the lower jaw came upwards, the upper jaw moved outwards so that the sloping surfaces on the upper and lower teeth ripped past one another, producing the grinding action. The fact that the rows of teeth were set in from the side of the skull shows that these animals had cheeks. The cheeks formed pouches that held the food as it was being ground. A sharp beak at the front of the mouth, supported by a bone that the theropods and sauropods did not possess, created an efficient, food-gathering mouth. With such a complex chewing mechanism there was no need for stomach stones to aid digestion. The armoured and horned dinosaurs also had these features.

Hollow nasal tubes, forming crest

Growing teeth

Lower jaw

Grinding surface

Teeth (above): Although the teeth of a duckbill numbered many hundred, only a few of them were working at any one time. They were tightly-packed, like dates in a box, and continually growing. Those at the grinding edge were constantly being worn away and replaced by those growing from below.

Tall spines on backbone
These are lashed together with tendons, giving a stiff tail.

Ischium

Swept-back pubis

Hip (right): The ornithopod hips are bird-like.

Ilium

Pubis

Ischium

Tail end

Ilium

Prepubis

Ischium

Pubis

Tail end

Hypsilophodon
A typical small ornithopod. Its plump body was supported on its hind legs, balanced by the tail. Its head had the chewing teeth, the cheeks and the beak. Its long hind legs show that this was a running animal, able to flee from danger.

Below: Hypsilophodon.

Ouranosaurus
One of the larger ornithopods. It is so heavy that it spent most of its time on all fours. A tall array of spines along its back may have supported a sail, or some kind of fatty hump for use as a food store, like a modern camel.

Below: Ouranosaurus.

Tenontosaurus
A particularly common dinosaur on the early Cretaceous plains of North America. It was intermediate in size between *Hypsilophodon* and *Ouranosaurus*, and had a particularly long tail.

Below: Tenontosaurus.

ARMOURED DINOSAURS

The basic ornithopod dinosaur, with a heavy plant-eating body and bird-shaped hips was the shape from which the armoured dinosaurs developed. The stegosaurs had plates, the ankylosaurs had armour, and the ceratopsians had horns.

Three groups of dinosaurs developed from the basic ornithopod pattern, all bearing armour of one kind or another: these were the stegosaurs, ankylosaurs and ceratopsians. Various lines of ornithischian dinosaur developed armour and, because of the added weight, reverted from the two-footed ornithopod-like stance to a four-footed mode of life. The first of these to evolve were the stegosaurs, in which the armour consisted of a double row of plates and spines. Later came the ankylosaurs, with armour that formed a mosaic across the back. Last to appear were the ceratopsians, where the armour was confined to the head.

Stegosaurs

The stegosaurs, or the plated dinosaurs, are found in Jurassic and early Cretaceous times. They had plates that ran down their backs, and often spines on the tail. These tail spines were used as weapons when swung at an attacker. Sometimes there were spines on the shoulder as well.

There has always been disagreement regarding the function of the

Left: With spines on the back and neck, plates down the tail, and an armoured shield across the lower back *Polacanthus was one of the more lightly armoured ankylosaurs.*

plates. The traditional view is that they were a defence mechanism, no doubt being covered in horn, and probably had pointed corners and sharp edges. They protected the backbone from an attack by a tall predator, but that meant the flanks were unprotected. A second theory is that the plates were part of a heat-exchange mechanism. If they were covered in skin rich in blood vessels, they would absorb the sun's warmth in the morning and shed heat with the midday wind. This seems an attractive theory to explain

the broad plates like those of *Stegosaurus*, but it is not so logical when dealing with the narrower plates of the smaller and more primitive forms like *Kentrosaurus*.

Ankylosaurs

The ankylosaurs developed in Cretaceous times, as the stegosaurs declined. They were the armoured dinosaurs, with the armour formed from bony studs set in the skin and covered in horn.

There were several different types of ankylosaur. Some, such as *Polacanthus*, had a thick rigid mosaic of tiny studs over the hips and spines sticking out at the side. Others, such as *Edmontonia*, had a fierce array of spikes sticking sideways and forward from the shoulders. Others, such as *Euoplocephalus*,

Right: The broad plates of Stegosaurus *would have been brightly coloured for signalling.*

Right: The plates of Kentrosaurus *were smaller and less prominent.*

Right: Edmontonia *was an ankylosaur with defensive spines on its shoulders.*

Below: Euoplocephalus *had armour in plates all over the head, back and down the tail, finishing with a massive tail club.*

had the defence concentrated around the tail where the bones at the tip were fused into a monstrous club, and the tail vertebrae were fused together to make a rigid shaft.

The ankylosaurs had very stout bodies that contained very sophisticated digestive systems. Probably like modern cows they carried fermenting guts to allow bacterial action to break down tough plant material before it could be fully digested.

Ceratopsians

The last dinosaur group to evolve was the ceratopsians, or the horned dinosaurs. Ceratopsians were distinguished by having all their armour on their heads. It may be that the head armour evolved from ridges at the back of the skull. The ridges anchored the very strong jaw muscles which these animals needed to chew the tough cycad fronds on which they fed. *Psittacosaurus* was typical of these early ceratopsians. All these dinosaurs had the same beak and cheek arrangement as the ornithopods. However, in the ceratopsians, the teeth were evolved for chopping rather than chewing.

The most spectacular ceratopsians, like *Triceratops*, lived at the end of the Cretaceous period, when herds of them roamed the plains of North America. Each kind had a different set of horns and a different neck shield, allowing them to tell each other apart.

A side-branch of the ceratopsian family was a group we call the boneheads. Their armour consisted of a solid dome of bone on top of the skull, possibly covered in horn. They were mostly quite small goat-sized animals, such as *Tylocephale*, but others, such as *Pachycephalosaurus*, grew to the size of its *Triceratops* cousin.

Below: The armoured neck shields of ceratopsians like Chasmosaurus *were for signalling as well as defence. They would have been brightly coloured like flags.*

Above: In plan view, the ceratopsian skeleton shows how the head shield covers the neck and protects the shoulders.

COPROLITES AND DIET

*We have plenty of information that shows what the dinosaurs ate and how they ate it. The evidence
includes close examination of their fossilized jaws and teeth, fossilized stomach contents and also
of fossil dung, known as coprolites.*

We know that the carnivorous
theropods had a particular kind of
tooth and jaw apparatus for shearing
meat. The teeth of sauropods were
designed for raking leaves from trees.
The teeth and jaws of ornithopods and
the armoured dinosaurs were built for
chewing, chopping and munching, and
processing plant food before
swallowing it. Teeth and jaws
represent physical evidence of the type
of diet that the dinosaurs had. But
there are other lines of investigation.

Palaeontologists describe fossil dung
as "coprolite". A coprolite can be a very
valuable tool to work out the diet of
something that is dead and fossilized.
Dinosaur coprolites are rare. Most
coprolites we find come from marine
animals, such as fish. As with all other
fossils, marine conditions provide a far
better preservation medium than any
land habitat. A piece of dung deposited
on land will be trampled on,
decomposed or eaten by bacteria, fungi
or other organisms. In fact, one of the
ways of identifying a structure as a
coprolite of a land-living animal is by
the presence of dung-beetle burrows
through it. Dung is a remarkably
nutritious substance to certain
creatures. Yet for all that, there have

Above: A Tyrannosaurus upper jaw.

been a fair number of dinosaur
coprolites discovered, and they give a
good insight into their diets.

Theropod coprolites seem to be
more common than those of plant-
eating dinosaurs. This is probably
because they contain a high proportion
of bone material that makes them
more robust than the stuff produced
by a plant-eater. Studies have been
done to try to determine what that
bone material might be. The chemistry
of a coprolite reflects the chemistry of
the meal, and chemical studies on
coprolites from a *Tyrannosaurus* or
one of its relatives suggest a diet
predominantly of duckbilled
ornithopods.

The coprolites of herbivorous
dinosaurs are more problematic. They
are very difficult to identify and, when
they are found, it is almost impossible
to determine what dinosaur produced
them. Coprolites in Jurassic rocks in
Yorkshire, England, are identified as
dinosaur droppings simply because of
their size – no other plant-eating
animals about at that time could
produce such a volume of dung. These
coprolites consist of a mass of pellets
like deer droppings, and contain the

partly digested remains of cycad-like
plants. Coprolites have been found
close to duckbill nesting sites in
Montana, and they contain shredded
conifer stem material. Duckbills had
powerful enough teeth to allow them
to chew woody twigs to extract the
nourishment. Grass structures in
sauropod coprolites from India show
that grass existed in late Cretaceous
times, much earlier than first thought.

*Below: Some dinosaurs swallowed stomach
stones to help grind up food once in the
stomach. These polished stones would have
been regurgitated and swapped for stones
with a sharper edge.*

*Below: Fossilized dung, shaped like huge
droppings, can provide a wealth of information
about what dinosaurs ate.*

Cololites are similar to coprolites. These are fossilized stomach contents that we are sometimes lucky enough to find in the body cavity of a fully preserved dinosaur skeleton. The best examples of stomach contents are those that consist of a meal that has not been digested. The excellent skeleton of *Compsognathus* found in the lagoonal limestones of Solnhöfen, in southern Germany, contains the bones of a little lizard in its stomach area, its last meal. The skeleton of one of the *Coelophysis* that died of hunger as a water hole dried up contains the bones of one of its youngsters; it had been driven to cannibalism by hunger.

Most cololites are in less good condition. A cololite found in the skeleton of the Australian ankylosaur *Minmi* contains the seeds of flowering plants and the spores of ferns. The concentration of seeds suggests that the ankylosaur went for those in particular, and may have been important in spreading the seeds around through its faeces. The chopped up state of the material indicates

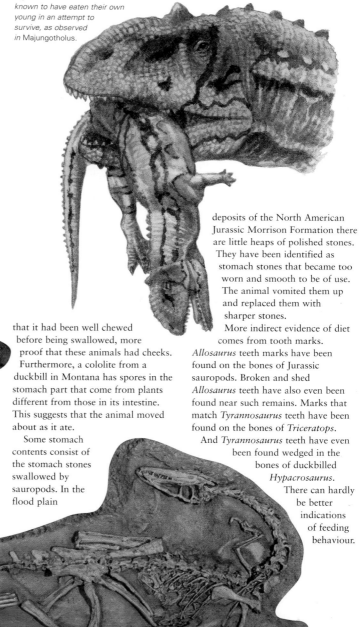

Right: In extreme times dinosaurs are known to have eaten their own young in an attempt to survive, as observed in Majungotholus.

that it had been well chewed before being swallowed, more proof that these animals had cheeks.

Furthermore, a cololite from a duckbill in Montana has spores in the stomach part that come from plants different from those in its intestine. This suggests that the animal moved about as it ate.

Some stomach contents consist of the stomach stones swallowed by sauropods. In the flood plain deposits of the North American Jurassic Morrison Formation there are little heaps of polished stones. They have been identified as stomach stones that became too worn and smooth to be of use. The animal vomited them up and replaced them with sharper stones.

More indirect evidence of diet comes from tooth marks. *Allosaurus* teeth marks have been found on the bones of Jurassic sauropods. Broken and shed *Allosaurus* teeth have also even been found near such remains. Marks that match *Tyrannosaurus* teeth have been found on the bones of *Triceratops*. And *Tyrannosaurus* teeth have even been found wedged in the bones of duckbilled *Hypacrosaurus*. There can hardly be better indications of feeding behaviour.

Below: The tiny bones in the stomach of this Coelophysis are those of a young member of its own species – clear evidence of cannibalism.

DINOSAUR FOOTPRINTS

Footprints are the best tools we have to tell how an animal lived. The study of footprints is known as ichnology, and the study of fossil footprints, is known as palaeoichnology. The science involves an understanding of how sediments and animals behave.

A dead dinosaur may have left a skeleton, but in all likelihood there will only be a few bones left for us to study once the rest of it has been eaten or eroded. It is even more likely that there will be no surviving physical remains. At best, there can only be one skeleton. However, that single animal will have made millions of footprints during its lifetime, and these footprints, if preserved, can tell us all sorts of things about the animal's lifestyle.

The specialist study of fossil footprints is known as palaeo-ichnology. It has a long pedigree. In the early nineteenth century, three-toed fossil footprints were found in the Triassic sandstones of Connecticut, USA. They were studied by local naturalist Edward Hitchcock, who thought they were the footprints of giant birds (there was no concept of dinosaurs at that time). Even today it is almost impossible to match a footprint to the animal that made it. Palaeoichnologists overcome this problem by attributing 'ichnospecies' names to them. This allows each footprint or footprint track to be catalogued and studied without

Below: Three-toed theropod footprints found in early Jurassic rocks in Spain.

Above and right: Fossil footprints can tell us whether an animal travelled alone or in groups, whether there were different sizes of animal passing by and the speed at which the animal travelled. Here are several Apatosaurus *trackways from the Morrison Formation, USA.*

referring to the animal that may have made it. The name of an ichnospecies often ends with the suffix *opus*. For example, *Anomoepus* is a footprint that may or may not have been made by a small ornithopod, *Tetrasauropus* is a footprint that was probably made by a prosauropod, and *Brontopodus* is almost certainly the footprint of a big sauropod, but which one? Most footprints are so vague that we can only tell that an animal of some kind passed this way.

If we find a series of footprints (or a trackway), made as the animal was moving along, we can tell if it had been travelling alone, in a pair, or in a larger group. Sometimes we find different sizes of the same kind of footprint, indicating that a family group of old and young had passed. Some trackways can be traced over tens or hundreds of kilometres, usually in separate outcrops. They are known as megatracksites.

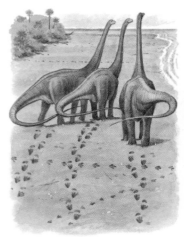

Below: A true print in the top layer usually has the mud squeezed up around it, and may preserve the skin texture of the foot.

Dinosaur trackways

The dimensions measured when studying fossil tracks are:
- Foot length (a).
- Foot width (b).
- Step length (the distance between successive prints) (c).
- Stride length (the distance between successive prints made by the same foot) (d).
- Pace angulation (the angle between three successive prints) (e).
- Angle of rotation (the angle that the direction of the foot makes with the direction of travel).
- Various equations using these parameters can be used to estimate a dinosaur's speed.

Below and left: Paleoichnology can be a highly mathematical science. Using the values measured here, and the dimensions of a dinosaur's leg (if known), we can calculate its walking speed.

However, gauging the size of the footprint can be tricky. The print that fossilized in the rock may not be the actual print impressed by the dinosaur's foot. When an animal leaves a footprint in the top layer of sediment, another less distinct print is impressed into the layer beneath. Often the top layer is washed off and only the impression in the lower layer (the underprint) is preserved. It is essential to recognize underprints so that we do not make erroneous measurements; they tend to be smaller than true footprints and, if confused, would make calculations of speed unreliable. Fortunately, a true print can usually be recognized by its detail. If we find a print that actually preserves the skin

Below: Many prints are preserved as casts. A footprint may be filled by later sediments, and when these sediments become rock they preserve a three-dimensional impression of the print that stands proud of the bedding plane.

texture of the underside of the foot, then we can be confident that we are looking at a true print.

Trackways

Sometimes, tracks can be so spectacular that it is easy to jump to unwarranted interpretations in the excitement of discovery. A track that was clearly made by a herd of sauropods may have a theropod track running parallel to it. This would appear to show a sauropod herd being stalked by a theropod that is waiting to attack. Or the theropod may have followed the same route days later – it is difficult to tell. And a track of sauropod footprints that only show the tips of the front toes and nothing else, may be interpreted as the marks made by the animal swimming and poling

Below: The shapes of footprints can be misleading. The very long print (top), was made in soft mud, where the toes dragged a blob of mud along (bottom).

Left: Footprint in top layer of earth.

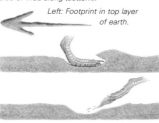

itself through the water using its front feet. More likely it is a set of underprints, with the narrower front feet penetrating deeper into the sediment and forming underprints, while the broader hind feet spread the weight and remain on the surface.

Below top: A dinosaur makes a footprint, with the true print impressed into the top layer of sediment and an underprint pressed into the layer below.

Below middle: The top layer of sediment is washed away.

Below bottom: Only the underprint is preserved.

DINOSAUR EGGS

We know that dinosaurs reproduced by laying eggs. However it is difficult to match the fossil eggs found with a particular species of dinosaur – the first dinosaur eggs were misidentified for 70 years. The study of these fossils show that dinosaurs laid eggs that were more like those of birds than of other reptiles.

It has always been assumed that dinosaurs laid eggs, as other large reptiles do. However, it was in the 1920s that the first dinosaur eggs were actually found. A series of expeditions into the Gobi Desert, led by Roy Chapman Andrews of the American Museum of Natural History, USA, uncovered several nests containing eggs and a large number of skeletons of the ceratopsian *Protoceratops*. For 70 years these "*Protoceratops* eggs" were the most famous and important dinosaur eggs found. At the site, the structure of the nest, the arrangement of eggs within it, and the proximity of the nests to one another were all clearly visible. It was suggested that there was evidence that the nests had been attacked. A small theropod named *Oviraptor*, or egg-stealer, was found close to one of the nests, having been overwhelmed by a sandstorm while trying to rob it. Then, in the 1980s, more of the same nests were found, again by an expedition from the American Museum of Natural History, but this time one of the nests had the skeleton of an *Oviraptor* actually sitting on it. The nests and eggs are now believed to be those of the *Oviraptor*.

Above: Sauropod eggs are almost spherical in shape.

At another site known as Egg Mountain in Montana, USA, nests with eggs and remains of the ornithopod *Orodromeus* were found, with remains of the theropod *Troodon*. The *Troodon* was first believed to be at the site intending to rob the nests, but the nests are now believed to be those of the *Troodon*. The *Orodromeus* remains are thought to be from corpses that the parent theropods

had brought back to feed their young. The provenance of the *Troodon* eggs was confirmed when some of the eggs were dissected and baby *Troodon* were found inside. However it is unusual to be able to dissect a dinosaur egg. A fossilized egg is an egg that died before it had a chance to hatch. When the egg dies the embryo inside is usually destroyed. Beetles burrow through the shell and lay eggs, and when the larvae

Left: There are skeletons of Oviraptor *parents, in Mongolia, actually lying over their eggs in the nest, wings spread out over them to keep them warm while they incubated.*

Left and right: Skeletons of baby Troodon *have been found inside eggs. As in other animals, the babies have proportionally larger heads and feet than the adults.*

hatch they eat the dinosaur egg's contents. Most fossilized eggs contain only scraps of bone that are so jumbled that they are unidentifiable.

Shell fragments are often found associated with nesting sites, where the eggs came to maturity and hatched. Modern microscopy techniques have been used on shell fragments so that dinosaur egg shells can be compared with those of modern animals. It was once assumed that dinosaur eggs had soft or flexible shells, like modern reptiles. However, study of the crystalline structure of fossil egg shells shows that the majority of them were hard and rigid, like those of birds.

Egg identification

As with footprints, it is difficult to assign a fossil egg to a dinosaur species, and it can only be successfully accomplished in rare occurrences when an identifiable embryo can be found inside the egg, or if a nest has the skeleton of a parent nearby.

The south German Solnhöfen skeleton of *Compsognathus* has a number of spherical objects associated with it. They are thought to be unlaid eggs that burst out of the body cavity after the animal died. Such an association of eggs and skeleton is very rare, though.

As with footprints, palaeontologists give names to particular fossil eggs – names that do not imply

Above: Maiasaura *nests show all stages of nesting behaviour – complete eggs, broken eggshell, bones of hatchlings and skeletons of well-grown youngsters that have not yet left the nest.*

identification of the egg layer. Such a classification is an "oospecies".

Most dinosaur eggs are found in nests. Occasionally they look as if they have been laid in holes in the ground without much preparation – a spiral of eggs found in France is thought to have been laid by a stegosaur that just deposited them and left. Most of the nests that we know about are quite complex structures, and are found together in rookeries or nesting sites. The nests laid by the duckbill ornithopod *Maiasaura* in Montana are the best studied. A typical nest consists of a mount of mud or soil, with a depression in the top. That depression is filled with ferns

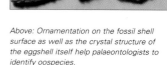

Above: Ornamentation on the fossil shell surface as well as the crystal structure of the eggshell itself help palaeontologists to identify oospecies.

and twigs, providing insulation for the eggs that are laid inside it. Each nest is positioned about an adult dinosaur length from the next, so that they do not disturb one another.

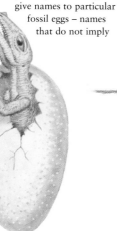

Right: Egg nests have been found in Montana, USA, at a site known as Egg Mountain.

DINOSAUR BABIES AND FAMILY LIFE

Fossils found at nesting sites can tell us how quickly dinosaurs grew, for example, by looking at the structure of the individual bones and analyzing their growth lines. We can also tell that growing dinosaurs were subject to diseases and injuries throughout life.

The *Maiasaura* nests in Montana, USA, have been the subject of the most studies on dinosaur family life. The Montana nesting site was by a lake in the uplands. Besides nesting behaviour, this site also provided all sorts of information about how the animals lived and grew. Once out of the egg, a baby *Maiasaura* was a small, vulnerable animal, about 45cm (18in) long. It remained in the nest and was nurtured by its parents, who brought food in the form of leaves and fruits to the nest. Nests were protected against raids by large lizards and theropod dinosaurs, such as *Troodon*. Seasonal climates meant that once in a while the *Maiasaura* herd had to leave the upland nesting site and migrate to the coastal plains, where food was still plentiful. The nestling grew to twice its birth length in about five months. At this juvenile stage the young dinosaur

Below: The quite recent find of a parent Psittacosaurus *with 34 young demonstrates that some dinosaurs did nurture their young in the nest.*

Above: Evidence suggests that some genera of dinosaurs looked after their young, nurturing the hatchlings and bringing back food until they were five months old.

would have left the nest, and followed the parents, learning how to find food for itself. Juveniles continued to grow, and after about a year reached a length of about 3.5m (11½ft).

The microscopic structure of the bones show us that the full adult length of about 7m (23ft) was reached in about six years, after which time the growth rate became very much slower. The conclusion is that dinosaurs, or at least duckbill dinosaurs, grew extremely quickly in the early part of their lives, and that the growth rate then slowed. This is the same growth pattern that we see in modern mammals and birds, but not in reptiles. The *Maiasaura* herds returned to their nesting area when it was time to breed, a round trip of around 300km (185 miles).

Sauropods

Similar evidence for the family life of sauropod dinosaurs comes from a site called Auca Mahuevo, in Argentina. This late Cretaceous site has a vast array of nests from a titanosaurid sauropod, probably *Saltasaurus*, made on a flood plain. During the breeding season herds of pregnant females descended on the area – a region that was just out of reach of most floodwaters from nearby streams. Their nests were simple compared with the nest structures of *Maiasaura* – they merely scooped out a shallow basin, about 1m (3ft) in diameter, with their clawed front feet, and laid their eggs in this. Each female laid 15–40 eggs in nests positioned 1.5–5m (5–16½ft) apart. Having laid their eggs, the females abandoned them, without any intention of looking after the nestlings once they hatched. It is possible that they remained in the area to discourage the big meat-eating dinosaurs, such as *Aucasaurus*, that roamed the area.

Below: Young Maiasaura *left their nests after about five months of growing and being tended by their parents.*

The young sauropods, on hatching, would have been about 30cm (12in) long – little animals compared with the adults who were 30 times their length – but they were equipped to find their own food immediately. Youngsters able to look after themselves immediately, are known as "precocious hatchlings" by biologists, and this is true of most reptiles today. Once out of the nest, the perils of the outside world awaited the young dinosaurs.

Dinosaur remains have been found showing signs of sickness and injury that would have prevented many from reaching old age. Palaeopathology, the study of ancient diseases, has shown that a range of traumas and diseases afflicted the dinosaurs.

Many ceratopsian skeletons have healed fractures to their rib-cages, suggesting injuries due to fighting. One ceratopsian skull has been found with a hole in it that matches the dimensions of a ceratopsian horn,

Above: As Maiasaura *youngsters grew, they joined the herd in their annual migrations to the productive feeding grounds.*

suggesting a head-on fight between the two animals. Tyrannosaurs suffered from gout. There are at least two examples of this disease that would have been brought on by a surfeit of red meat in the diet. Arthritis has also been identified in quite a few dinosaurs, including *Iguanodon*, where foot bones have been fused together. Additionally, a number of the bigger ornithopods have fractures to the upward-projecting spines from the tail vertebrae close to the hips. It has been suggested that these breakages were caused by excessive violence during mating.

DEATH AND TAPHONOMY

The study of what happens to a body between its death and its fossilization is known as taphonomy. The taphonomy of an individual fossil organism is extremely important to the scientist who wishes to find out about that animal, and the conditions under which it lived and died.

Fossil-bearing rocks are usually full of bivalves, or sea urchins, or corals, or even fish. Fossils of sea-living things are common because when they die, creatures that live in the sea are likely to sink to the bottom and be covered by sand, mud and other marine sediments. As these sediments solidify, the dead creatures become fossils. An animal that lives and dies on land, on the other hand, may have been killed by a hunting animal and may be eaten on the spot. Once the hunter has eaten its fill, scavenging animals will take

their pick of the prey. The bones are broken up and carried away, and what is left is eaten by insects and broken down by fungi and bacteria. After a few weeks there is nothing left of the dead animal but a smear on the ground – certainly nothing left to fossilize.

The vast majority of dinosaurs have left nothing to tell us of their existence. However if a dinosaur's body falls into a river and is buried in flood deposits, or if it dies in a toxic lake where no scavengers can live, or if it is engulfed

and buried in a sandstorm, there is a chance that the remains will survive in fossil form.

Fossil finds

Dinosaur fossils are found in a number of forms. The most spectacular is the "articulated skeleton". This is the ultimate prize because it has all the bones still joined together as when the dinosaur died. However, the temptation is to leave it as found, and so information is lost if the skeleton is not dissected and studied minutely.

The fate of a *Stegosaurus*

A dinosaur skeleton is rarely found complete. Many things can have happened to the dead body before it became buried and fossilized.

1 Above: A *Stegosaurus* is killed by a hunting *Allosaurus*. The big meat-eater eats its fill. Inevitably much of the body is left uneaten.

2 Above: Smaller meat-eaters, such as *Ceratosaurus*, wait until the *Allosaurus* has fed and departed and then scavenge most of the flesh left behind.

3 Above: The scraps that remain are eaten by even smaller scavengers, and the rest is broken down by insects and fungi. By this time the skeleton is scattered.

4 Above: Eventually the scattered bones are buried by river deposits during the wet season as the plain floods. Now the process of fossilization can begin.

Above: The skin impression of a Hadrosaur *was created as mud solidified around a corpse before it decayed – a rare occurence.*

The next best thing is an "associated skeleton". This is a jumble of bones that obviously comes from the same animal, but which have been broken up and scattered. Usually something is missing, carried away by the forces that pulled it apart. More common is an "isolated bone". Sometimes its origin is obvious, but not always, and scientific errors have been perpetrated by the misidentification of isolated bones. *Dravidosaurus* was thought to be a stegosaur, living in India where no

Below: A partly excavated isolated femur (thigh bone) that belonged to a sauropod dinosaur, preserved in Upper Jurassic redbed mudstones of the Upper Shaximiao Formation, in Sichuan, China.

other stegosaurs had been found, and existing tens of millions of years after the stegosaurs were supposed to have died out. The isolated bones on which the identification was based were actually parts of a plesiosaur, a marine reptile.

Bits of bone that have no scientific value whatsoever are termed "float". Some dinosaur excavations map and catalogue every bit of float found, in the hope that they might one day yield some information. Most excavations just ignore them.

Above: The associated skeleton of a sauropod in Jurassic sediment in the Dinosaur National Monument, Utah, USA.

Finally there are trace fossils, footprints, droppings, eggs and other lines of evidence that a dinosaur existed, but without any physical remains. Paradoxically it may be that trace fossils tell us more about the animal's life than the body fossils.

Below: An articulated skeleton of Lambeosaurus *where the components have been kept together as in life.*

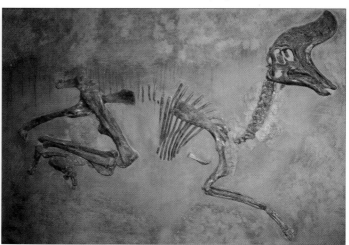

DESERTS AND ARID HABITATS

We often associate dinosaur remains with water, and animals living near it are more likely to be buried in sediment once they have died than those that lived far away from it. However, there are a number of occurrences where dinosaurs of very dry habitats have been found fossilized.

The Gobi Desert was a desert even in Cretaceous times. Ephemeral lakes surrounded by plains of scrubby vegetation were separated by vast areas of shifting sand dunes. Here roamed herds of the small ceratopsian *Protoceratops*, in such numbers that one palaeontologist dubbed them 'the sheep of the Cretaceous period'. Occasionally, the harsh, dry winds of the desert overwhelmed these herds in a sandstorm or a dust storm, and sometimes they were engulfed by sliding sand as an unstable dune collapsed. Either occurrence had the effect of killing the animals where they stood, and preserving them in a tomb of dry sand that eventually solidified into desert sandstone.

A specimen that shows just how quickly such an event occurred was found in the 1970s. The skeletons of a *Protoceratops* and a fierce little *Velociraptor* were found wrapped around one another. The meat-eater's arm was seized in the ceratopsian's beak, while the long claws of the former were clutched tightly to the latter's head shield. The two had been buried and died while in the middle of their struggle. Elsewhere in the desert, dinosaur nests and eggs have been preserved in sandstone. One nest, belonging to the theropod *Oviraptor*, even had the mother dinosaur sitting across it, in a vain attempt to incubate her eggs as the sandstorm swept down and engulfed her.

In the Triassic period, back at the beginning of the Age of Dinosaurs and a whole supercontinent away from the Gobi Desert, southern England consisted of an arid, limestone plateau where more fossils have been found. Early prosauropods, such as *Thecodontosaurus*, eked out a living here on the scrappy vegetation. The plateau was riddled with gullies and caves, and moist underground air seeped to the surface producing a slightly lusher vegetation at the cave mouths. Plant-eaters were

Below: The shifting sands and airborne dust clouds of the desert were constant threats to the dinosaurs of the late Cretaceous Gobi Desert. Often they were buried in hollows or on the flanks of collapsing sand dunes.

tempted to these areas, and occasionally lost their footing and fell to their deaths. There they were devoured by cave-dwelling animals, or lay until they were covered by cave debris and fossilized. This leads us to the anomaly in which Triassic dinosaur remains are found in much older Carboniferous rocks. The Carboniferous limestones formed the uplands and the Triassic animals were fossilized in them.

We can often tell if a dinosaur died in an arid environment. As a dead body dries out in the heat, the tendons that link the bones shrink. In most dinosaurs tendons lash together the bones of the backbone and help to support the weight of the neck and tail. When these tendons dry and

contract, they pull the tail upwards and the neck backwards, drawing the skull over the shoulders and back. An articulated skeleton found in this position shows that the body dried out before burial. Many articulated skeletons are found like this.

Above: A sandstorm could mean death for a large animal like a dinosaur. It could also mean that its skeleton was preserved entirely and articulated. An animal engulfed in tonnes of sand suffocated quickly, but the sheer weight of the sediment kept it in place and once the soft tissues decayed the bones remained undisturbed.

Below: An arid limestone plateau habitat is the setting for a swallowhole into which lizards and a Thecodontosaurus *have fallen. Eventually the swallowhole is filled with debris and the bones of the dead animals fossilize.*

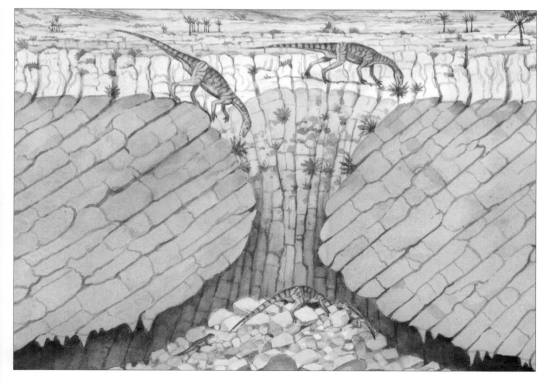

OASES AND DESERT STREAMS

In an arid landscape, the presence of water is a great attraction to life. Where desert depressions reach down to the moist rocks of an aquifer and water bubbles to the surface in cool pools, or where streams flow down from mountains to parched lowlands, vegetation grows and animals live.

Rivers and streams flowing away from rapidly eroding uplands can carry vast quantities of broken rock fragments in the form of grains of sand and silt. They are spread across the lowlands as the current slackens. The result may be accumulating beds of fertile soil in which plants grow. So, in desert areas, the river supports the riverside vegetation and the seasonal floods spread layers of silt over the landscape, fertilizing it. The River Nile, with its annual floods, keeps the Egyptian desert alive and was responsible for producing one of the earliest prolific farming civilizations.

In Triassic times the vast supercontinent of Pangaea was beginning to break up. The interior was still arid, being so far from the sea, and life could only really flourish in the flood plains of rivers. It was the time of the prosauropods – the first big plant-eating dinosaurs – the early relatives of the huge long-necked sauropods. They fed on the cycad-like plants that grew close to the ground, and also reached up to browse the coniferous trees that lined the water courses. This heavy browsing led to the evolution of trees and foliage that had strong, sharp, sword-like leaves as a defence, such as the monkey puzzle tree that still exists in the mountains of South America and in ornamental gardens across Europe. Among the predators, the various crocodile relatives, such as the phytosaurs, were giving way to the first of the theropods. When the plant-eaters were attracted to the water and became trapped in quicksands, their great

weight preventing them from escaping, the meat-eaters would take advantage of their helplessness. Remains of all these animals are found in the rocks formed from the river sandstones and the silty flood deposits of the time. Occasionally their fossils tell the story of trapped animals devoured by predators, or even of animals killed by the changing seasonal conditions.

Below: A prosauropod, attracted to the water, is trapped in quicksand. Small theropods and crocodile-like predators close in for the kill.

Petrified Forest National Park
The most famous fossil occurrence of a Triassic desert waterside habitat is Petrified Forest National Park in Arizona, USA. Close by, at Ghost Ranch, a pack of meat-eating theropods, *Coelophysis*, was discovered as articulated skeletons. The evidence shows that they died from dehydration around a drying water hole. The best evidence for prosauropods being caught in quicksand and killed by predators comes from the Frick Brick Quarry, Switzerland. Almost identical circumstances are known from the Molento and Lower Elliot Formations of South Africa, showing how widespread these habitats were on Pangaea.

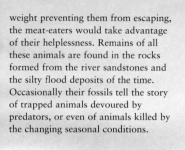

1 *Plateosaurus* – prosauropod dinosaur.
2 *Liliensternus* – theropod dinosaur.
3 *Rutiodon* – crocodile-like phytosaur.
4 Sauropod leg bones set vertically in river sandstones showing where they were caught in quicksands.
5 The rest of the skeleton torn apart and scattered.
6 Footprints of theropods.
7 Scattered teeth of theropods and phytosaurs.
8 Articulated skeletons of dinosaurs killed by drought.
9 Mud cracks revealing a drying water hole.
10 Burrows of worms and arthropods showing the presence of water.

CALM LAGOONS

Seawater trapped behind reefs forms shallow lagoons that evaporate slowly under a hot, tropical sun.
The salt concentration in the water increases and the water becomes toxic, killing anything that enters it.
Detailed evidence of life in dinosaur times comes from deposits formed under these conditions.

In late Jurassic times the break up of Pangaea was well under way. A huge embayment, known as the Tethys Sea, separated the continent of Europe and Asia from the landmasses of the south; those that became Africa, South America and Australia.

Along the northern flanks of the Tethys a deep-water reef developed, formed from sponges. The remains of this reef can be found today in rocks from Spain to Romania. As the reef grew, and as continental movements raised the sea floor, the reef approached the surface where the deep-water sponges died. The reef growth was continued by corals building on the sponge-formed

structures. The reefs reached the surface where they cut off lagoons between the deep Tethys waters and the shoreline at the continent to the north. Debris from the reefs filled the lagoons and the water became shallow. The heat of the sun evaporated the water from these shallows, and salt and other minerals settled on the lagoon floor. The water was constantly replenished from the ocean beyond the

Solnhöfen

There is a concentration of lagoon deposits in Solnhöfen, southern Germany. These so-called lithographic limestone quarries have yielded famous fossils of flying animals and land-living creatures. There are fossils of ammonites that have plunged into the limy mud, horseshoe crabs that have dropped dead at the end of their tracks, and floating animals, such as brittle stars and jellyfish, that drifted into the poisonous waters.

Right: This pterosaur fossil is from Solnhöfen, Germany. The details of its anatomy are clearly defined.

Below: The lagoon environment is best known from Solnhöfen, Germany, where some of the finest fossils have been uncovered, but lagoons of this kind probably extended across the whole of southern Europe.

reefs, but the lower layer of the lagoon water became poisonous with the concentration of minerals. Any fish that swam in it died and sank to the bottom. Any arthropod that crawled in was poisoned and died. The bodies lay undisturbed as the water was poisonous for scavengers too.

Islands were scattered across the lagoons which, with the arid shoreline, were formed from the stumps of the sponge reefs that had emerged from the water as the land rose. The animals of these dry lands consisted of pterosaurs, the first-known bird *Archaeopteryx*, little lizards and small dinosaurs, such as the chicken-sized theropod *Compsognathus*. All of these animals have been found as articulated skeletons in the deposits formed in the lagoons.

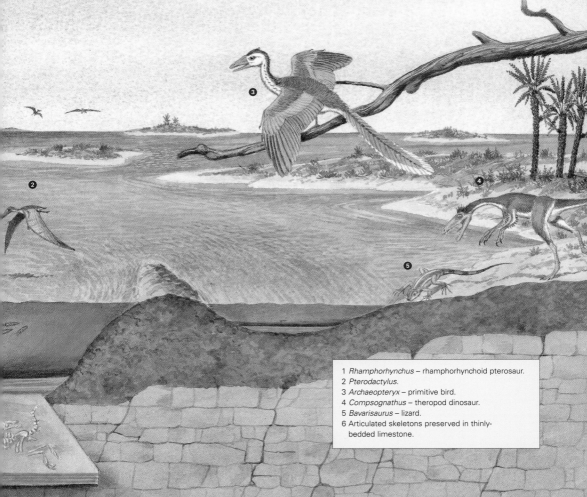

1 *Rhamphorhynchus* – rhamphorhynchoid pterosaur.
2 *Pterodactylus.*
3 *Archaeopteryx* – primitive bird.
4 *Compsognathus* – theropod dinosaur.
5 *Bavarisaurus* – lizard.
6 Articulated skeletons preserved in thinly-bedded limestone.

RIPARIAN FORESTS

Many dinosaur fossils from North America come from the late Jurassic Morrison Formation that stretches across much of the Midwest. It consists of river- and flood-deposits. Under seasonally dry conditions, vegetation was mostly restricted to river banks and only flourished in the wet season.

When the pioneers headed west across North America in the late nineteenth century, they found the bones of strange animals. The finds drew the attention of scientists who, within 30 years, discovered more than 100 previously unknown dinosaur types. Most came from a sequence of rocks called the Morrison Formation, which was formed from river sediments in late Jurassic times. The landscape was once that of a riparian forest. In the late Jurassic a shallow seaway stretched north-south along the length of the North American continent. To the west the ancestral Rocky

Mountains arose along the edge of the ocean. Between the two stretched a plain built up of sediment washed down from the mountains. Rivers meandered across the plain, flooding frequently and depositing sediment. In times of flood, the sediment built up on the riverbanks, forming levees, until the surface of the river became higher than the elevation of the surrounding plain. The water frequently broke through the levees spreading sediment in a fan-shaped deposit, and river water often seeped through the levees as springs filled freshwater ponds and lakes across the plain. Other lakes,

formed at times of flood, dried out and became poisonous with alkaline minerals leached from the soil.

Below: During the late Jurassic, in North America, in the wet seasons the rains were plentiful and floods spread across the plains. Flood sediment was deposited over the landscape causing prolific plant growth on which animals browsed.

The original flood deposits were disturbed by the trampling feet of the animals – called "bioturbation". The alkaline lakes formed beds of limestones with ribbons of river sediments running through them. Dinosaur fossils are found as isolated bones on the flood deposits or sometimes as associated skeletons in the river sediments.

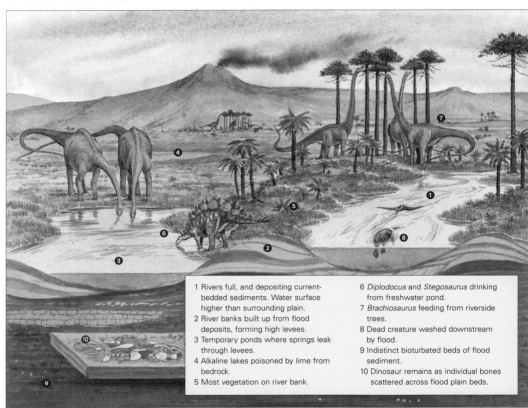

1 Rivers full, and depositing current-bedded sediments. Water surface higher than surrounding plain.
2 River banks built up from flood deposits, forming high levees.
3 Temporary ponds where springs leak through levees.
4 Alkaline lakes poisoned by lime from bedrock.
5 Most vegetation on river bank.
6 *Diplodocus* and *Stegosaurus* drinking from freshwater pond.
7 *Brachiosaurus* feeding from riverside trees.
8 Dead creature washed downstream by flood.
9 Indistinct bioturbated beds of flood sediment.
10 Dinosaur remains as individual bones scattered across flood plain beds.

Morrison Formation

The Morrison Formation, named after the Colorado town of Morrison, USA, consists of 30–275m (100–900ft) of shale, siltstone and sandstone, and stretches from Montana to New Mexico. Its landscape was once that of a riparian forest, or forestry at the side of riverbanks. The most famous outcrops of the Morrison Formation are in Dinosaur National Monument in Utah, and the Fruita Palaeontological Area and Dry Mesa Quarry, both in Colorado. At Tendaguru in Tanzania, a similar environment existed at exactly that time and almost the same range of animals lived there.

Above: The Morrison Formation is seen most obviously as a ridge along the foothills of the modern Rocky Mountains.

On this topographic surface the plants were confined largely to the river banks and around the freshwater pools. The plants formed forests and isolated stands of primitive conifers and ginkgos, with a lower-storey of cycad-like plants and tree ferns, and an undergrowth of ferns and beds of horsetails close to the water. The open plain had a scrappy growth of ferns. The alkaline pools were barren. The climate was seasonal with dry periods interrupted by times of abundant rain.

Herds of sauropods migrated across this landscape. *Diplodocus,* *Apatosaurus, Camarasaurus, Brachiosaurus* and many others, moved from thicket to thicket wherever there was food. The plated *Stegosaurus* fed from the trees and undergrowth. The main ornithopod was *Camptosaurus*. There were theropods aplenty, ranging from the enormous *Allosaurus,* through the medium-size *Ceratosaurus* to smaller animals like *Ornitholestes* and *Coelurus*. The abundant fossil remains of these dinosaurs are found mostly as isolated bones in flood deposits, but there are also associated skeletons and articulated skeletons in river channel deposits.

Below: In the dry season rivers dried to a trickle, and ponds became hard-packed mud. Water evaporation brought up the mineral calcite and left it as lumpy beds of limestone beneath the soil surface. Animals migrated into areas where there was still food to be had.

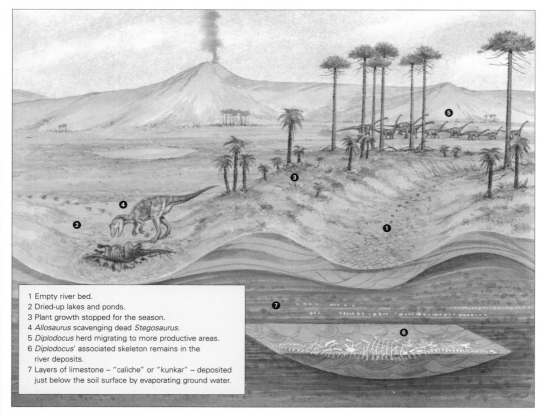

1 Empty river bed.
2 Dried-up lakes and ponds.
3 Plant growth stopped for the season.
4 *Allosaurus* scavenging dead *Stegosaurus*.
5 *Diplodocus* herd migrating to more productive areas.
6 *Diplodocus'* associated skeleton remains in the river deposits.
7 Layers of limestone – "caliche" or "kunkar" – deposited just below the soil surface by evaporating ground water.

LAKE ENVIRONMENTS

*Lakes tend to be rather ephemeral. Water-filled hollows left by glaciers or landslide-blocked valleys soon
fill with sediment and disappear. However, lakes formed in rift valleys, such as those in modern East
Africa may be long-lasting landscape features. Good fossil remains are found in rift valley lake deposits.*

At the end of the Jurassic period and the beginning of the Cretaceous, the eastern part of the northern continent, in the area where China now lies, was beginning to break up. The continental surface was split by fault lines, running north-east to south-west, which produced rift valleys. These rift valleys contained lakes. The water of these lakes was clear, and the only sediment was from very fine particles. In quiet times the valleys were filled with lush forests of long-needled conifers and ginkgoes. Horse-tails, ferns and mosses flourished around the lake edges. This environment supported an astonishing assortment of animals. The biggest were the dinosaurs, such as the

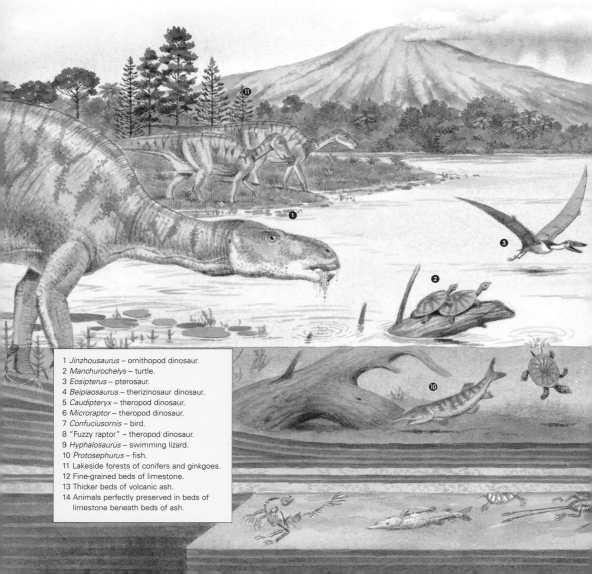

1 *Jinzhousaurus* – ornithopod dinosaur.
2 *Manchurochelys* – turtle.
3 *Eosipterus* – pterosaur.
4 *Beipiaosaurus* – therizinosaur dinosaur.
5 *Caudipteryx* – theropod dinosaur.
6 *Microraptor* – theropod dinosaur.
7 *Confuciusornis* – bird.
8 "Fuzzy raptor" – theropod dinosaur.
9 *Hyphalosaurus* – swimming lizard.
10 *Protosephurus* – fish.
11 Lakeside forests of conifers and ginkgoes.
12 Fine-grained beds of limestone.
13 Thicker beds of volcanic ash.
14 Animals perfectly preserved in beds of
   limestone beneath beds of ash.

horse-sized ornithopod *Jinzhousaurus*, which browsed the low plants of the shorelines. The hairy therizinosaur *Beipiaosaurus* hunted small animals on the banks. The smaller dinosaurs were theropods, many of which seemed to show transitional features between conventional dinosaurs and birds. *Sinosauropteryx* was like *Compsognathus* but covered in fur or feathers. *Caudipteryx* and

Below: Lake environments existed throughout the changing landscape of the Age of Dinosaurs. The lake deposits found in Liaoning, China, are beds of fine limestone interspersed with beds of volcanic ash that formed in early Cretaceous times.

Protarchaeopteryx had stubby wing-like forelimbs with feathers, and a fan of feathers on the tail. *Microraptor*'s fore and hind limbs even evolved into wings. There were also true birds, some of which still had primitive claws on the wings. These smaller creatures hunted the swarms of insects that lived close to the water.

We know of the existence of these animals because the mode of preservation is so fine that even the finest structures are visible. Every now and again the volcanoes that formed along the fault lines would erupt, engulfing the valleys in poisonous gas and showering the lakes with a thick deposit of ash. Gassed by volcanic

fumes and buried in fine volcanic ash, the smallest details of skin and feathers have been preserved around the almost perfectly articulated skeletons. We also have the fossils of their insect prey, and all sorts of other animals, such as turtles and other swimming reptiles, and also early mammals, that lived in the area.

WETLANDS

Wetlands tend to be most extensive in moist, tropical climates, close to upland areas. Heavy erosion from exposed hills brings masses of sediment down to low-lying regions. The sediment spreads out to form mud flats through which streams meander.

The most extensive, modern, tropical wetlands are found in the papyrus-choked Nile drainage area in Sudan and Uganda. Wetlands also appear further south in the Okovango flood plain at the edge of the Kalahari Desert. The mud flats support vegetation peculiarly adapted to the conditions. In these places the dominant plants are water-loving grasses and reeds, which developed in the Tertiary period. In the Mesozoic,

the water courses and mud banks supported thick beds of horsetails. Like grasses, horsetails spread by means of underground stems which held the sediments together, and provided a firm surface on which more permanent vegetation could grow. Like reeds, the horsetails could grow in shallow water. On stabilized sandbanks, thickets of ferns such as *Weichselia* grew. Stable ground developed and on it thrived thickets of conifers and cycads.

During the Cretaceous period, herds of *Iguanodon* roamed the wetlands of northern Europe, grazing on horsetail beds. Smaller herbivores, such as the fleet-footed *Hypsilophodon*, also scampered there. In the more permanent thickets, armoured dinosaurs such as *Polacanthus* grazed. Where plant-eating animals flourished,

Below: Dinosaur bones may be quite abundant and diverse in wetland deposits, but they tend to consist of isolated bones or, at best, associated skeletons.

European wetlands

The dinosaur wetlands of northern Europe are best represented by the Wealden and Wessex formations, laid down during the early Cretaceous. They outcrop along southern England and the Isle of Wight, and across the Paris Basin. At the time, mountains of limestone and metamorphic rock to the north supplied the sediment that washed down to the lowlands. The result is a thick sequence of sandstone, mudstone and clay, with frequent beds of plant-rich material. Dinosaur remains mostly occur as isolated bones, some showing signs of gnawing. They are often found with the teeth of freshwater fish, and are sometimes encrusted with the eggs of freshwater snails, showing that they were deposited in stream beds or in the backwaters of rivers.

meat-eaters that preyed on them followed, and in the horsetail swamps of northern Europe they included *Megalosaurus* and *Neovenator*. Wading in the waters were fish-eating theropods, such as *Baryonyx*.

Dinosaurs were not the only animals that existed here. In the water were crocodiles, including the dwarf form, *Bernissartia*, no bigger than a domestic cat, and turtles such as *Chitracephalus*, about the size of a dinner plate. In the skies there wheeled pterosaurs, such as the condor-sized *Ornithodesmus*. Disarticulated bones of all these animals have been found in the swamp deposits laid down at that time. The isolated fossil bones in these deposits may be worn and polished,

showing that they were washed about for a while before settling. They are sometimes encrusted with the eggs of water snails, suggesting that they lay on the stream bed for some time before becoming buried.

1 *Oviraptor* – theropod dinosaur.
2 *Eotyrannus* – theropod dinosaur.
3 *Neovenator* – theropod dinosaur.
4 *Bernissartia* – crocodile.
5 *Pelorosaurus* – sauropod dinosaur.
6 *Polacanthus* – armoured dinosaur.
7 *Baryonyx* – theropod dinosaur.
8 *Iguanodon* – ornithopod dinosaur.
9 *Hypsilophodon* – ornithopod dinosaur.
10 River and flood deposits.
11 Lens-shaped stream deposits.
12 Fossils appear as isolated bones in river deposits.

SWAMP FORESTS

By the end of the Cretaceous period the vegetation of the world was taking on an appearance familiar to humans. There were still no grasses, but flowering plants had appeared in the undergrowth, and deciduous trees were beginning to take over from conifers as the main woodland flora.

It is easy to imagine dinosaurs as being the inhabitants of deep, dark jungles. Where there were deep forests and woodlands, there were dinosaurs well adapted to living there. As the advanced ornithopods, especially the duckbills, evolved and diversified during the Cretaceous period, plant-life evolved with them. The broad mouths and low-slung necks of these ornithopods showed that they fed close to the ground. The evolutionary response would have been for plants, having been grazed, to develop survival strategies that would allow them to rebuild populations quickly. Flowers and enclosed seeds do this, allowing the main part of the germination process to take place after the parent has been destroyed. Flowering plants in deciduous woodlands evolved with the low-feeding habits of the later plant-eaters such as the broad-mouthed duckbills. Many of these dinosaurs had extravagant head structures, linked to their nasal passages. They made grunts and trumpet noises, producing sounds that would penetrate dense forest undergrowth so that the animals could communicate with one another.

The deciduous woodlands spread across the broad plains and deltas, between the newly arisen Rocky Mountains and the spreading inland sea, that now reached from the Arctic Ocean down across the middle of North America. The deposits formed the Hell Creek Formation. On higher ground the forests consisted of vegetation such as tall stands of primitive conifers with an understorey of cycads and ferns. These forests were inhabited by the last of the long-necked sauropods that had been the most important plant-eaters since the beginning of the Jurassic period.

Fossil survival

Duckbilled dinosaurs are sometimes found as 'mummies', with the skin still fossilized around them. This occurred when the animal died and was stranded in the open, possibly on a sandbank in a delta. One flank of the dead animal would be pressed down into the mud, impressing the skin texture into the mud. The insides would have shrivelled away and the skin would have dried to leather, shrinking around the bones of the leg and the rib cage. The exposed part of the skeleton would have deteriorated quickly, the bones carried off by scavengers or washed away in the

river. The next flood would have filled its insides with sand swirled along by the floodwaters. Fish have been found fossilized inside the rib cages of such dinosaurs. The skin would not have survived the subsequent fossilization process but by then the impression in the surrounding sediment would have solidified. The result is an articulated skeleton that is surrounded by the impression of the skin, giving a valuable insight into what the outer covering of a dinosaur was like.

Below: All manner of creatures from the tiniest insects to the largest dinosaurs have been found fossilized in areas that were once river beds and swamps.

1 *Kritosaurus* – ornithopod dinosaur.
2 *Troodon* – theropod dinosaur.
3 *Tyrannosaurus* – theropod dinosaur.
4 *Ceramornis* – modern-type bird.
5 Skeleton of *Kritosaurus* drying out, with underside buried in mud.
6 Deciduous trees.
7 Flowering herbaceous undergrowth.
8 Swamp deposits.
9 Channel deposits.
10 Dinosaur remains appear as articulated or associated skeletons.

OPEN PLAINS

Herd-living animals have always lived on wide open plains. This was also true in dinosaur times. The late Cretaceous rocks formed on the plains of North America have ample evidence of herding behaviour in the horned dinosaurs – including the fossilization of entire herds in deposits known as "bonebeds".

Modern open landscapes are dominated by herds of plant-eating animals. Look at the high Serengeti plain of Tanzania today, with its herds of wildebeest, impala, elands, gazelles, zebras and so on. So it was with the open plains of the Cretaceous period.

The low vegetation that clothed the open landscapes in the Cretaceous period consisted largely of ferns but no grass. This was where herds of ceratopsians roamed. Like the modern herds that consist largely of different types of antelope, with different horn arrangements, the Cretaceous herds consisted of a number of different types of ceratopsian. They differed from one another by the arrangement of horns and frills, and other head ornamentation. The youngsters all looked the same – when they were young they were sheltered by the rest of the herd and had little contact with other ceratopsians. As adults, however, the range of horn and frill types was very marked. This suggests that the ceratopsians lived in tightly organized herds, moving from place to place as a unit, and keeping away from herds of other types, as happens in modern

grassland animals. Additionally, like the ceratopsians, modern grassland animals migrated with the seasons to where food was most abundant.

We can see evidence of herd behaviour in the fossil occurrences known as bonebeds. They consist of a mass of bones, usually of one species of ceratopsian. The bones lie on the bottom of stream channel deposits and tend to be of the same size, suggesting that they have been sorted out by flowing water. They may consist of the remains of more than 1,000 individuals of the same species. It is

Below: Herds of various species of ceratopsian roamed the plains of North America in the late Cretaceous, harassed by the big meat-eaters of the time.

easy to visualize the scenario. A herd of ceratopsians was overcome by water while crossing a river during migration, and the bodies were washed downstream. The bodies washed ashore on a river beach, where they decayed and were scavenged by other dinosaurs and pterosaurs. Then the remains were picked up by the flooding river and deposited on the channel bottom, along with shed teeth from the scavengers.

As modern plant-eaters have predators like lions to harass them, so the ceratopsians had the tyrannosaurs.

Dinosaur Provincial Park

The best-studied ceratopsian bone beds are in the late Cretaceous Dinosaur Park Formation, in Dinosaur Provincial Park, in Alberta, Canada, where there are at least eight occurrences, one stretching for almost 10km (6 miles). Others occur in Montana, USA, and as far north as the North Slope of Alaska.

Below: Dinosaur Provincial Park, Alberta.

1 *Chasmosaurus* in defensive circle.
2 *Albertosaurus*.
3 *Quetzalcoatlus* – pterosaur.
4 *Centrosaurus* in migrating herd.
5 *Styracosaurus* displaying to other ceratopsians.
6 Plain formed of flood sediments.
7 Lens-shaped channel deposits.
8 Bonebeds at the bases of channel deposits.

Formation of a bone bed
A Herd of ceratopsians crosses a river.
B Herd panics and individuals drown.
C Bodies are washed up on a sandbar.
D Bodies are scavenged by meat-eaters.
E Remains are washed into the river and settle on the bed.

SHORELINES AND ISLANDS

The edge of the sea is a popular habitat for modern birds. It was probably the same for their ancestors, the dinosaurs, though evidence is sparse. Shoreline deposits tend to be sparse and rather ephemeral since they are constantly disrupted by waves and tides.

In Cretaceous times the seaway that spread down the length of the North American continent stretched from Alaska to the Gulf of Mexico, splitting the dry land in two. In some areas the hinterland was thickly forested, and migrating dinosaurs travelled along the beaches where the walking was easier. North-south trackways have been found in Oklahoma, Colorado and New Mexico, USA. These discoveries have led to the concept of the "dinosaur freeway" – a beach migration route indicated by the consistent direction of Cretaceous dinosaur footprints. There are also others.

Island inhabitants

Crocodiles inhabited the rivers that opened out into the continental sea, and their toe marks have been found where they clawed their way across the river sediments. Birds found plenty to eat on the tidemarks, and their three-toed footprints abound. They can be distinguished from the footprints of small dinosaurs by their more splayed toes.

Out at sea, islands, as always, can support a variety of specialized life forms. Animals reach newly formed

Below: The sequence of early Cretaceous rocks known as the Dakota Group was formed as seashore deposits along the interior sea of contemporary North America. It is famous for its footprints and trackways.

Dinosaur Ridge

The most publicized exposure of the dinosaur freeway is Dinosaur Ridge, in the Rocky Mountain foothills, just a few kilometres west of Denver, Colorado.

the mainland's animal life. In either scenario, if the island becomes a permanent geographical feature the animal life will adapt and evolve to survive there. One such adaptation is the development of dwarf forms. Small animals need less food to survive, and islands have limited natural resources. In the Ice Age there were elephants the size of pigs on the Mediterranean islands, and giant ground sloths on the Caribbean islands that were much smaller than their mainland relatives. A modern example is the Shetland pony, a breed well adapted to the bleak habitat of the Scottish islands. The northern edge of the Tethys Sea in Europe had a whole archipelago of islands, thrown up by

earth movements as the northern and southern continents approached one another. Bones of dinosaurs unearthed in Romania show that duckbills, only one-third of the size of their relatives, and ankylosaurs the size of sheep, existed there.

islands by being rafted there on logs or other land debris. Also, an island that formed when an area of land was cut off by the sea usually retains some of

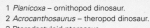

1 *Planicoxa* – ornithopod dinosaur.
2 *Acrocanthosaurus* – theropod dinosaur.
3 Pterodactyloid pterosaur.
4 Birds.
5 Alternating beds of beach sands and shallow sea sediments.
6 Sand ripples.
7 Ornithopod trackways following the shoreline as herds migrated.
8 Theropod trackways at right angles to the shoreline, as individuals came out of the forest to scavenge or to attack the migrating herds.
9 Bird trackways in random feeding pattern.
10 Individual dinosaur bones in sea sediments – badly worn and encrusted with sea life.

MOUNTAINS

Mountainous habitats in Mesozoic times would have had specific flora and fauna, adapted to high altitude environments with thin air, dryness and cold. Creatures from such habitats are rarely found as fossils, and so we can only speculate as to the animals and plants that lived in these regions at that time.

The Mesozoic world was continuously on the move. The shifting plates that tore Pangaea apart and made the continents drift crumpled up mountain ranges along the edge. The debris that formed the sediment of the sedimentary rocks of the time was worn from these Mesozoic mountains. These mountains would have had their share of animal life, as they do in the present day, and they must have had some dinosaurs.

What do we know of animals that lived in areas of erosion, rather than of deposition? What of animals that lived in mountains where no sediment was accumulating, and no rocks were being formed?

In modern times we can see how mountains produce particular habitats. Most rain falls on low ground, particularly on the windward sides of mountains. High up the rainfall becomes less and the temperatures drop. This affects the kinds of plants that grow and in turn the animal life. Mountain animals must be specifically adapted to the extremes of cold, and to clambering about on exposed rocks – the modern ibex is a good contemporary example. We expect that the dinosaurs living in the mountains would have been quite different from the well-known ones that lived in the lowlands, but there has been speculation.

We know the bonehead dinosaurs by the fossil skulls that have been found. This is unusual – the skulls of dinosaurs are usually the first to break down and disappear when the animal fossilizes. The boneheads had skulls thick enough to withstand all sorts of taphonomic punishment. Many of them are found badly worn, as if they

had been rolled along river beds for a long time before coming to rest and being buried. It is possible that the home ranges of the boneheads were well inland, and perhaps in mountainous areas where the streams had their sources.

Ankylosaur skeletons are often found lying on their backs in marine sediments. It seems likely that their dead bodies had been drifting down rivers for a long time – long enough for decay to begin. When the gases of putrefaction expanded in the body cavity the floating animal would have rolled on its back, the armour acting as a keel. Eventually it would have sunk to the bottom in this attitude. Again it is possible that the dead animals were carried downstream from mountainous areas where they lived. All this, however, is mere speculation. It is one of the reasons why we will only find the remains of a fraction of the dinosaurs that ever lived.

Below: Mountains are areas of erosion rather than of deposition, and so no sedimentary rocks would have accumulated there, and hence no fossils would have been preserved. Herds of boneheads and solitary ankylosaurs may have roamed the slopes, while pterosaurs would have soared overhead.

1 Mountain conifer.
2 Meadow of heath-like or moor-like vegetation.
3 *Quetzalcoatlus* – pterosaur.
4 *Stegoceras* – bonehead dinosaur.
5 *Edmontonia* – ankylosaur dinosaur.
6 Scree slope – rubble formed as erosive forces break down the mountain rocks and carry them away.

EXTINCTION

What happened 65 million years ago, at the end of the Cretaceous period has always been a mystery. After 155 million years as the most prominent animal life on Earth, the dinosaurs suddenly died out. With them, flying and swimming reptiles, a large number of fish and other animals disappeared.

There have been many theories regarding the reasons behind the extinction of the dinosaurs. Mostly the theorists have been divided into two camps – those who favour gradual extinction as an explanation and those who favour sudden catastrophe, although "sudden", in geological terms, can actually cover half a million years, looking instantaneous in the geological record.

The gradualists stress that for most of the late Cretaceous period, conditions had been very stable, with not much variation in climate for tens of millions of years. It may be that the specialized animal life of the time had become too specialized and too adapted to these stable conditions. A slight change in the environment, such as the raising of the temperature or the cooling of the atmosphere, may have put intolerable stresses on the dinosaurs so that they could not cope.

The corresponding change of vegetation that would have accompanied a climatic change would have resulted in a change in dinosaur food stocks and as a result the dinosaurs could have starved.

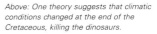

Above: One theory suggests that climatic conditions changed at the end of the Cretaceous, killing the dinosaurs.

Another theory suggests that the gradual movement of continents brought landmasses into close proximity to one another. As a result animals would have been able to migrate and carry with them, into new geographical regions, diseases to which the endemic population would have been vulnerable.

Below and left: Another theory backed by scientific understanding is that a meteor hit the Earth's surface, resulting in huge changes to the Earth's atmosphere and the extinction of the dinosaurs. The meteor is said to have caused the Chicuxlub Crater, at the edge of the Gulf of Mexico.

Above: A third theory suggests that volcanic activity caused the death of the dinosaurs.

Above: The eruptions that caused vast lava flows in Cretaceous India would have altered the climate considerably.

However, some view the demise of the dinosaurs as a sudden occurrence. They envisioned cataclysms brought about by violent volcanic eruptions, or earthquakes producing such extreme conditions that the dinosaurs could not have coped. However, then, in the 1970s, a discovery was made that changed this discussion for ever. It was found that the end of the Cretaceous and the beginning of the Tertiary (the K/T boundary in geological parlance) was marked by a bed that was rich in the element iridium. Iridium is rare at the surface of the Earth but quite common in meteorites. So a new theory was put forward that a massive meteorite had struck the Earth 65 million years ago. The immediate result would have been shock waves and fires. The explosion would have sent masses of dust and steam into the atmosphere, blanketing the Earth and causing temperatures to fall as the sunlight was blocked out. Plants would have died. Plant-eating dinosaurs would have starved. Then meat-eating dinosaurs, denied their plant-eating prey, would have perished. After months or years the skies would have cleared and plants would have started to grow. By that time all the dinosaurs would be dead.

A huge buried structure, on the coast of Yucatan in Mexico, was identified as a meteorite crater dating from the end of the Cretaceous period. Signs of shattered rocks and sea wave damage were identified around the Caribbean. Corroborative evidence was being amassed.

Other theories were put forward too. Iridium is also found beneath the Earth's crust, and could be brought out by volcanic activity. There was indeed a great deal of volcanic activity at that time – half of the continent of India is made up of lava flows that erupted just then. Volcanic activity as intense as that could have had exactly the same effect as a meteorite strike, blanketing the Earth with smoke and fumes. India and Yucatan were on exactly opposite sides of the globe 65 million years ago. A massive meteorite impact in one could have sent shock waves through the crust and instigated volcanic activity in the other. Or the meteorite could have split into two, one part falling in India, the other in Yucatan. In any case, the theories of those who believe in catastrophe are popular.

However, there were other subtleties. Statistical analysis of fossil finds suggested that the dinosaurs had been dying out for several million years before the end of the Cretaceous.

Perhaps they were already on their way out when an impact just finished them off. Whatever the cause, the reign of the dinosaurs came to an end. The land was swept clear, to be colonized by something else. The something else turned out to be the mammals – small, insignificant creatures while the dinosaurs were alive, but adaptable enough to take over once they had gone. As a result they evolved and spread and took over every niche that had been previously occupied by the dinosaurs.

Below: Volcanic eruptions occurred throughout the Age of Dinosaurs, for example, producing the Triassic ash beds seen here. However, none was so great as that which occurred at the end.

THE FOSSILIZATION PROCESS

After an animal dies and its body is subjected to the different destructive forces of taphonomy, what is left may be fossilized. This process involves the burial and preservation of the remains. The sediments in which the remains are buried become sedimentary rock, and the original remains are turned into mineral.

Taphonomy is the study of what happens to an organism after its recent death and before it becomes a fossil. Diagenesis is what happens thereafter to turn what remains of the body into something that will withstand the passage of time.

The fossils that we pick up from the rocks, or that we see in museums, have changed a great deal in substance and appearance since they were parts of living organisms.

Fossil formation

The fossil is formed at the same time as the sediment in which the organism lies is turned to sedimentary rock. This diagenesis usually involves two processes – compression and cementation. Compression is produced by the sheer weight of the overlying sediments, and acts to compact the grains of the sediment into a more coherent mass. In cementation, the ground water percolating through the rock deposits mineral, often calcite (the same as builders' cement), between the grains gluing them together as a solid mass. Further compaction and the input of heat from the Earth's interior may alter the mineral content of the rock. If this

happens it will take the rock from the realms of a sedimentary rock to a metamorphic rock, a new type in which usually all the fossils are destroyed.

In the rare occurrences in which an embedded organism becomes a fossil in a sedimentary rock, one of a number of different processes may be involved. An organism, or part of an organism, may remain unaltered. This is a very rare form of fossilization, but we sometimes see it in an insect preserved in a piece of amber. The procedure is simple. An insect settles on a glob of resin seeping from a tree and becomes trapped. It is engulfed and preserved before it has a chance to decay. Later the resin becomes buried and turned into amber through the normal processes that turn sediment to rock, and the insect is preserved within. Sometimes, though, this perfect preservation is illusory for the insects innards may have decayed through the action of its own bacteria. There are no dinosaur fossils that have been

preserved like this, and no dinosaur blood in preserved mosquito stomachs.

The hard parts of an organism may, however, remain unaltered. We find this in the teeth of sharks from the last few tens of millions of years. The teeth are so much tougher than the rest of the skeleton that they survive for a long period. The bones of Ice Age mammals trapped in the tar pools of Los Angeles are another example. Again, no dinosaurs are preserved like this. Sometimes only the original carbon of the organic substance remains. This is seen in the black shapes of leaves that are sometimes seen in plant fossils. Taken to an extreme, this process gives us coal.

Below: Over subsequent years mud and sand are deposited by flooding. The movement of the sediment means that the skeleton begins to break up and be dispersed. The skeleton is still white, fresh bone.

Below: In a dry season the following year, the vegetation has not survived, the river has dried up and mud cracks appear in the mudbanks, and the dinosaur's body is now a skeleton. The neck and tail are pulled back, as the tendons have dried and shrunk. Other scavenging dinosaurs take the last bits of nutrition from the skeleton.

Below: A dinosaur's body is washed up on a river sandbank. It has been dead a few days, and its stomach is bloated by decomposition. Below its body are the river sands.

Over long periods of time, the cellular structure of the original may be replaced by a totally different substance. Silica in ground-water passing through rocks may replace the original carbon, molecule by molecule, and give a fossil that shows the

Below: Eventually the skeleton is buried so deep below the surface that the sediments become rock and the skeleton becomes a fossil. The previously white skeleton turns black as the bone becomes mineralized. Above the dinosaur may be river sands and muds, beds of conglomerate (fossilized shingle), and marine limestone.

original microscopic structure, but made of silica. Petrified wood is a good example. It is also seen in some Australian plesiosaurs, in which the bone has been replaced by opal.

Groundwater percolating through the rock may also dissolve away all traces of the original organism. The result is a hole in the rock called a "mould", in exactly the same shape as the original. Permian reptiles from the desert sandstones of Elgin, in Scotland,

Below: The mountain is rising. The sedimentary rocks are uplifted and distorted. The skeleton is distorted too.

occur in this way, and pouring latex into the moulds produces casts of the original bones. This casting is sometimes done naturally, with dissolved groundwater minerals filling the moulds. Sea urchins in chalk are sometimes found replaced by flint. Then there are fossils which contain no part or impression of the original animal at all. They are called trace fossils and encompass footprints and trackways, coprolites and eggs.

All these processes mean nothing to us unless the fossil is returned to the surface. This only happens when the rocks containing the fossil are uplifted through earth movements, usually mountain building associated with the movement of the tectonic plates. Then erosion has to wear away all the rocks above so that the fossil is exposed. If this erosion is too vigorous, the exposed fossil will not last for long as it will be eroded away as well. All in all, even if a dinosaur does become fossilized, the odds are very much against our finding and excavating it. There are good reasons why dinosaur fossils are rare.

Below: Tens of millions of years later the rocks erode, revealing the fossilized skeleton.

EXCAVATION IN THE FIELD

Once the fossil dinosaur is discovered, the job of excavation can take on military-style logistics. Not only does the skeleton have to be extracted without damage, but the setting and the surrounding rocks must be analysed as well to give as full a picture as possible.

Usually a dinosaur skeleton is found by chance, by somebody out walking, or by a quarry worker turning over a rock. A *Stegosaurus* in Colorado was found recently when a palaeontologist working on another dig threw his hammer at random into a cliff. Another, in Montana, was found by a farmer digging a hole for a fence post. Once found, the skeleton is reported to a museum, a university or another institution that has the means to excavate it. Planning can take months or even years, and much of this involves finding the money to do the work – because practical palaeontology can be a very expensive business.

On site, the first thing is to find out how much of the skeleton there is. The overburden – that is the rock directly over the skeleton – has to be removed. This can be done with earth movers. Then, when the rock is down to a few centimetres of the bed that contains

Below: A palaeontologist marks out a grid over the bones, providing a reference to all finds in preparation for drawing a site map.

the skeleton, the last of the overburden is removed carefully by hand, usually with fine tools and brushes.

Once the skeleton has been exposed, the next phase is to catalogue what is there. A site map is drawn. This is a plan of the site showing where each

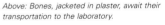

Above: Bones, jacketed in plaster, await their transportation to the laboratory.

bone lies, and the presence of anything else that may be of interest. Only when all this has been done can the excavation begin. Fossils that are newly exposed to the air may be very fragile. They may be unstable, and chemicals in them may react with the atmosphere and cause the fossil to decay quickly. Exposed bones are treated as quickly as possible with a chemical or varnish to seal them, and stop the air causing deterioration. The exact nature of the chemical or varnish has to be recorded for the technicians at the laboratory or the museum.

Fossil bones can be so fragile that they crumble away if they are lifted. To avoid this, they are encased in a jacket. It consists of a layer of moist paper, then a layer of plaster bandages – just like the plaster cast used in medicine for repairing a broken limb. (In the days of the "bone wars", this technique was pioneered by a fossil hunter, using rice, the staple food of

the expedition.) The exposed part of the bone is covered in this plaster jacket. Then the rock surrounding and underlying it is dug away, the bone turned over, and the rest of the jacket applied to the other side. How many jackets, and how big a part is jacketed at one time, will depend on the state of the skeleton. In an associated skeleton nearly every individual bone needs to be jacketed. If the skeleton is articulated and there is access to heavy lifting equipment, the whole skeleton can be jacketed in one go. The complete Colorado *Stegosaurus* mentioned above was airlifted from the site by helicopter, but it is rare for such expensive resources to be put at the disposal of a dinosaur excavation.

Once the skeleton has been removed the adjacent area is sifted for other specimens. Palaeontologists will look for the teeth of animals that may have scavenged the skeleton, the seeds of the plants that lived at the same time, and all sorts of other information that can be used to build up a picture of the animal's life.

Back in the laboratory, preparators, the technicians skilled in handling fossil material, prepare the fossil so

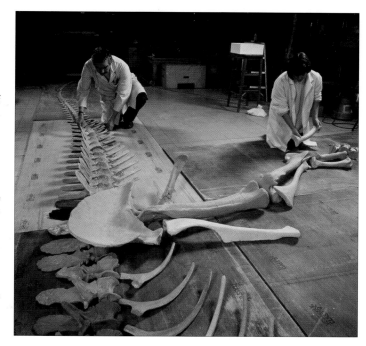

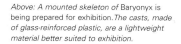

Below: The mounted skeleton of Triceratops is exhibited in life-like pose. It is made from a cast of the fossils.

Above: A mounted skeleton of Baryonyx is being prepared for exhibition. The casts, made of glass-reinforced plastic, are a lightweight material better suited to exhibition.

that it can be studied by the palaeontologist. Jackets are removed from the skeleton, and any stabilizing chemical is removed or replaced with one that is more appropriate for laboratory study.

In the past, the ultimate result was a mounted skeleton, with the fossil bones clamped to a welded steel frame erected in the public display area of a museum. Today a display skeleton is formed using casts of the skeleton rather than the actual fossils. Technicians make moulds from the fossils and cast reproductions of the bones in a lightweight material that is easier to handle, and which can be mounted more efficiently. This allows for the fossils to be stored under controlled conditions and kept available for study. Missing bones are replaced by casts from other skeletons, or sculpted by artists.

THE PALAEONTOLOGISTS

There are many hundreds of palaeontologists, the scientists who have discovered and studied the fossils, who should receive a mention in this list, all having pushed forward the frontiers of the science or still actively doing so. The following is merely a selection of those who have contributed to the science.

Camarasaurus *discovered by Cope.*

Florentino Ameghino
(1857–1911)
More famous for his work on the unique fossil mammals of South America, Ameghino pioneered the study and excavation of dinosaurs and other extinct vertebrates in Argentina in the late nineteenth century. Most of his discoveries are now housed in the La Plata Museum, Argentina.

Roy Chapman Andrews
(1884–1960)
Andrews led several expeditions from the American Museum of Natural History into the Gobi desert in the 1920s. The intention was to find the earliest remains of human ancestors. Instead he found a vast array of new dinosaurs in Cretaceous rocks. Perhaps the most important find was the first example of a dinosaur nest with its eggs. He pioneered the use of motorized transport to reach fossil sites.

Robert T. Bakker (1945–)
The concept of warm-bloodedness in dinosaurs is associated more with the charismatic Bakker than anyone else. Since the 1960s he has maintained that dinosaurs were lively active animals, using many lines of evidence, including comparative anatomy and fossil population studies. He has named several new dinosaur genera.

Rinchen Barsbold (1935–)
Barsbold is a Mongolian palaeontologist who, since the 1980s has worked with the Palaeontological Centre, Mongolian Academy

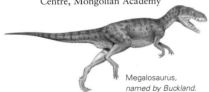

Megalosaurus,
named by Buckland.

of Sciences, Ulan Baatar, and has done a great deal to uncover and name the central Asian dinosaurs.

José F. Bonaparte (1928–)
The most famous contemporary Argentinean palaeontologist, Bonaparte has added hugely to the understanding of dinosaurs in South America. His work includes studies of the late Cretaceous armoured titanosaurs and an investigation of South American pterosaurs that led to a renaissance of the subject.

William Buckland (1784–1856)
The first professor of geology at Oxford University, Buckland was the first to publish a scientific description of a dinosaur – *Megalosaurus* in 1824. Dinosaur bones had been noted before but never studied seriously. He recognized that the fossil jawbone and teeth that had been brought to him had come from a giant reptile of some kind. This was before the concept of a dinosaur had been established.

Edwin Colbert (1905–2001)
This curator of the American Museum of Natural History and later the Museum of Northern Arizona is famous for his discovery, in the 1960s, of the fossil of a mammal-like reptile, *Lystrosaurus*, in Antarctica. This discovery helped to confirm the

understanding of the movements of the continents caused by plate tectonics. He famously studied the bonebed of *Coelophysis* found in New Mexico.

Edward Drinker Cope (1840–97)
Cope was one of the two figures in the nineteenth century "bone wars" – a long-lasting rivalry with Othniel Charles Marsh to find and describe as many dinosaur specimens from the newly-opened Midwest of North America as possible. He worked from Philadelphia, USA.

Earl Douglass (1862–1931)
Douglass worked for the Carnegie Museum, in Pittsburgh, USA, (set up by the Scottish industrialist and philanthropist Andrew Carnegie to accommodate the dinosaur remains discovered in the Midwest at the end of the nineteenth century) and opened up the dinosaur beds of Utah. The site that he studied eventually became Dinosaur National Monument.

Eberhard Fraas (1862–1915)
Most dinosaur discoveries had been made in Europe and in North America. Fraas, a German authority on dinosaurs sent an expedition, led by Werner Janensch, to German East Africa (now Tanzania) in 1907 and uncovered the Jurassic dinosaur deposits at Tendaguru, including the

huge *Brachiosaurus* (now renamed *Giraffatitan*) that for years stood in the Humboldt Museum in Berlin as the biggest mounted skeleton in the world.

John R. (Jack) Horner (1946–)

As state geologist for Montana, USA, Horner was the principal investigator of the dinosaur nesting sites of Egg Mountain and Egg Island, and named *Maiasaura*. The result of his work has been a new understanding of dinosaur family and social life.

Joseph Leidy (1823–91)

The first dinosaur to be studied in North America, was found in New Jersey, and named by Leidy. *Hadrosaurus* is now regarded as a *nomen dubium* as there was no skull to identify it. Leidy was based in Philadelphia and went on to name more dinosaurs. He is best known for his work on fossil Tertiary mammals.

Gideon Mantell (1790–1856)

As a country doctor in south-east England, Mantell collected fossils in his spare time. He and his wife Mary Ann found the bones of *Iguanodon* and named it in 1825. As time went on he spent less time as a doctor and devoted his time to amassing fossil collections, which he established in Brighton.

Othniel Charles Marsh (1831–99)

Marsh was Cope's opponent in the "bone wars", working from Yale University. They sent rival teams to sites to try to outdo one another. Before this time there had been only six genera of dinosaurs described. By the time the frenzy of the bone wars was over there were more than 130.

Herman von Meyer (1801–1896)

The first German palaeontologist, von Meyer described and named the first bird *Archaeopteryx*, as well as some of the

Maiasaura was named by John R Horner.

pterosaurs from the same area of southern Germany. He also discovered *Plateosaurus*, and pioneered the study of dinosaurs as well as invertebrate fossils in Germany and northern Europe.

John H. Ostrom (1928–2005)

With his discovery and description of *Deinonychus* in 1969, American palaeontologist Ostrom established the evolutionary connection between birds and dinosaurs. This also gave rise to the theory that the dinosaurs, at least the theropods, were warm-blooded like birds. The theory was eventually vindicated by the discoveries of the feathered dinosaurs and dinosaur-like birds in China.

Richard Owen (1804–92)

It was Sir Richard Owen, a British anatomist working at the British Museum (Natural History) – now the Natural History Museum – who is credited with creating the concept of dinosaurs by announcing at a meeting of the British Association for the Advancement of Science, in 1841, a new group of animals named the Dinosauria – based on *Megalosaurus*, *Iguanodon* and the ankylosaur *Hylaeosaurus*, which was also found by Mantell.

Ernst Stromer von Reichenbach (1870–1952)

This German palaeontologist from Munich was the first to excavate the dinosaur sites of Egypt. He did so for 30 years discovering such dinosaurs as *Aegyptosaurus*, *Carcharodontosaurus* and *Spinosaurus*. His specimens were lost during World War II when the museum in Munich was bombed. His excavation sites were forgotten too, until rediscovered by a team from Washington University in 2000.

Harry Govier Seeley (1839–1909)

A British palaeontologist, his main contribution to palaeontology was to divide the dinosaurs into the orders Saurischia and Ornithischia based on their hip structures. He named a number of dinosaurs, but most were so fragmentary that they have been renamed or declared *nomen dubia*. He also published an early pioneering account of pterosaurs.

Paul Sereno (1957–)

Perhaps the most prolific dinosaur hunter today, Sereno, based in Chicago, USA, has discovered new dinosaurs in North Africa and in central Asia. In 1986 he advanced the understanding of dinosaurs by reclassifying the ornithischians in a system that is still used today.

Charles H. Sternberg (1850–1943) and his sons, Charles M., George and Levi

Charles senior contributed to the bone wars by excavating for Cope. Between 1912 and 1917, the family pioneered the Canadian boom in dinosaur discoveries. The skeletons that they extracted are displayed in museums all around the world.

Dong Zhiming (1937–)

The most famous modern Chinese palaeontologist, Dong opened up the vast dinosaur beds of Sichuan and in north-west China. Working with the Institute of Vertebrate Palaeontology and Palaeoanthropology in Beijing, China, he named about 20 new dinosaur genera in as many years. Among his wealth of discoveries he established the homalocephalid family of boneheaded dinosaurs.

Coelphysis, named by Cope but extensively studied by Colbert.

Anserimimus discovered by Barsbold.

1 *Iguanodon.*
2 *Istiodactylus.*
3 *Hypsolophodon.*
4 *Neovenator.*
5 *Bernissartia.*
6 *Eotyrannus.*

THE WORLD OF DINOSAURS

For 155 million years dinosaurs evolved and became one of the most successful groups of animals the world has ever known. They changed as the world around them changed.

At the beginning of the Age of Dinosaurs, during the late Triassic period, all the land areas of the Earth were fused together as a single landmass that we term the supercontinent of Pangaea. The centre of this supercontinent was searing desert, and the only habitable zones were around the edges. It is against this background that the dinosaurs developed. Since there was only one landmass, the same dinosaurs lived in all of the habitable areas – the prosauropods whose skeletons we find in Triassic Germany are largely the same as those we find in South Africa; the small theropods that existed in Arizona are almost identical to those in Zimbabwe.

Then, with the beginning of the Jurassic, the supercontinent began to split, opening up rift valleys that let the ocean into the heart of the landmass. Shallow seas spread over the margins giving broad continental shelves. The climate became moist. New types of dinosaur evolved in these conditions.

Finally, in the Cretaceous period, Pangaea had split into individual continents, most of which we would recognize today. As a result, different dinosaurs evolved in different areas, with the giant theropods of North America being of quite different families from the giant theropods of South America, and the long-necked sauropods of the southern continents being replaced by the duckbills as the main plant-eaters in the north. The biodiversity of the world was much greater at the end of the Age of Dinosaurs than it was at the beginning.

Against this changing background we can apply all the lines of evidence we have seen in the introductory chapters and study the dinosaurs that lived in different places at different times. The directory that follows depicts most of the dinosaurs known. However, new dinosaurs are being found all the time, and a catalogue such as this will require constant updating. There are many, many more dinosaurs that are still to be found.

Left: Early Cretaceous Europe had herds of large and small plant-eaters, menaced by large and small meat-eaters, along with their relatives, the crocodiles and the pterosaurs.

THE
TRIASSIC
PERIOD

The Triassic period is regarded as the beginning of the Age of

Dinosaurs, although the dinosaurs did not appear until its end.

At that time the landmass of the world was joined together in a

supercontinent known as Pangaea. Its centre was so far from the sea

that desert conditions prevailed, and the only habitable areas were

around the coastal rim. Here the dinosaurs evolved, alongside the

mammals that were later destined to replace them. The dinosaurs

evolved along with the vegetation that was to support them (from seed

ferns in the early Triassic to conifers in the late Triassic), and later they

evolved to prey upon other creatures which ate the plantlife.

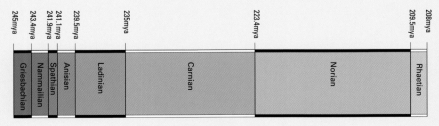

Top: The Earth as it would have looked in Triassic times.

*Above: The timeline shows the different chronological stages of the Triassic
period of Earth's history. (mya = million years ago)*

1 *Eudimorphodon.*
2 *Plateosaurus.*
3 *Placerias.*
4 *Liliensternus.*
5 *Proganochelys.*

PLACODONTS

The placodonts were a strange group of swimming reptiles that appeared in the middle Triassic and became extinct at the Triassic/Jurassic boundary. They were heavy-bodied swimmers that mostly fed on shellfish: their teeth were specialized for picking shells from rocks and crushing them between their jaws. In the Triassic seas they would have been the equivalent of modern walruses.

Paraplacodus

The placodonts consisted of several families, including the unarmoured placodontoids, which must have resembled gigantic newts, and the armoured cyamodontids, which evolved into turtle shapes. *Paraplacodus* was typical of the former. The species lived in shallow seas and lagoons at the edge of the Tethys Ocean and fed from the banks of shellfish that grew there.

Features: The jaws of *Paraplacodus* are uniquely adapted to picking up shellfish, with three pairs of protruding teeth in the top and two in the bottom. The teeth project from the front of the mouth. They have a series of rounded crushing teeth in the upper and lower jaws. The thick ribs produce a distinctly box-like body with an almost square cross-section, a strong set of belly ribs forming the flat floor of the body – a heavy design that kept it close to the seabed.

Distribution: Northern Italy.
Classification: Sauropterygia, Placodontia, Placodontoidea.
Meaning of name: Almost *Placodus* (see below).
Named by: Peyer, 1931.
Time: Anisian to Ladinian stages of the middle Triassic.
Size: 1.5m (6ft).
Lifestyle: Shellfish-eater.
Species: *P. broilii*.

Left: A specimen of a similar placodont called Saurosphargis *was destroyed during World War II; this may have been a close relative.*

Placodus

Aquatic animals adopt several methods of dealing with buoyancy. The placodonts used what is known as pachystasis: the development of particularly thick and heavy bones to keep the body submerged. Voluminous lungs allowed them to adjust their buoyancy. This technique is often used by animals that feed while walking along the sea bed, such as modern dugongs and sea-otters.

Features: Like *Paraplacodus*, *Placodus* has protruding teeth at the front. However they are shorter, thicker and more spoon-shaped. The crushing teeth are not confined to the edges of the mouth but form a broad pavement across the palate, and the skull is particularly strong to withstand the stresses of crushing seashells. A row of bony scutes forms a jagged ridge along the back. A gap at the top of the skull may have held a light-sensitive organ.

Distribution: Germany.
Classification: Sauropterygia, Placodontia, Placodontoidea.
Meaning of name: Flat toothed.
Named by: Agassiz, 1833.
Time: Anisian to Ladinian stages of the Triassic.
Size: 3m (10ft).
Lifestyle: Shellfish-eater.
Species: *P. gigas*.

Left: When the teeth were first found in 1830 the ichthyologist Louis Agassiz misidentified them as fish teeth, and gave the name Placodus. *Richard Owen, in 1858, recognized them as reptile teeth.*

Cyamodus

Although some of the placodontids had armour plates embedded in the skin, it was the cyamodontids that took this to an extreme. The bodies became broad and flat, almost turtle-like, and they probably spent much of their time close to the shallow sea bed, pulling themselves over the sand and searching for shellfish, very much as modern rays do.

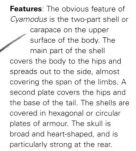

Features: The obvious feature of *Cyamodus* is the two-part shell or carapace on the upper surface of the body. The main part of the shell covers the body to the hips and spreads out to the side, almost covering the span of the limbs. A second plate covers the hips and the base of the tail. The shells are covered in hexagonal or circular plates of armour. The skull is broad and heart-shaped, and is particularly strong at the rear.

Left: Juvenile specimens of Cyamodus *appear to have an extra tooth on the roof of the mouth, compared with adults. It seems possible that they reduced the number of teeth as they grew to maturity.*

Distribution: Germany.
Classification: Sauropterygia, Placodontia, Cyamodontidae.
Named by: Meyer, 1863.
Time: Anisian to Ladinian stages of the Triassic.
Size: 1.3m (4ft).
Lifestyle: Shellfish-eater.
Species: *C. rostratus, C. hildegardis, C. kuhnschneyderi.*

OTHER PLACODONTS

We do not know the ancestors or the origin of the placodonts. The latest understanding is that they may be diapsids (a group of reptiles with two pairs of holes in the skull behind the eyes) and may be related to the ancestors of the plesiosaurs, although their limbs were not so highly adapted for swimming.
Helveticosaurus zollingeri from the Anisian stage of the Triassic in Switzerland was once thought to have been a primitive, possibly ancestral, placodont. It resembles *Placodus* in appearance. However it now seems to be totally unrelated.
Psephoderma alpinum from several localities in Europe resembled *Cyamodus* with its two-part carapace. Its shell has three distinct ridges running longitudinally along it.
Placochelys placodonta from the Landinian stage of the Triassic in Hungary has jaws that were pointed at the front, and no front teeth. It is possible that it possessed a horny beak to peck the shellfish from the reef.
Protenodontosaurus italicus appears to be intermediate between *Cyamodus* and the more advanced turtle-like placochelyids.

Right: The skulls of Paraplacodus (top) and Placodus (bottom). Though the two creatures have a similar appearance they are quite different.

Henodus

We can imagine *Henodus* like some kind of reptilian ray, paddling its broad flat body across the bed of a shallow lagoon, foraging in the rippled sand with its broad mouth. Its plate-like body would have made it better adapted to searching along flat sea beds than to the shellfish-encrusted reefs frequented by its relatives. Its weak limbs suggest that it did not spend much time on land.

Features: This is the placodont that most resembles the turtles. There is a carapace over the whole of the body that stretches out well beyond the span of the limbs, and as in the other cyamodonts this is matched by a plastron, a lower shell, which covers the undersurface. Both carapace and plastron are made up of a geometric array of individual plates. The head is squared off at the front and is shortened in front of the eyes.

Distribution: Southern Germany.
Classification: Sauropterygia, Placodontia, Cyamodontidae.
Meaning of name: Armoured.
Named by: Huene, 1936.
Time: Carnian stage of the late Triassic.
Size: 1m (3ft).
Lifestyle: Seafloor-forager.
Species: *H. chelyops.*

Left: Henodus *is the only placodont so far found in non-marine deposits. It seems to have lived in brackish or freshwater lagoons. Its broad mouth was adapted for lifting shellfish from soft sandy beds.*

ICHTHYOSAURS

Perhaps the best known of fossil sea reptiles are the fish-lizards, the ichthyosaurs. The best are known from fossils in Jurassic rocks, but their history reaches back into Triassic times. There are several distinct early shapes of ichthyosaurs, before the classic dolphin-like appearance was standardized. Some are eel-like and others are enormous and whale-shaped.

Cymbospondylus

Although *Cymbospondylus* is formally regarded as part of the shastasaurid group of primitive ichthyosaurs, recent studies suggest that it may be more primitive than originally thought – maybe too primitive to be regarded as an ichthyosaur. Indeed, in appearance it does not seem to have the physical features, such as the dorsal fin and fish-like tail, which are so distinctive of the later members of the group. *Cymbospondylus* is the state fossil of Nevada.

Features: *Cymbospondylus* is more eel-shaped than the other ichthyosaurs, with a narrow body and a long flexible tail that takes up about half the length of the animal. The legs have already evolved into paddles but these were probably used more for stabilization, while the swimming action was produced by undulations of the body. The head is quite small in relation to the body but is quite typical of the ichthyosaurs, with the long snout and the small, sharp fish-catching teeth already present.

Distribution: Nevada, USA, and Germany.
Classification: Ichthyosauria, Shastasauridae.
Named by: Leidy, 1868.
Time: Middle Triassic.
Size: 6m (18ft).
Lifestyle: Fish-hunter.
Species: C. natans, C germanicus, C. nevadanus, C. parvus, C. piscosus, C. grandis, C. petrinus, C. buchseri.

Mixosaurus

In appearance *Mixosaurus* seems intermediary between the eel-like forms typified by *Cymbospondylus*, and the more familiar dolphin-shaped forms such as the later *Ichthyosaurus*, hence the name. The long tail with its low fin suggests that it was a slow swimmer, moving forward by undulation of the tail.

Features: The body and tail are long, with the tail having a low fin that is not as well-developed as the shark-like fin of the later types, and there is a stabilizing dorsal fin on the back. The paddle-like limbs are still made up of five toes each, unlike the multi-toed structure found in later ichthyosaurs. However, each toe has more individual bones than is usual (polyphalangy) and the front limbs are longer than the rear.

Distribution: China, Timor, Indonesia, Italy, Alaska, Canada, Nevada, USA, and Spitsbergen, Svalbard.
Classification: Ichthyosauria, Mixosauridae.
Meaning of name: Mixed lizard.
Named by: Baur, 1887.
Time: Middle Triassic.
Size: 1m (3ft).
Lifestyle: Fish-hunter.
Species: M. atavus, M. kuhnschnyderi, M. cornalianus.

Left: Mixosaurus *lived at the same time as* Cymbospondylus.

Shonisaurus

In the 1920s miners found a deposit of huge bones in Nevada. When they were excavated 30 years later they were found to be the remains of 37 individual enormous ichthyosaurs, and were named *Shonisaurus*. They may have been a shoal that was beached, or cut off from the sea. More recently, the scientific view is that they died and sank to the sea bed.

Features: This ichthyosaur has a whale-like shape and long narrow paddles. It has teeth only at the front of the jaws. *S. popularis* held the record as the biggest of the ichthyosaurs, but the skeleton of an even bigger species, *S. sikanniensis*, was found in the 1990s in British Columbia. With an estimated length of 21m (70ft), this was so big it could only have been seen in its entirety from the air. Studies of this new species suggest that *Shonisaurus* may not have been as deep-bodied as originally thought.

Distribution: Nevada, USA, to British Columbia, Canada.
Classification: Ichthyosauria, Shastasauridae.
Meaning of name: Lizard from the Shoshone Mountains.
Named by: Camp, 1976.
Time: Norian stage of the late Triassic.
Size: 15m (50ft).
Lifestyle: Ocean hunter.
Species: *S. popularis*, *S. sikanniensis*.

Right: Of the skulls found, only the smaller ones had teeth. Shonisaurus *may have become toothless as it grew.*

ICTHYOSAUR CLASSIFICATION

The exact classification of the ichthyosaurs has always been a mystery. For most of the history of their study it was thought that they belonged to an evolutionary line that was totally distinct from that of any other reptile. Then there was the puzzle of the origin of their appearance and lifestyle. How did such fish-like animals evolve from land-living creatures? Indeed, were their ancestors ever land-living? For a while there was a theory that sometime in the late Palaeozoic they evolved directly from swimming amphibians without going through a land-living phase at all.

This intriguing suggestion was laid to rest in 1998 with the re-examination of the original *Utatsusaurus* skeleton by Ryosuke Motani and Nachio Minoura from Berkeley and Hokkaido Universities. By using computer imagery, they were able to reverse the distortion of the original specimen and study it carefully. They found that *Utatsusaurus*, despite its fishy shape, was actually quite closely related to the lizard-like diapsid reptiles such as *Petrolacosaurus*, making it distantly related to the ancestors of the lizards and snakes, and even the dinosaurs. They were more closely related to these than to the other main line of sea-going reptiles, the turtles.

Utatsusaurus

This, the earliest-known ichthyosaur, was found in 1982 in Japan, but it was not until a study of the skeleton was published in 1998 that its significance as the most primitive of the group was noted. Like most primitive ichthyosaurs it swam by undulations of its body rather than by sweeps of the tail, and it probably inhabited the shallow waters of the continental shelf.

Features: The skull is rather broad and tapers gradually towards the snout, unlike the narrow jaws of most other ichthyosaurs. The teeth are small for the size of skull and arranged in a groove – a primitive feature. Its paddles are small and unusually, the hindlimb is bigger than the forelimb. The paddle is made up of four fingers, unlike the usual five found in other ichthyosaurs. Its undulating swimming action is shown by the large number of very thin narrow vertebrae, making for a very flexible body.

Distribution: Japan and Canada.
Classification: Ichthyosauria.
Meaning of name: Lizard from Utatsugyoryu in Japan.
Named by: Shikama, Kamei and Murata, 1998.
Time: Spathian stage of the early Triassic.
Size: 3m (10ft).
Lifestyle: Swimming hunter.
Species: *U. hataii*.

Left: The side-to-side swimming action of Utatsusaurus *would have been fairly inefficient and would have restricted it to life in shallow seas on continental shelves.*

NOTHOSAURS

If the ichthyosaurs are the whales and dolphins of the Mesozoic world, then the plesiosaurs can be regarded as the seals and sea lions. Before the plesiosaurs were fully established, a more primitive offshoot from their line, the nothosaurs, were the most abundant fish-catchers of the time. The nothosaurs had the same long necks as their plesiosaur cousins but were not so well adapted to water.

Ceresiosaurus

With a long body and powerful tail *Ceresiosaurus* has features usually indicative of an animal that swims by undulations of its body. However the bone structure, especially in the thick tail and strong hips, suggests that it was a pursuit diver and hunted underwater like a penguin, manoeuvring with its strong paddles. Analysis of its stomach contents has shown it to be a hunter of marine reptiles.

Features: *Ceresiosaurus* has much longer toes than other nothosaurs. The additional length was gained by an increase in the number of bones in each toe (polyphalangy). The toes are fused into a swimming paddle. The tail bones are thick and evolved to support powerful muscles. The skull is short, much shorter than that of any other nothosaur, and looks very much like that of a plesiosaur with the nostrils placed well forward. The front paddles are bigger than the hind, suggesting that they played a bigger part in the animal's locomotion.

Left: The preserved stomach contents of Ceresiosaurus contain the remains of pachypleurosaurs. Ceresiosaurus must have been a fast hunter to catch such agile prey.

Distribution: Europe.
Classification: Nothosauridae.
Meaning of name: Lizard of Ceres.
Named by: Peyer, 1931.
Time: Anisian stage of the middle Triassic.
Size: 4m (13ft).
Lifestyle: Swimming hunter.
Species: *C. calcagnii*, *C. russelli*.

Nothosaurus

The main fish-eaters on the northern shores of the Tethys Ocean in Triassic times were the nothosaurs, and *Nothosaurus* itself was typical. It swam in the surf and shallow inshore waters but rested and bred on the beaches and rock caves of the shoreline. It may have laid eggs in the sand as modern turtles do, or it may have given birth at sea.

Features: The skull is long and flat, and the long jaws are equipped with sharp interlocking teeth, some in the form of paired fangs, making a formidable fish trap. Each foot is webbed and has five long toes, and can be used for walking on land and for swimming. The body, neck and tail are long and flexible. A muscular tail helped in swimming.

Distribution: Europe, North Africa, Russia and China.
Classification: Nothosauridae.
Meaning of name: False lizard.
Named by: Munster, 1834.
Time: Anisian and Ladinian stages of the middle Triassic.
Size: 3m (10ft), although a newly found species, *N. giganteus*, reached 6m (20ft).
Lifestyle: Swimming hunter.
Species: *N. mirabilis*, *N. giganteus*, *N. procerus*.

Pachypleurosaurus

The pachypleurosaurs were not true nothosaurs but were closely related. They were quite small and probably lived close to shore or in lagoons, rather like the modern marine iguana. They may have originated in China and migrated to Europe along the northern shores of the Tethys Ocean. The small head suggests that they hunted small fish or shellfish.

Right: The pachypleurosaurs appear to be the link between the placodonts and the group comprising the nothosaurs and plesiosaurs. However, the group has such extreme skeletal specializations that it is difficult to be sure.

Features: The tail is deep and obviously used as a swimming organ. The hips and shoulders, although adapted for swimming, are still strong enough to support the animal on land. The head is very small and the structure of the ear suggests that it was sensitive to sounds above the sea surface rather than underwater. Some of the bones are thick, an adaptation associated with buoyancy control in some aquatic animals.

Distribution: Italy, Romania, Switzerland.
Classification: Pachypleurosauridae.
Meaning of name: Thick rib lizard.
Named by: Cornalia, 1854.
Time: Anisian to Ladinian stages of the middle Triassic.
Size: 1m (3ft).
Lifestyle: Swimming shellfish-eater.
Species: *P. edwardsi.*

SEA-LIVING CREATURES

It is a strange fact that as soon as vertebrates evolved to live on land, away from the water, casting off the need to return to the water to breed as amphibians do, there was an evolutionary movement that took a certain number back to the sea. Adaptations for living in the sea – such as streamlined body shape, limbs that worked as paddles, a body density that allowed floating or sinking – which had all been lost began to evolve once more.

In some groups, such as that of the ichthyosaurs, this re-adaptation became almost total. In others, such as the nothosaurs, it was partial. It is as if the nothosaurs adopted a kind of half-way stage so that they could exploit both the marine environment and that on the land. Such adaptations included legs that were webbed for swimming but still retained the bone structure and musculature for moving about on land, and the specialist teeth that would catch slippery prey like fish.

In common with all other secondarily marine vertebrates, even those as highly adapted as the ichthyosaurs, they still needed to breathe air. Gills that would have allowed underwater breathing never re-evolved once they had been lost by their ancestors.

Below: Nothosaurus breathing out of the water.

Lariosaurus

It appears that at least this genus of nothosaur was viviparous (able to bear live young). Several skeletons have been found associated with embryos indicating that they carried their young to maturity in their bodies. In one specimen two juvenile placodonts of the genus *Cyamodus* have been found in the stomach area, a clue to the diet of this nothosaur. Much of our modern knowledge of nothosaurs comes from the work of Dr Olivier Rieppel of the Field Museum in Chicago, the present day specialist in these uniquely Triassic marine reptiles.

Features: The primitive features of this small nothosaur include the short neck and toes. The back legs are five-toed with claws, and slightly webbed. The front legs are adapted into paddles. Both pairs of legs are quite short and do not give the impression of powerful swimming structures. The front legs are stronger than the hind, suggesting that they were the main swimming organs, unlike in the pachypleurosaurs. Fangs at the front of the broad head interlock as the jaws are closed, forming a vicious trap for catching fish and other aquatic animals.

Distribution: Spain, France, Italy, Germany, Switzerland and China.
Classification: Nothosauridae.
Meaning of name: Lizard from Lake Lario.
Named by: Curioni, 1847.
Time: Anisian to Landinian stages of the middle Triassic.
Size: 60cm (2ft).
Lifestyle: Fish- and crustacean-eater.
Species: *L. valceresii, L. balsamii, L. curioni, L. xingyiensis.*

Left: The front legs are stronger than the hind suggesting that they were the main swimming organs, unlike in the pachypleurosaurs.

PTEROSAURS

As soon as reptiles evolved there was a certain number that could exploit the possibilities of a life of flight. During the Permian and Triassic periods there were several reptiles that could glide on flaps of skin. However, it was only with the evolution of the pterosaurs in the late Triassic that true flight developed among the reptiles.

Sharovipteryx

Although not a pterosaur, *Sharovipteryx* shows what might be regarded as a primitive equivalent of the pterosaur flying membrane. Its long hind legs suggest that it jumped like a grasshopper, and once airborne the gliding

membrane would then catch the air and allow the animal to glide. The membrane was possibly also used as a display device.

Below: Sharov originally named this pterosaur Podopteryx in 1971, but that name was found to have been taken, or pre-occupied, by a beetle.

Features: The legs of *Sharovipteryx* are extremely long and support a panel of skin – a patagium. This stretches from the hind legs to the tail and from the hind legs to the little front limbs, undoubtedly forming a gliding surface. The leg bones are too lightly built for powered flight. The forelimbs are tiny and possibly only used for adjusting the trim of the patagium while in flight.

Distribution: Madygen, Kyrgyzstan.
Classification: Archosauromorpha, Ornithodira.
Meaning of name: Sharov's wing.
Named by: Cowen, 1981.
Time: Triassic.
Size: 30cm (1ft) long.
Lifestyle: Probably a gliding insectivore.
Species: *S. mirabilis*.

Eudimorphodon

Distribution: Northern Italy.
Classification: Archosauromorpha, Ornithodira, Pterosauria, Rhamphorhynchoidea.
Meaning of name: The true two-formed tooth.
Named by: Sambelli, 1973.
Time: Norian stage of the late Triassic.
Size: 1m (3ft) wingspan.
Lifestyle: Fish-eater.
Species: *E. ranzii*.

This is one of the earliest pterosaurs known, and it was a fully formed flying animal. It is known from several articulated skeletons, including those of juveniles. Its claws show that it lived on steep surfaces such as cliffs or tree trunks, from which it could launch its fish-hunting flights over water. Scales of fish have been found in the stomach regions.

Features: The teeth mark this animal out as a fish-eater, since they are mostly small and densely packed in the jaw, with more than 100 in the whole mouth. Some have several points. Some are larger than others, forming big fangs near the front of the mouth. This is an ideal arrangement for dealing with slippery prey. No later pterosaur had as elaborate an arrangement of teeth as this. The hind legs are quite sturdy for the size of animal and the tail is stiff and straight.

Peteinosaurus

This was a contemporary of *Eudimorphodon*, a little smaller and probably an insect-eater rather than a fish-eater. It shows that the newly-evolved pterosaurs had already diversified into a variety of lifestyles and pursued several different types of food in the variety of habitats that existed around the shores of the Tethys Ocean. It may have been the ancestor of the better-known early Jurassic *Dimorphodon*.

Features: *Peteinosaurus* is a little more primitive in construction than *Eudimorphodon*; the teeth are mostly the same size, apart from two pairs of small fangs in the front lower jaw, and all have single points. The wings are about twice as long as the hind legs (this is short for a primitive pterosaur). Most rhamphorhynchoids (a main group of the pterosaurs) had wings that were at least three times as long as the hind legs.

Distribution: Northern Italy.
Classification: Archosauromorpha, Ornithodira, Pterosauria, Rhamphorhynchoidea.
Meaning of name: Winged lizard.
Named by: Wild, 1978.
Time: Norian stage of the late Triassic.
Size: 60cm (2ft) wingspan.
Lifestyle: Flying insectivore.
Species: *P. zambelli*.

Above: Peteinosaurus *probably weighed as little as 100g (3½oz).*

EVOLUTION OF THE PTEROSAURS

Above: Heleosaurus, *a primitive archosaur and possible pterosaur ancestor.*

spring fully evolved into the fossil record in the rocks formed in late Triassic times. There is no record of any undisputed ancestral forms. The evolution of their powers of flight is unknown. This is not really surprising. The odds are against any animal becoming fossilized, and we are lucky to have any examples, especially of lightweight flying animals whose delicate bones would rarely fossilize anyway.

It is possible that their ancestors lie among the primitive archosaurs, probably lizard-like animals that may have run on hind legs. Some of these may have taken to trees in pursuit of tree-living insects, and developed long limbs and lightweight bones as an adaptation to tree-living life. The next stage may have been the development of gliding patagia between the limbs, to allow travel from tree to tree. How such a form developed into the warm-blooded active flying animal, with muscular manoeuvrable wings, is still to be discovered.

Preondactylus

This pterosaur is known from two partial skeletons, and some bits of forelimbs. One interesting fossil is jumbled up and packed together. The inference is that this was a fish pellet. A dead *Preondactylus* was swallowed by a massive fish and after it was digested the hard parts were regurgitated in a lump that was subsequently fossilized.

Features: This is the earliest pterosaur known. Only part of the skull has been found, but what there is shows the animal to have been very similar to the Jurassic *Dorygnathus*. The tail is well preserved and is similar to that of all other rhamphorhynchoids, in that it consists of long vertebrae lashed together into a stiff column by tendons that have become solidified into bone. This forms a balancing and steering organ.

Distribution: Northern Italy.
Classification: Archosauromorpha, Ornithodira, Pterosauria, Rhamphorhynchoidea.
Meaning of name: Finger from the Preone Valley.
Named by: Wild, 1983.
Time: Norian stage of the late Triassic.
Size: 1.5m (6ft) wingspan.
Lifestyle: Fish-eater.
Species: *P. buffarinii*.

Left: Dorygnathus *for comparison.*

Above left: Computer imagery shows the partial skull of Preondactylus *to have straight jaws with different-sized teeth, big fangs at the front and also about halfway along, and a sclerotic ring (a strengthening ring of fine bones), in the eye socket.*

EARLY MEAT-EATERS

What were the first dinosaurs? The truth is that we are not quite sure. We can only look at the earliest remains found, and speculate that these must be the earliest and most primitive of the dinosaur group. As far as we know the dinosaur dynasty began with the small meat-eaters in South America, but they may have existed in other places as well.

Eoraptor

The Valley of the Moon in north-western Argentina consists of dusty outcrops of sandstone and mudstone laid down in lushly forested river valleys in late Triassic times. These river banks were prowled by the first dinosaurs, including fox-sized *Eoraptor*, and various other reptiles that it hunted for food.

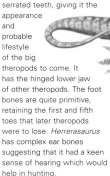

Right: Even the most perfect dinosaur skeleton can say little about the skin or the coloration. As in most restorations, skin colour and pattern are conjectural.

Features: *Eoraptor* is known from a complete skeleton that is lacking only the tail, and in shape and size *Eoraptor* conforms to every idea of the primitive dinosaur. Its lower jaw lacks the bone joint behind the tooth row that is seen in every other meat-eater, and there is more than one kind of tooth, something unusual in a meat-eating dinosaur. However, all other skeletal features such as the shape of the hips, the upright stance and a reduction of the number of fingers on the hand show that this dinosaur is definitely an early theropod.

Distribution: North-western Argentina.
Classification: Saurischia, Theropoda.
Meaning of name: Dawn plunderer.
Named by: Sereno, 1993.
Time: Carnian stage of the late Triassic.
Size: 1m (3ft).
Lifestyle: Hunter.
Species: *E. lunensis*.

Herrerasaurus

Distribution: North-western Argentina.
Classification: Saurischia, Theropoda, Herrerasauridae.
Meaning of name: From Victorio Herrera, its discoverer.
Named by: Sereno, 1988.
Time: Carnian stage of the late Triassic.
Size: 5m (16½ft).
Lifestyle: Hunter.
Species: *H. ischigualestensis*.

A contemporary of *Eoraptor* on the late Triassic riverbanks of Argentina, *Herrerasaurus* was a much bigger and more advanced theropod. Because of the difference in size, it must have hunted different prey from its smaller relative. Its skeleton was found in 1959, although it was several decades before it was scientifically studied. The complete skull was not found until 1988.

Features: A big animal with heavy jaws and 5cm- (2in-) long serrated teeth, giving it the appearance and probable lifestyle of the big theropods to come. It has the hinged lower jaw of other theropods. The foot bones are quite primitive, retaining the first and fifth toes that later theropods were to lose. *Herrerasaurus* has complex ear bones suggesting that it had a keen sense of hearing which would help in hunting.

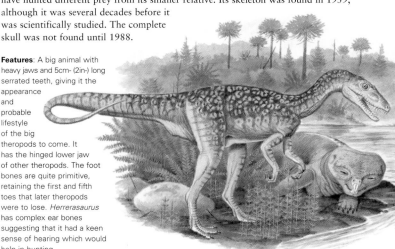

EARLY HISTORY OF MEAT-EATING DINOSAURS

It is possible that the dinosaurs first evolved in South America. Most of the early dinosaur fossils are found in the Santa Maria Formation in Brazil and the Ischigualasto Formation in Argentina, both dating from the Carnian stage of the Late Triassic. Certainly the most complete early dinosaurs have been found here, although scattered remains have cropped up in other parts of the world. South America was also home to the lagosuchids, a group of archosaurs very closely related to the dinosaurs. It is likely that their common ancestor lived in this area.

Most of the animals on these two pages were once regarded as too primitive to be allocated to either of the major dinosaur groups – saurischia or ornithischia. However, research in the 1990s, mostly by Paul Sereno of Chicago, shows that they were primitive saurischians, as shown by the structure of the skull and the bones of the hind foot.

These early meat-eaters had certain interesting features. There was a joint in the lower jaw, just behind the tooth row, that absorbed the stresses involved in biting struggling prey, curved claws on the first three fingers for grasping, and hollow limb bones for speed – all features possessed by the later meat-eating dinosaurs.

Below: Lagosuchus was an anchosaur close to the ancestry of the dinosaurs.

Staurikosaurus

In 1936, an expedition from the Museum of Comparative Zoology at Harvard, USA, looked at the Santa Maria Formation in southern Brazil. They found the skeleton of a big meat-eating dinosaur, but it was not until 1970 that it was studied scientifically and named by the great American palaeontologist Edwin H. Colbert. Since then opinions have differed widely about the classification of *Staurikosaurus*.

Features: *Staurikosaurus* is very similar to its Argentinian contemporary, *Herrerasaurus*, but more lightly built and with a longer, more slender neck. The head is quite large and constructed for tackling large prey. Like *Herrerasaurus* it has a primitive five-toed foot, but the arm and hand bones are missing from the only known skeleton. Some scientists believe that it is the same animal as *Herrerasaurus*, while others think it is so primitive that it cannot be classed as either a theropod or a sauropod. The classification given here is therefore tentative.

Distribution: South-eastern Brazil.
Classification: Saurischia, Theropoda, Herrerasauridae.
Meaning of name: From the Southern Cross constellation.
Named by: Colbert, 1970.
Time: Carnian stage of the late Triassic.
Size: 2m (6½ft).
Lifestyle: Hunter.
Species: *S. pricei*.

Left: Although it must have had the appearance of a theropod, Staurikosaurus may have belonged to a completely different group. It may not even have been a saurischian but something ancestral to both the saurischian and ornithischian lines.

Chindesaurus

The story of *Chindesaurus* shows the confusion caused by the primitiveness of the early dinosaurs. It was unearthed in the 1980s by Bryan Small of the University of California, Berkeley, USA, and was regarded as a prosauropod, an early plant-eater. The media dubbed it "Gertie" after a cartoon dinosaur from the early days of the cinema. In 1985 it was found to be a herrerasaurid theropod.

Right: This animal lived in North America in the Carnian stage of the late Triassic. Unfortunately it is difficult to equate the age with that of similar rocks in South America, and so it is unclear whether Chindesaurus *predated the South American forms or was later.*

Features: *Chindesaurus* is similar to the other early dinosaurs, but its legs are particularly long and its tail very whip-like. It is important because it is the first herrerasaurid found in the Northern Hemisphere. Its presence in the Chinle Formation of Petrified Forest National Park, with the remains of other meat-eaters, shows that a variety of prey animals existed in the river-bank forests of the time.

Distribution: Arizona and New Mexico, USA.
Classification: Saurischia, Theropoda, Herrerasauridae.
Meaning of name: From Chinde Point.
Named by: Murray and Long, 1985.
Time: Carnian stage of the late Triassic.
Size: 3m (10ft).
Lifestyle: Hunter.
Species: *C. bryansmalli*.

LITHE MEAT-EATERS

The classic shape of a meat-eating dinosaur lent itself rather well to a fast-moving lifestyle. The powerful hind legs propelled the animal forward into an attack, and the teeth and claws would be used as weapons. Some of the smaller dinosaurs became quite lightweight and were obviously built for speed, developing slim bodies and long, agile legs.

Aliwalia

Distribution: South Africa.
Classification: Saurischia, Theropoda.
Meaning of name: From Aliwalia in South Africa.
Named by: Galton, 1985.
Time: Carnian to Norian stages of the late Triassic.
Size: 8m (26ft).
Lifestyle: Hunter.
Species: *A. rex.*

Features: Hardly anything is known about this animal except that it is a very large meat-eater, probably the first of the really big flesh-eating dinosaurs. The only remains that we know are parts of the hind leg bones and a jaw fragment. We often think of the very big meat-eating dinosaurs as coming later in the Mesozoic, when they would have had time to evolve from the earlier small dinosaurs. However, the presence of animals such as *Aliwalia* shows that they existed very early. They probably preyed on prosauropods or any of the other big plant-eaters, such as the pick-toothed rhynchosaurs or the last of the herbivorous mammal-like reptiles.

The few remains that we have of *Aliwalia* were found in a shipment of prosauropod fossils sent from South Africa to a museum in Austria, in 1873. For a century they were thought to have come from a prosauropod, and led to the idea that prosauropods may have been meat-eating animals as well as the plant-eaters we now know them to have been.

Left: Life restorations of animals based on only a handful of bones are difficult. However, we know enough about the general appearance of the early meat-eating dinosaurs to make an educated guess as to the probable appearance of Aliwalia.

Shuvosaurus

Distribution: Texas, USA.
Classification: Saurischia, Theropoda.
Meaning of name: From Shuvo, the son of the finder.
Named by: Chatterjee, 1993.
Time: Norian stage of the late Triassic.
Size: About 3m (10ft).
Lifestyle: Possibly omnivorous or an egg-stealer.
Species: *S. inexpectatus.*

When the skull of *Shuvosaurus* was discovered by the son of palaeontologist Sankhar Chatterjee in the early 1990s, it caused a surprise. It was suggested that it was the skull of an ostrich-mimic because of its lack of teeth. However, ostrich mimics are only known from Cretaceous rocks, and the Triassic would have been far too early for this group to have lived. The unexpected discovery is the reason for the specific name of the only species found to date.

Above: The toothless jaws for Shuvosaurus suggest it had a very specialized diet, however there is no evidence as to what that diet was.

Features: Nothing is known about this dinosaur except a toothless skull. Its original identification as an ostrich-mimic did not hold water, and soon it was reclassified as a rauisuchian, one of the land-living, crocodile-like predators that shared the world with the early dinosaurs. On this basis it was assigned to the already known genus *Chatterjeea*, which lacked the skull. Vertebrae and other isolated bones found in the formation known as the Upper Dockum Group of the Norian of Texas may belong to this animal. Current thinking suggests that it is a very specialized coelophysoid, but the matter is not settled.

Gojirasaurus

The dinosaur *Gojirasaurus* was found by Kenneth Carpenter of Denver Museum of Nature and Science. As a child he was fascinated by the 1954 Japanese science-fiction film *Godzilla*, which later sparked his interest in palaeontology and dinosaurs. He named his newly found fossil *Gojirasaurus* in honour of his childhood passion.

Features: This dinosaur is only known from an associated skeleton, consisting of ribs, bits of backbone, the shoulder girdle, part of the hips and hind leg, and a tooth. What is more, the skeleton is of an individual that was not fully grown, and so we have no good idea of the size of the adult animal. Consequently the restoration here is highly speculative. What we can say is that *Gojirasaurus* was an early meat-eater that attained a respectable size.

Left: The original Japanese name for Godzilla is Gojira, which is itself an amalgam of the Japanese words kujira, meaning whale, and gorira, meaning gorilla.

Distribution: New Mexico, USA.
Classification: Saurischia, Theropoda, Ceratosauria.
Meaning of name: From Gojira.
Named by: Carpenter, 1991.
Time: Carnian to Norian stages of the late Triassic.
Size: 5.5m (18ft), from an immature specimen.
Lifestyle: Hunter.
Species: *G. quayi*.

THE SHAPE OF MEAT-EATERS

The basic shape of the meat-eating dinosaur became established early in the history of the group. The upright posture, with the long hind legs and tail, is familiar to anyone with any knowledge of dinosaurs. Originally scientists believed that a meat-eating dinosaur stood with its back at an angle of 45 degrees with the tail dragging on the ground – the stance of the kangaroo, which is the only large, familiar, tailed bipedal animal of today. No doubt it was the kangaroo that inspired this vision.

About the 1970s the more modern version of the meat-eating dinosaur came to be accepted. The back is held horizontally, the jaws and claws are held out at the front where they can do the most damage, and the whole body is balanced by the heavy tail.

One of the theories for the evolution of this stance was that the dinosaurs' immediate ancestors were crocodile-like semi-aquatic forms, with strong swimming back legs and powerful tails. When they took to the land they walked naturally on their strong hind legs, with their shorter front legs clear of the ground. The heavy muscular tail then balanced the rest of the body quite naturally.

Below: The dinosaur skeleton on the left shows the contemporary understanding of how theropods stood. The skeleton on the right shows the old-fashioned view.

Liliensternus

In Triassic Europe *Liliensternus* was one of the largest meat-eating dinosaurs. In size and appearance it was somewhere between *Coelophysis* and *Dilophosaurus*. It lived in the forests that grew on the banks of the rivers that crossed an otherwise barren continent. It may have hunted the big plant-eating prosauropods that lived in the region.

Features: *Liliensternus*, in common with the other early meat-eaters, has five fingers on the hand. However, the fourth and fifth are much reduced – seemingly a step on the way to the established three-finger pattern of the vast majority of theropods. It is a very slim animal with a particularly long neck and tail. The crest on the head would probably have been brightly coloured, and would have been used for signalling and communicating with other dinosaurs.

Distribution: Germany.
Classification: Saurischia, Theropoda, Ceratosauria.
Meaning of name: From Ruhle von Lilienstern.
Named by: von Huene, 1934.
Time: Carnian to Norian stages of the late Triassic.
Size: 5m (16½ft).
Lifestyle: Hunter.
Species: *L. orbitoangulatus*, *L. airelensis*, *L. liliensterni*.

COELOPHYSIDS

The coelophysids seem to have been the most widespread of the meat-eaters in late Triassic times. So successful were they that members of the group survived into the beginning of the Jurassic. They were small, fast-running dinosaurs and probably preyed on the smaller animals of the time, leaving larger prey to the big crocodile-like animals that still existed.

Procompsognathus

One of the earliest, European, meat-eating dinosaurs was *Procompsognathus*. What we know of its skeleton shows that it was an active hunter, but it was not the major hunter of the time because it is clear that much bigger herrerasaurids lived alongside it. Despite the fact that little is known about this animal, and we are uncertain of its appearance or classification, it became widely known as a principal dinosaur in Michael Crichton's 1995 novel *The Lost World: Jurassic Park*, and the subsequent film.

Features: We only know of a single skeleton, and that is badly crushed, with most of the neck, tail, arms and hips missing. The classification, putting it among the Coelophysoidea, is largely based on the long, narrow skull and the teeth. These, however, now seem to have belonged to a different animal altogether, possibly a crocodile. As a result, the restoration shown here and the classification given are rather spurious.

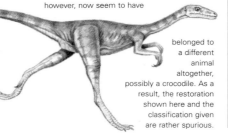

Distribution: Germany.
Classification: Theropoda, Coelophysoidea (unproven).
Meaning of name: Before *Compsognathus*.
Named by: Fraas, 1913.
Time: Norian stage.
Size: 1.2m (4ft).
Lifestyle: Hunter.
Species: *P. traissicus*.

Camposaurus

Charles Lewis Camp (1893–1975) was an American vertebrate palaeontologist who did a great deal of work on the sites that produced most of our knowledge of the animal life of the Triassic Petrified Forest fossil site in Arizona. One of his finds consisted of the partial remains of this small meat-eater that was named after him long after his death.

Distribution: Arizona, USA.
Classification: Theropoda, Coelophysoidea.
Meaning of name: From Charles Lewis Camp.
Named by: Hunt, Lucas, Heckert and Lockley, 1998.
Time: Carnian stage of the late Triassic.
Size: 1m (3ft).
Lifestyle: Hunter.
Species: *C. arizonensis*.

Features: We only know of the limb bones and a few vertebrae of this dinosaur. They show it to be very similar to *Coelophysis* in having long running legs and clawed, three-toed feet. However, there are enough differences in the shape and arrangement of bones to show that, although closely related, it is a totally different animal. It would have hunted the smaller animals that lived in the fern thickets and coniferous forests of the Petrified Forest National Park, Arizona.

Right: The few bones known of Camposaurus are just different enough to distinguish it from the slightly later Coelophysis, but the two must have been very similar.

COELOPHYSIS DISCOVERY

The discovery of a pack of *Coelophysis* at the evocatively named Ghost Ranch in New Mexico, in 1947, gives an astonishing insight into the lives and deaths of the early meat-eating dinosaurs. Some were complete and fully articulated. They all had a typical death pose, with the head and neck pulled back, indicating that the body twisted as the tendons dried out in the open air. Inside the rib cages of two were the bones of *Coelophysis* youngsters. The interpretation is that a pack of *Coelophysis* had gathered near a drying waterhole and had perished as the water disappeared. They were driven to such extremes of hunger that they were forced to eat their young to survive, but clearly it did not work. *Coelophysis* had already been named by Cope in 1889, and the Ghost Ranch skeletons were identified as the same animal. Later work, however, showed that they were different enough to be a whole new genus, and they were renamed *Rioarribasaurus*. By this time the site was famous, and the dinosaurs found there had been accepted as *Coelophysis*, resulting in the International Commission of Zoological Nomenclature (ICZN) – the body that regulates the naming of new animals – ruling that the name *Coelophysis* be applied to the Ghost Ranch animals rather then Cope's original find. It is unusual for the ICZN to do this.

Coelophysis

This is without doubt the best known of the coelophysid group and hundreds of skeletons, some complete and articulated, provide our knowledge of the group in general. A dozen skeletons were found at Ghost Ranch in New Mexico, in 1947, indicating that they died of starvation around a drying water hole. In 1998 the space shuttle 'Endeavor' took a skull of *Coelophysis* to the Mir space station, making it the first dinosaur in space.

Features: *Coelophysis* is a slimly built running hunter with a long skull and neck, a lightweight body and a long tail. The premaxilla – the front bone of the skull – is loosely articulated to the rest of the jaw mechanism, suggesting that it could be moved to manipulate small prey. Its fossil remains show that it hunted in packs. Two different weights of skeleton suggest that both males and females travelled together.

Distribution: Arizona, New Mexico and Utah, USA.
Classification: Theropoda, Coelophysoidea.
Meaning of name: Hollow form.
Named by: Cope, 1889.
Time: Carnian and Norian stages of the late Triassic.
Size: 2.7m (9ft).
Lifestyle: Hunter.
Species: *C. bauri*. Classification is in dispute, but some regard *Podokesaurus* and *Syntarsus* as species of *Coelophysis*, and give the name *Rioarribasaurus* to the Ghost Ranch remains.

Left: Coelophysis is named after its hollow bones, a feature shared with modern birds. They made for a very light, fast-moving animal.

Eucoelophysis

It used to be thought that there were very few species of hunting dinosaurs living during Triassic times. The discovery of *Eucoelophysis* and *Camposaurus* shows a range of animals all closely related in the coelophysoid group, living in the same area at more or less the same time. *Eucoelophysis* was found in the Petrified Forest beds, slightly older than those containing the remains of *Coelophysis*.

Features: What we know of this animal comes from the discovery of leg bones and other parts of the skeleton in New Mexico in the early 1980s. Their shape shows that it was closely related to *Coelophysis* but different enough to be regarded as a separate genus. Based on this it seems that some of the original *Coelophysis* material found by amateur fossil collector David Baldwin in the 1880s actually belongs to this genus.

Distribution: New Mexico, USA.
Classification: Theropoda, Coelophysoidea.
Meaning of name: True *Coelophysis*.
Named by: Sullivan and Lucas, 1999.
Time: Carnian to Norian stages of the late Triassic.
Size: 3m (10ft).
Lifestyle: Hunter.
Species: *E. baldwini*.

Right: The original remains of Coelophysis found by Cope, along with some other putative Coelophysis species, may actually belong to the genus Eucoelophysis.

PRIMITIVE SAUROPODOMORPHS

The sauropodomorphs were the first plant-eating dinosaurs. The earliest, probably evolving from the primitive meat-eaters, were small, rabbit-sized beasts but, as the Mesozoic advanced, they evolved into the biggest land animals that have ever existed. They can be divided into two main groups, the earlier prosauropods and the sauropods. There were also animals that were too primitive to be classed in either.

Saturnalia

As an early dinosaur species *Saturnalia* gives us a taste of things to come in the dinosaur world. It is the earliest known plant-eating dinosaur, and was originally thought to have been an early prosauropod. It is now thought to have been more primitive than the prosauropods, and we can only classify it as an early member of the Sauropodomorpha.

Features: From three partial skeletons we can construct the appearance of this elegant, rabbit-sized animal, with a long neck and tail. The head is small and the teeth coarsely serrated, in keeping with a vegetable diet. The body and legs are quite slender. The hip bones are quite primitive, making it a rather borderline dinosaur, but its ankle bones are similar to those of the contemporary meat-eating dinosaurs. It was very close to the ancestry of the prosauropods and the sauropods.

Distribution: Brazil.
Classification: Sauropodomorpha.
Meaning of name: After the Roman solstice festival.
Named by: Langer, Abdala, Richter and Benton, 1999.
Time: Carnian stage of the late Triassic.
Size: 1.5m (5ft).
Lifestyle: Low browser.
Species: *S. tipiniquim*.

Right: The Saturnalia was the Roman winter solstice festival, and the name was given to this dinosaur as that was the time of year when it was found. The species name means "native" in the local Portuguese.

Thecodontosaurus

An early discovery, *Thecodontosaurus* was only the fourth dinosaur to have been named. It was named from a jaw bone with teeth, which was found in limestone quarries near Bristol, England. It was not originally included in Owen's classification of the Dinosauria, possibly because it was so small compared with *Iguanodon* and *Megalosaurus*. It was only drawn into the dinosaur fold by Thomas Huxley in 1870.

Features: *Thecodontosaurus* is known from hundreds of fossils, both juvenile and adult, including partial skeletons. Unfortunately, the earliest specimens were lost in the bombing of Bristol City Museum, England, during World War II. As with all other prosauropods, it has a small head with plant-shearing teeth, and front legs that are shorter than the hind. It probably moved about mostly on all fours, but could spend much of its time on its hind legs.

Distribution: England and Wales.
Classification: Sauropodomorpha, Prosauropoda.
Meaning of name: Socket-toothed lizard.
Named by: Riley and Stutchbury, 1836.
Time: Norian to Rhaetian stage of the late Triassic.
Size: 2m (6½ft).
Lifestyle: Low browser.
Species: *T. antiquus, T. minor, T. caducus.*

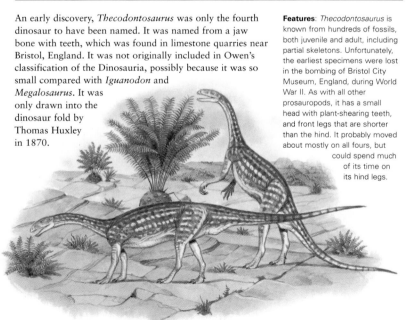

Left: Thecodontosaurus probably ran on two legs and browsed on four.

Efraasia

The only remains of this animal were discovered in 1909 by German dinosaur pioneer Eberhard Fraas. The animal was named after him in 1973. The skeleton shows the distinctive nature of the prosauropod's hand. This dinosaur is now thought by some to be the juvenile form of the larger *Sellosaurus*, although there is a suggestion that it may be a very primitive sauropodomorph, like *Saturnalia*.

Features: As with other prosauropods, *Efraasia* has multipurpose hands with long, grasping fingers and mobile thumbs. The wrist joint is well-developed and flexible, and the palm could be turned towards the ground allowing the animal to walk on all fours. There are only two sacral vertebrae joining the backbone to the hips, however, which is quite a primitive feature because most other saurischian dinosaurs have three. This may be the sign of a young animal.

Distribution: Germany.
Classification: Sauropodomorpha, Prosauropoda.
Meaning of name: After Eberhard Fraas, the German dinosaur pioneer.
Named by: Galton, 1973.
Time: Norian stage of the late Triassic.
Size: 2.4m (8ft).
Lifestyle: Low browser.
Species: E. diagnostica, E. minor.

Right: The original skeleton of Efraasia was associated with a set of crocodile jaws, and the whole thing was restored with a meat-eater's head and a plant-eater's body. Peter Galton's research in 1973 established the current rendering of the dinosaur.

PLACING DINOSAURS IN THEIR CORRECT PERIOD

Left: The skull of Thecodontosaurus.

The prosauropods, a group of sauropodomorphs, comprise one of the first dinosaur groups to be studied, but we still do not know a great deal about them compared with other dinosaurs. This is probably because they lived at the beginning of the Age of Dinosaurs, and their fossil record is so old that it is fragmentary and incomplete.

The first prosauropod remains to be found were those of *Thecodontosaurus*. Its fragmentary fossils were found in Triassic infillings of Carboniferous limestones near Bristol in southern England. The seeming paradox of Triassic fossils found in Carboniferous rocks can be explained by the geography of the time. In the Triassic period this area was an arid limestone upland, formed from rock laid down during the Carboniferous period. It was riddled with caves and chasms which may have provided shelter for animals. Swallow holes would have supported vegetation around the rim, encouraged by moist air rising from the caves below. Plant-eating animals such as *Thecodontosaurus* would have been attracted to this richer vegetation, and inevitably some fell to their deaths in the caves. The rubble and bones on the cave floor eventually solidified, resulting in fossils of Triassic animals surrounded by Carboniferous rock.

Sellosaurus

One of the prosauropods of the European Triassic deserts, *Sellosaurus* would have fed on the primitive monkey puzzle-like conifers that grew in the moister areas. The presence of stomach stones inside some of the skeletons suggests that they had quite a sophisticated digestive system, possibly with a big gut full of fermenting bacteria to help break down the tough plant material.

Features: *Sellosaurus* is a typical, medium-sized prosauropod. It has a heavy body, a long neck and a small head. Its hind legs are bigger and stronger than its arms, which it would use to pull down twigs and leaves from trees. It differs from other prosauropods by the shape of the tail vertebrae – the neural spine on each has a saddle-shaped structure, hence the name. It may have been ancestral to the later *Anchisaurus* but this would be difficult to prove. It is known from over 20 partial skeletons, three of them with skulls. Gastroliths, or stomach stones, have been found associated with some of the skeletons.

Distribution: Germany.
Classification: Sauropodomorpha, Prosauropoda, Plateosauridae.
Meaning of name: Graceful saddle lizard.
Named by: von Huene, 1908.
Time: Norian stage of the late Triassic.
Size: 6.5m (21ft).
Lifestyle: Low browser.
Species: S. gracilis.

Left: Originally the name Sellosaurus encompassed two other animals, Efraasia diagnostica and Plateosaurus gracilis. The research of Adam Yates, an Australian palaeontologist, in 2002 indicated that they were three separate genera.

TRUE PROSAUROPODS

The prosauropods, a group of sauropodomorphs, rapidly developed into the main plant-eaters of late Triassic times, when the landmass of the world was still joined together. Consequently, the remains of similar animals have been found all over the world. Although Europe seems to have been the centre for prosauropod discoveries, they have also been found in North and South America, Africa and China.

Euskelosaurus

The first dinosaur to be discovered in Africa was *Euskelosaurus*. It was known from a set of leg bones found in South Africa in 1866, and was sent to London for study. Since then, bones of this animal have been found all over southern Africa, suggesting that it had been a very common plant-eater of the time. It appears to have lived at a time when dry climates were spreading across the southern continents, and many remains have been found in the Lower Elliot formation in South Africa, entombed in sandstone formed from soft river sands in which they had become mired – rather like the Swiss occurrences of *Plateosaurus*.

Features: This is one of the biggest of the prosauropods, and as such it looks very much like one of the true sauropods that succeeded the group. It supports its bulky body by moving on all fours. While we have many bones of this animal, we do not have the skull, and its relationship with the rest of the prosauropod group is unclear.

Distribution: Lesotho, South Africa, Zimbabwe.
Classification: Sauropodomorpha, Prosauropoda.
Meaning of name: Well-limbed lizard.
Named by: Huxley, 1866.
Time: Carnian to Norian stages of the late Triassic.
Size: 9–12m (30–39ft).
Lifestyle: High browser.
Species: *E. browni, E. africanus, E. capensis, E. fortis, E. molengraafi.*

Right: Euskelosaurus *resembles both the advanced prosauropods such as* Melanorosaurus, *and the primitive sauropods like* Antetonitrus. *However, it is the shape of the thigh bone that curves out that shows it is a traditional prosauropod.*

Blikanasaurus

Distribution: Lesotho.
Classification: Sauropodomorpha, Prosauropoda.
Meaning of name: Lizard from Blikana.
Named by: Galton and van Heerden, 1985.
Time: Carnian to Norian stages of the late Triassic.
Size: 5m (16½ft).
Lifestyle: High browser.
Species: *B. comptoni.*

Evolution is always developing new animal shapes as a response to new conditions. The variation seen in the prosauropods includes a movement towards bigger, heavier animals. One of these lines developed into the sauropods, but some became merely evolutionary sidelines that happened to look like sauropods but developed no further. *Blikanasaurus* was one of these.

Features: All that we know of *Blikanasaurus* is the stoutly built hind limb. It shows the very short foot of an animal that would have spent all its time on all fours, and as such it would have resembled the later sauropods. However, its fifth toe (that is present in all sauropods) is tiny, showing that *Blikanasaurus* represented a side-branch of the prosauropods rather than being part of the evolutionary line to the sauropods themselves.

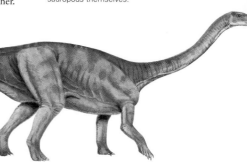

Right: Blikanasaurus *was, as far as we know, the first dinosaur to spend all its time on all fours. The rest of the contemporary early prosauropods would have had the ability to go about on their hind legs from time to time.*

Plateosaurus

Distribution: Germany, Switzerland and France.
Classification: Sauropodomorpha, Prosauropoda, Plateosauridae.
Meaning of name: Broad lizard.
Named by: von Meyer, 1837.
Time: Norian stage of the late Triassic.
Size: 8m (26ft).
Lifestyle: High browser.
Species: *P. engelhardti*, *P. gracilis*.

Below: The teeth of Plateosaurus were leaf-shaped and coarsely serrated. They were well suited to shredding the tough leaves of the tree ferns and primitive conifers.

One of the best known of the prosauropods is *Plateosaurus*. It was one of the most common of all European dinosaurs with fossils found in over 50 localities. The fact that so many skeletons have been found together suggests that they may have been herding animals that migrated across the arid plateaux and basins that formed the landscape of Europe in late Triassic times. The concentration of *Plateosaurus* remains in a few areas has led palaeontologists to suggest that frequent disasters overtook the herds, but it is just as likely that these accumulations of bones took place over long periods of time, highlighting the fact that the herds gathered at specific spots at particular times of the year.

Interesting remains have been found in Switzerland, and consist of the leg bones of *Plateosaurus* standing vertically in river sandstones, while the rest of the skeleton is scattered over a wide area, mixed with the teeth of meat-eating dinosaurs and crocodile-like animals. The interpretation is that the *Plateosaurus* became bogged down in quicksands and was torn apart by hunting animals. *Euskelosaurus* remains showing the same mode of preservation have been found in South Africa.

PROSAUROPOD LINEAGE

The prosauropods are sometimes given the name palaeopods. They evolved from two-footed meat-eating ancestors, and their early forms evolved into the later sauropods. The smaller prosauropods were active little two-footed animals which perhaps were not quite able to run. Even so, they would have been faster than any four-footed animal at the time. The biggest were totally quadrupedal animals. Those in between would have spent most of their time on all fours, but would have raised themselves up to reach the conifer foliage and fruits of the cycad-like plants on which they fed. The hind legs had heavy, five-toed feet, while the front legs were relatively short. They had five fingers and, a characteristic of the group, a big claw on the thumb. This may have been used as a weapon or to help them dig up roots or tear down branches. When the prosauropod walked on all fours, a specialized joint held the thumb claw up and out of the way.

So distinctive are the feet that the footprints in Triassic and early Jurassic sandstone can be readily identified as those of prosauropods. A set of footprints of the ichnogenus *Navajopus* has been linked to the prosauropod *Ammosaurus* – one of the few instances when this can be done.

Right: A prosauropod hand and clawed thumb.

Features: The general understanding of prosauropods is derived from the many remains of *Plateosaurus*. However, new analyses are constantly being made of the material. The traditional view is of an animal walking in a digitigrade stance – on the tips of its toes like a bird. More recent studies, however, suggest that it would have spent most time in a flat-footed plantigrade stance with the weight on the whole of the foot, like a bear.

MORE PROSAUROPODS

As the Triassic period progressed, the prosauropods evolved from small and medium-sized animals into big beasts that took on the appearance of the later sauropods with their four-footed stance, massive bodies, long necks and tiny heads. Most of the big ones fall into the melanorosaurid group of prosauropods.

Lessemsaurus

Similar prosauropods lived in South Africa and South America. In late Triassic times the Atlantic Ocean did not exist, and South America and Africa were part of one continent. Since environmental and climatic conditions were continuous across the two, it is not surprising that the remains of similar animals are found on both continents. Prosauropods were the major plant-eating animals there.

Features: The only thing known of *Lessemsaurus* is the partially articulated spinal column. This shows tall spines sticking upwards that would have produced a prominent ridge along the back. Perhaps this was some kind of signalling device and

Right: This prosauropod is named after Don Lessem, a prolific American writer who popularized dinosaurs.

was brightly coloured, or maybe it was something to do with temperature regulation. From the structure of the backbone it is assumed that the rest of the animal conformed to the typical prosauropod shape.

Distribution: Argentina.
Classification: Sauropodomorpha, Prosauropoda.
Meaning of name: Lessem's lizard.
Named by: Bonaparte, 1999.
Time: Norian stage of the late Triassic.
Size: 9m (30ft).
Lifestyle: High browser.
Species: *L. sauropoides*.

Riojasaurus

The best known South American prosauropod is *Riojasaurus*. However, although the skeleton has been known from the 1960s, the skull has only recently been discovered. It was known that the neck was long and slender, and this suggested that the head was small like that of other prosauropods. The discovery of the skull confirmed this.

Features: *Riojasaurus* is a member of the melanorosaurid family – the big prosauropods that tended to walk on all fours. More than 20 skeletons at different stages of growth give us a good idea of what this animal looked like. The back legs are only slightly larger than the front legs, suggesting a quadrupedal stance. The limb bones are thick and solid, but the backbone is hollow. *Riojasaurus* probably weighed something in the region of one tonne.

Distribution: Argentina.
Classification: Sauropodomorpha, Prosauropoda.
Meaning of name: Lizard from Rioja province.
Named by: Bonaparte, 1969.
Time: Norian stage of the late Triassic.
Size: 11m (36ft).
Lifestyle: High browser.
Species: *R. incertus*.

Left: Riojasaurus *was probably the heaviest land animal that existed before the evolution of the sauropods. All the skeletons have been found in the foothills of the Andes.*

Camelotia

There are not many differences between the big, advanced prosauropods and the more primitive of the sauropods, and those differences lie mostly in the arrangement of foot bones and the curvature of the leg bones. This can lead to confusion when only part of a skeleton is found. *Camelotia* is a dinosaur that seems to combine features of both groups. It was first found by Seeley in the 1890s and given the name *Avalonia*. There were some teeth associated with the skeleton that did not seem to fit

Features: The vertebrae, hip bones and hind leg bones – the only elements that are known for *Camelotia* – pose a bit of a mystery. The curve of the thigh bone and the areas of leg muscle attachment are similar to those of a sauropod, which has led to the suggestion that *Camelotia* is actually a primitive sauropod. However, the rest of the bones indicate that it is a member of the melanorosaurid prosauropods.

Distribution: England.
Classification: Sauropodomorpha, Prosauropoda, Melanorosauridae.
Meaning of name: From Camelot, the legendary castle of King Arthur.
Named by: Galton, 1985.
Time: Rhaetian stage of the Triassic and Jurassic boundary.
Size: 9m (30ft).
Lifestyle: High browser.
Species: *C. borealis*.

with the accepted idea of prosauropod teeth. These have since been identified as the teeth of an ornithopod, called *Avalonianus*.

Left: The name Avalonia *was pre-occupied and Peter Galton renamed this dinosaur when he redescribed it a century after it was found and first named, separating the spurious ornithopod teeth from the rest of the skeleton.*

PROSAUROPOD EVOLUTION
The classification of the prosauropods, as with all other dinosaurs, is continually undergoing revision. The main primitive groups are the Plateosauridae and the Melanorosauridae. The line leading to the sauropods branched off early, and the prosauropods themselves continued to evolve into the more advanced forms. The Massospondylidae then continued into the Jurassic.

Below: An advanced prosauropod above and a primitive prosauropod below.

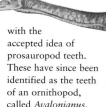

Melanorosaurus

The big prosauropods from the southern continents, such as *Melanorosaurus* and *Euskelosaurus* from South Africa, and *Riojasaurus* from South America, were very similar in appearance, leading to the suggestion that they are all members of the same genus. Certainly they all belonged to the melanorosaurid family, and it would probably be difficult to tell apart in life.

Features: This was a large prosauropod that established the melanorosaurid family. It is the heaviest known, with massive front and hind legs, and was probably the biggest land animal at the time. Unlike some of its relatives, this prosauropod was unlikely to have been able to move on its hind legs. Its shape and lifestyle, using its neck to reach the branches of trees in which to browse, anticipated the big sauropods to come.

Distribution: South Africa.
Classification: Sauropodomorpha, Prosauropoda, Melanorosauridae.
Meaning of name: Black Mountain lizard (from the site of discovery).
Named by: Haughton, 1924.
Time: Carnian to Norian stages of the late Triassic.
Size: 15m (50ft).
Lifestyle: High browser.
Species: *M. readi*, *M. thabanensis*.

Left: The huge size of Melanorosaurus *and its relatives may have been an early defensive adaptation. The sheer size of such an animal would have discouraged attacks by any of the smaller meat-eaters that existed at the time.*

HERBIVOROUS MISCELLANY

At the end of the Triassic period a totally new group of plant-eating dinosaurs, the ornithischians, which were distinguished from the saurischians by their bird-like hips, began to roam the earth. The most widespread ornithischians were the ornithopods, those with bird-like feet, which first appeared in Triassic times as small, two-footed running animals.

Pisanosaurus

Pisanosaurus is a dinosaur oddity. The ornithopods were thought to have evolved from animals like *Lesothosaurus*, a primitive group from the early Jurassic period in South Africa. They had not evolved the chewing mechanisms and the cheeks that are so distinctive of the group as a whole. Yet *Pisanosaurus* had these features, 25 million years before *Lesothosaurus* existed.

Features: The details of the pelvis and the ankle joint are more like those of a saurischian than an ornithischian, and this puts an animal with an advanced ornithischian feeding mechanism close to the ancestors of both the ornithischians and the saurischians. In brief, the head of *Pisanosaurus* is that of an ornithischian while the body is more like that of a saurischian. It is possible, however, that the specialist jaws evolved completely independently of those of the true ornithischians – an evolutionary 'one-off'.

Distribution: Argentina.
Classification: Ornithopoda (unproven).
Meaning of name: Pisano's lizard.
Named by: Casamiquela, 1976.
Time: Carnian stage of the late Triassic.
Size: 1m (3ft).
Lifestyle: Low browser.
Species: *P. merti*.

Technosaurus

Distribution: Texas, USA.
Classification: Ornithischia.
Meaning of name: Texas University lizard.
Named by: Chatterjee, 1984.
Time: Carnian stage of the late Triassic.
Size: 1m (3ft).
Lifestyle: Low browser.
Species: *T. smalli*.

The origin and evolution of the plant-eating ornithopod dinosaurs is unclear. The discovery of a single jaw bone and teeth in Triassic rocks of Texas, USA, suggests that they evolved quite early, but more evidence is needed before a worthwhile timetable can be established. *Technosaurus* would have foraged in the undergrowth of the coniferous forest that became Petrified Forest National Park, and was probably preyed upon by *Coelophysis* and its relatives.

Features: Palaeontologists only know of a single jawbone of this dinosaur, but it seems to pose the same kind of problems as *Pisanosaurus* – the presence of a sophisticated, plant-eating dinosaur long before the ornithischian dinosaurs are supposed to have evolved their advanced chewing mechanisms. The jawbone suggests that *Technosaurus* would have been a small, swift two-footed animal, walking on hind legs balanced by a heavy tail.

Above: The species name T. smalli *is in honour of Bryan Small, a palaeontologist who has done much work in the southern states of the USA.*

Antetonitrus

A specimen of the prosauropod *Euskelosaurus* lay on a shelf in the University of Witwatersrand after being unearthed near Bloemfontein, South Africa, in 1981. Then, in 2001, Australian palaeontologist Adam Yates looked closely at the bones and realized that they actually belonged to a sauropod – the earliest ever discovered. He and the original discoverer, James Kitching, re-examined the fossil and established the new genus.

Features: Though considered an early sauropod, *Antetonitrus* still has a prosauropod hand, with the ability to grasp with its fingers. However the hind legs, with the straight thigh bone and the short foot bones – adaptations to carrying heavy weights permanently on all fours – show that this is a primitive sauropod. It is the earliest true sauropod so far known. It was found in rocks only slightly older than those in which *Isanosaurus* (below) was found.

Distribution: South Africa.
Classification: Sauropodomorpha, Sauropoda.
Meaning of name: Before the thunder, referring to a later group of sauropods called the brontosaurs or 'thunder lizards'.
Named by: Yates and Kitching, 2003.
Time: Norian stage of the late Triassic.
Size: 10m (33ft).
Lifestyle: High browser.
Species: *A. ingenipes*.

Right: Although Antetonitrus is small in relation to the later sauropods (weighing about two tonnes) it is still bigger than any land-living animal that exists today.

DINOSAUR TEETH

Technosaurus is not the only dinosaur that is based on the fossils of puzzling teeth in the Triassic beds of North America. Carnian rocks reveal the teeth of other plant-eating dinosaurs.

There are primitive teeth that are curved and armed with coarse serrations, looking as if they have come from a meat-eating ornithischian – a dinosaurian absurdity (since ornithischians don't eat meat) but quite plausible considering the evolutionary types that were being 'tried out' in the early days. Some teeth have coarse serrations, which themselves have finer serrations, pointing to a complex feeding strategy.

The teeth of a dinosaur called *Pekinosaurus* have been found in New Jersey and New Mexico, USA, showing that these early plant-eaters were quite widespread in their range. *Tecovasaurus* is another dinosaur known from two kinds of teeth: one from the cheek and the other, longer and curved, from the front of the mouth. The teeth of *Tecovasaurus* are quite common, and useful as an index fossil providing proof that a particular geological bed is the same age as another with similar fossils.

All in all, the abundance and diversity of these teeth seem to suggest that ornithischians evolved in the early Carnian period, and had evolved rapidly into a number of different forms during late Carnian times. The ornithischians appear to have been more diverse in south-western North America than the prosauropods of the time. One day we may have the fossils of the rest of the skeletons. These may give us a better idea of the appearance of the animals that we now know from teeth.

Isanosaurus

For years it was suspected that the sauropods had a history that stretched back into the Triassic period. There has been footprint evidence to suggest this, and the complexity of sauropod types that suddenly appeared in the early Jurassic period indicated a long history. The discovery of *Isanosaurus* in Thailand, and then the discovery of *Antetonitrus* (above) in South Africa, prove that sauropods actually existed so early in history.

Features: *Isanosaurus* is known only from parts of a backbone, a shoulder blade, ribs and a femur. The femur is intermediate between the curved shape of a prosauropod femur and the straight sauropod femur, and the sites of muscle attachment are intermediate between the two. The tall backbones, however, mark this as a sauropod rather than a prosauropod. The lack of fusion between the bones of the backbone show that it was probably a youngster, only half-grown.

Distribution: Thailand.
Classification: Sauropodomorpha, Sauropoda.
Meaning of name: Lizard of Isan (in north-eastern Thailand).
Named by: Buffetaut, Suteethorn, Cuny, Tong, Le Loeuff, Khansubha and Jongauthariyakui, 2000.
Time: Norian and Rhaetian stages of the late Triassic to Jurassic boundary.
Size: 6m (18ft), but not fully grown.
Lifestyle: High browser.
Species: *I. attavipachi*, *I. russelli*.

Left: The species I. attavipachi honours P. Attavipach, a former director of the Thai Department of Mineral Resources, a supporter of palaeontological research.

THE
JURASSIC
PERIOD

During the Jurassic period the supercontinent of Pangaea began to break up. Rifts, similar to the Great Rift Valley of modern Africa, split across the landmass, and filled with seawater to form embryo oceans. Shallow seas spread across the continental margins and shelves. The result was moist equable climates, and the inland areas became habitable.

The plant-eating dinosaurs fed on conifer trees and in turn were hunted by fierce meat-eating dinosaurs. In the sky the pterosaurs flew, accompanied by the birds, which were just beginning to evolve. In the shallow shelf seas the unrelated sea reptiles fed on fish and invertebrates that inhabited the warm waters.

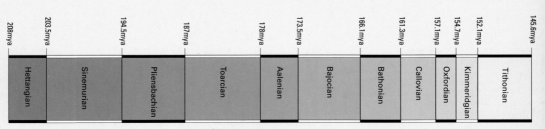

208mya	203.5mya	194.5mya	187mya	178mya	173.5mya	166.1mya	161.3mya	157.1mya	154.7mya	152.1mya	145.6mya
Hettangian	Sinemurian	Pliensbachian	Toarcian	Aalenian	Bajocian	Bathonian	Callovian	Oxfordian	Kimmeridgian	Tithonian	

Top: The earth as it looked in Jurassic times.

Above: The timeline shows the different chronological stages of the Jurassic period of Earth's history. (mya = million years ago)

1 Pterosaurs.
2 *Mamenchisaurus.*
3 *Omeisaurus.*
4 *Angustineriptus.*
5 *Tuoiangosaurus.*

ICHTHYOSAURS

The wide variety of forms of whale-like or eel-like ichthyosaur that existed in Triassic times were whittled down at the beginning of the Jurassic to a more dolphin-like shape. Most Jurassic ichthyosaurs lived in the early part of the period, with a few important genera remaining at the end. Only one, Platypterigius, is known to have survived into Cretaceous times.

Temnodontosaurus

The evolutionary trend among the Jurassic ichthyosaurs was towards a more compact, fish-shaped or dolphin-shaped type. However, the tradition of the whale-sized elongated beasts of the Triassic continued in the various species of *Temnodontosaurus*. The stomach contents show that they dined mainly on squid and the shelled cephalopods of the shallow continental seas.

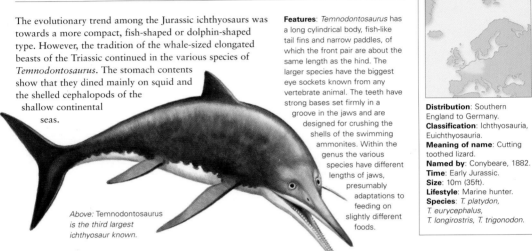

Above: Temnodontosaurus *is the third largest ichthyosaur known.*

Features: *Temnodontosaurus* has a long cylindrical body, fish-like tail fins and narrow paddles, of which the front pair are about the same length as the hind. The larger species have the biggest eye sockets known from any vertebrate animal. The teeth have strong bases set firmly in a groove in the jaws and are designed for crushing the shells of the swimming ammonites. Within the genus the various species have different lengths of jaws, presumably adaptations to feeding on slightly different foods.

Distribution: Southern England to Germany.
Classification: Ichthyosauria, Euichthyosauria.
Meaning of name: Cutting toothed lizard.
Named by: Conybeare, 1882.
Time: Early Jurassic.
Size: 10m (35ft).
Lifestyle: Marine hunter.
Species: *T. platydon, T. eurycephalus, T. longirostris, T. trigonodon.*

Ichthyosaurus

This is the creature that we think of when we hear the term ichthyosaur. Its dolphin-like appearance is familiar from all kinds of popular images. Its shape was a perfect adaptation to swift hunting in open waters. Propulsion was by powerful thrusts of the tail and the animal was stabilized by the paddle limbs and the dorsal fin.

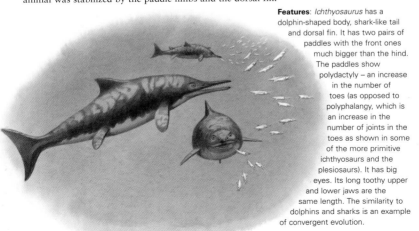

Features: *Ichthyosaurus* has a dolphin-shaped body, shark-like tail and dorsal fin. It has two pairs of paddles with the front ones much bigger than the hind. The paddles show polydactyly – an increase in the number of toes (as opposed to polyphalangy, which is an increase in the number of joints in the toes as shown in some of the more primitive ichthyosaurs and the plesiosaurs). It has big eyes. Its long toothy upper and lower jaws are the same length. The similarity to dolphins and sharks is an example of convergent evolution.

Distribution: Southern England.
Classification: Ichthyosauria, Thunnosauria.
Meaning of name: Fish lizard.
Named by: Conybeare, 1821.
Time: Sinemurian stage of the early Jurassic.
Size: 2m (6ft).
Lifestyle: Ocean hunter.
Species: *I. communis, I. intermedius, I. conybeari, I. breviceps.*

Excalibosaurus

A group of ichthyosaurs called the eurhinosaurids were the veritable swordfish of their day. The very long upper jaw, about four times the length of the lower, may have been used as a weapon, or it may have meant that, with the resulting downward-directed mouth, the animal foraged on the seabed probing in the sand for likely food.

Right: A specimen of Excalibosaurus was found and described in 2003. Rather than being a different species, the specimen found is probably an adult version of the already known species, and it is this find on which the 7m (23ft) length is based.

Features: The feature that distinguishes the members of the Eurhinosauria is, of course, the remarkable upper jaw. Where it projected beyond the mouth the teeth that were present extended sideways, rather like the teeth of a sawfish. *Excalibosaurus*, at 7m (23ft) long, is the giant of the group. Most others, including *Eurhinosaurus* itself – a similar animal after which the group is named – usually reached 3–4m (10–13ft). The paddles are long and narrow, suggesting manoeuvrability rather than fast swimming.

Distribution: Southern England.
Classification: Ichthyosauria, Eurhinosauria.
Meaning of name: Excalibur lizard.
Named by: McGowan, 1986.
Time: Sinemurian stage of the early Jurassic.
Size: 7m (23ft).
Lifestyle: Ocean hunter or scavenger.
Species: *E. costini*.

ICHTHYOSAUR FOSSILS

The ichthyosaurs were familiar to nineteenth-century naturalists long before dinosaurs were. This is understandable since sea-going animals were more likely to have their remains preserved as fossils. Their dead bodies would have sunk to the sea bed where sediments were gathering and would eventually, through a process of lithification, become sedimentary rock.

At first ichthyosaurs were thought of as some kind of ancient crocodile, and their remains were avidly collected in the quarries and coastal cliffs of southern England. The first professional woman fossil collector, Mary Anning, made important finds.

Complete skeletons all seemed to have had a kink in the tail, with the end of the tail bent downwards. This was thought to have been some kind of damage, and the first restorations of ichthyosaurs showed straightened crocodile-like tails. Full-sized statues of these animals built in the 1850s and still to be seen in the grounds of the Crystal Palace in south London, England, show this straightened tail.

In the 1880s ichthyosaur fossils were found in the slate quarries at Holzmaden in Germany. These were so well preserved that the soft tissues were still present as a thin film of carbon surrounding the skeletons. It was now obvious that the bent tail supported a fluke, like the tail of a shark but the other way up. A shark-like dorsal fin was also evident for the first time. Some of these skeletons also showed females in the act of giving birth, the first indication that ichthyosaurs were viviparous.

Opthalmosaurus

The big eyes, up to 10cm (4in) in diameter, suggest that *Opthalmosaurus* was a deep-water or night-time feeder. Its toothless jaws may indicate that it fed on soft-bodied prey such as squid. It has been estimated that *Opthalmosaurus* could dive to depths of about 500m (1,640ft). The disc-like vertebrae made a fairly rigid body shape.

Features: This is the most streamlined of the ichthyosaurs, with a body that was almost teardrop-shaped. The tail fin is big and half-moon-shaped, with muscles for fast propulsion. The eyes fill almost the whole of the side of the head and each contains a sclerotic ring (a ring of bone) to stop it from collapsing under pressure. The front paddles are bigger than the hind, suggesting that these did most of the steering.

Below: Opthalmosaurus seems to have been prone to the "bends", the non-technical name for the injury that divers sustain when they surface too quickly. The condition is caused by the formation of nitrogen bubbles in the bloodstream and can lead to bone damage. Such damage has been found in fossils of Opthalmosaurus.

Distribution: Southern England, northern France, western North America, Canada, Argentina.
Classification: Ichthyosauria, Opthalmosauria.
Meaning of name: Eye lizard.
Named by: Seely, 1874.
Time: Late Jurassic (but included here because of its importance).
Size: 3.5m (11ft).
Lifestyle: Fish hunter.
Species: *O. discus, O. icenius*.

PLESIOSAURS

The dawn of the Jurassic saw the dominance of two main lines of swimming reptiles, the ichthyosaurs and the plesiosaurs. Other swimming reptiles existed as well, such as marine crocodiles and turtles, but it was these two groups that were most abundant. Among the plesiosaurs there were two groups – the long-necked, small-headed plesiosauroids, and the short-necked, big-headed pliosauroids.

Macroplata

The plesiosaur *Macroplata* appears to have been quite a long-lived genus. The two species known span 15 million years at the beginning of the Jurassic, an unusual time span that has led some to believe that they are two different genera. The structure of the shoulder girdle indicates that the front paddles could produce a powerful forward stroke for fast swimming.

Features: As in all pliosaurs, *Macroplata* has a long head, in this case half the length of the neck. The neck vertebrae are numerous, about 29 of them, but short and compact, producing a rigid structure. The skull tapers to a needle-like snout with the front of the two lower jaws fused for most of their length, not just at the tip. The length of this symphysis is used by palaeontologists to determine the identity of pliosaurs.

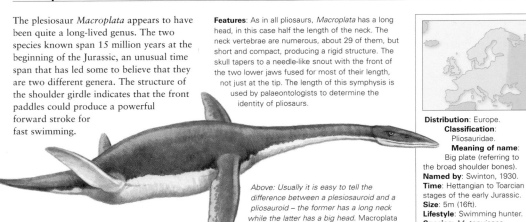

Distribution: Europe.
Classification: Pliosauridae.
Meaning of name: Big plate (referring to the broad shoulder bones).
Named by: Swinton, 1930.
Time: Hettangian to Toarcian stages of the early Jurassic.
Size: 5m (16ft).
Lifestyle: Swimming hunter.
Species: *M. tenuiceps*, *M. longirostirs*.

Above: Usually it is easy to tell the difference between a plesiosauroid and a pliosauroid – the former has a long neck while the latter has a big head. Macroplata appears to have possessed both.

Rhomaleosaurus

The nostrils of *Rhomaleosaurus* were not used for breathing (they were too small to allow for the passage of enough air), but for hunting prey. Water that passed into the open mouth while the animal was swimming was channelled out through the nostrils, which were situated on the top of the head. This water would have been tasted or smelled by *Rhomaleosaurus*, and used to analyze the surrounding environments.

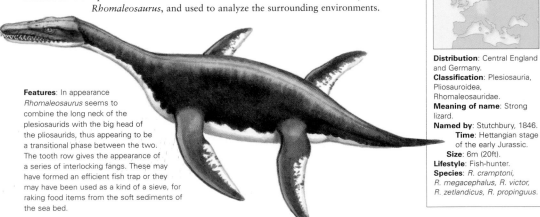

Features: In appearance *Rhomaleosaurus* seems to combine the long neck of the plesiosaurids with the big head of the pliosaurids, thus appearing to be a transitional phase between the two. The tooth row gives the appearance of a series of interlocking fangs. These may have formed an efficient fish trap or they may have been used as a kind of a sieve, for raking food items from the soft sediments of the sea bed.

Distribution: Central England and Germany.
Classification: Plesiosauria, Pliosauroidea, Rhomaleosauridae.
Meaning of name: Strong lizard.
Named by: Stutchbury, 1846.
Time: Hettangian stage of the early Jurassic.
Size: 6m (20ft).
Lifestyle: Fish-hunter.
Species: *R. cramptoni*, *R. megacephalus*, *R. victor*, *R. zetlandicus*, *R. propinguus*.

Attenborosaurus

The original specimen of *Attenborosaurus*, in Bristol City Museum, England, was regarded as a species of *Plesiosaurus* (*P. conybeari*), but it was destroyed during World War II. A good plastercast survived for later study to be carried out, and to allow for its description as a new genus. Unfortunately the important evidence of the skin was lost.

Features: The original fossil of *Attenborosaurus* carried traces of the skin as a thin brownish film. This showed the skin to have been a continuous membrane with no sign of scales. However there were small oblong bones in the hip region that may have been dermal plates. The skull is quite large and pliosaur-like, but the neck is very long, and the pelvic girdle is quite primitive for a plesiosaur of this type. The snout is tapered. *Attenborosaurus* has fewer teeth than normally found in plesiosaurs, and these have massive crowns.

Distribution: Europe.
Classification: Plesiosauria, Plesiosauroidea.
Meaning of name: Attenborough's lizard (after Sir David Attenborough, the natural historian).
Named by: Bakker, 1993.
Time: Sinemurian stage of the early Jurassic.
Size: 5m (16ft).
Lifestyle: Fish hunter.
Species: *A. conybeari*.

Above: The genus was named after the naturalist and journalist Sir David Attenborough, whose childhood interests in plesiosaurs eventually led him to a distinguished career in natural history journalism.

SWIMMING ACTION OF THE PLESIOSAUR

The paddles of a plesiosaur are stiff and relatively inflexible. The bones consist of five toes, which are very elongated. Each toe has more than the usual number of bones or phalanges, up to about 20, which compares with the four or five normally found in a tetrapod. This condition is known as polyphalangy. All these bones are tightly lashed together with gristle to produce a powerful swimming organ.

A plesiosaur swam by means of a flying action, with an undulating body movement very much like that of a modern penguin or sea lion. The hydrodynamics suggest that in most plesiosaurs the rear paddles were used for power with the downward stroke producing the propulsive force. The front paddles were used mainly for stabilization and to alter the angle and the direction of swimming. There may have been a vertical diamond-shaped fin on the tail for steering, but palaeontologists are not all in agreement about this. The body was solid and compact, with the shoulder and pelvic girdles occupying about half the area of the undersurface. This formed a firm base for the swimming muscles.

Gastroliths, or stomach stones, have been found in association with plesiosaur skeletons. Swallowing stones helped the animal to adjust its buoyancy; this is found in contemporary sea lions and penguins, modern marine animals that swim by using a similar flying action.

"Plesiosaurus"

This was the first of these plesiosaurs to have been described, and hence has given its name to the whole group. It is also something of a wastebasket taxon in that all kinds of specimens have been referred to it. "A snake threaded through the body of a turtle", was the description given by Dean Buckland, one of the Victorian naturalists who first studied it.

Features: The skull is small and the neck very long, about half as long again as the body, and composed of about 40 cervical vertebrae. The paddles are long, each consisting of five toes but with many more joints than usual. The forelimbs are slightly longer than the hind, although in the youngsters this is reversed. The teeth are long and sharp, and interlock when they close to form a fish trap.

Distribution: Europe.
Classification: Plesiosauria, Plesiosauroidea.
Meaning of name: Almost a lizard.
Named by: de la Beche and Conybeare, 1821.
Time: Sinemurian and Toarcian stages of the early Jurassic.
Size: 3.5m (11ft).
Lifestyle: Swimming hunter.
Species: *P. dolichodeirus*, *P. guilelmiimperatoris*, *P. macrocephalus*, *P. megdeirus*, *P. winspitensis*, *P. eurymerus*.

PTEROSAURS

Much of what we know about early Jurassic flying creatures, or pterosaurs, comes from the remains found in the shales of southern England and Germany. In early Jurassic times much of Europe was covered in a shallow shelf sea, and the fine mud of the sea bed resulted in deposits of heavy shale which were later quarried as building slate at Holzmaden, in southern Germany.

Dimorphodon

Distribution: Dorset, England.
Classification: Pterosauria, Rhamphorhynchoidea.
Meaning of name: Two types of teeth.
Named by: Owen, 1859.
Time: Liassic stage of the early Jurassic.
Size: 1.4m (4½ft) wingspan.
Lifestyle: Fish hunter.
Species: *D. macronyx*.

The first specimen of *Dimorphodon* was found by the famous fossil collector Mary Anning in the cliffs of Lyme Regis, in Dorset, in 1828. Only a few specimens have been found and these are all from the Dorset coast, although another possible specimen has been collected from the banks of the River Severn in Gloucestershire, England.

Features: The two types of teeth implied in the name consist in each jaw of a set of 30–40 tiny pointed teeth, with four large teeth at the front. They are set in a very deep, narrow skull, made up of flimsy struts of bone surrounding large cavities, keeping the body weight to a minimum. The legs are very long and powerful and the claws on both the feet and the hands are strong. This is seen as an adaptation to clinging to cliffs and rocks.

Below: As in all fossil animals, the coloration is speculative. It does seem possible that the deep face of such pterosaurs as Dimorphodon could have been brightly coloured for signalling, like modern deep-beaked birds.

Campylognathoides

This pterosaur is known from several skeletons, some complete, from the shale deposits of Holzmaden in Germany. One species, *C. indicus*, is known from India. The site at Holzmaden is famous for the complete ichthyosaur skeletons. This genus was originally named *Campylognathus* but the scientific description was not valid, and so it was redescribed and renamed later. The name refers to the slight upward hook to the front of the jaw.

Features: The head of *Campylognathoides* is large with a pointed snout. It has very large eye sockets, which suggests it had acute eyesight, and possibly a nocturnal habit. The hip bones of a pterosaur that may be *Campylognathoides* show that the legs were held sideways and upwards, suggesting an inability to walk on its hind legs while on the ground. It is not known if this applies to all others of the rhamphorhynchoid group.

Distribution: Germany and India.
Classification: Pterosauria, Rhamphorhynchoidea.
Meaning of name: Like the curved jaw (it was originally named *Campylognathus* or curved jaw).
Named by: Wild, 1971.
Time: Pliensbachian stage of the early Jurassic.
Size: 1.75m (5¾ft) wingspan.
Lifestyle: Fish-hunter.
Species: *C. liasicus*, *C. indicus*, *C. zitteli*.

Above: The quarries of Holzmaden in east Germany, where Campylognathoides was found, produced many fossils of early Jurassic animals, including the pterosaurs that hunted for fish in the original shallow sea.

Dorygnathus

Some very fine skeletons of *Dorygnathus* have been uncovered from the slate quarries of Holzmaden. In the early 19th century it was originally classed as a species of *Pterodactylus*, but that was before a distinction was made between the long-tailed rhamphorhynchoids and the short-tailed pterodactyloids.

Features: The wings are relatively short for a rhamphorhynchoid. The foot has a long, narrow fifth toe, which may have supported a web of skin for use in manoeuvring, possibly compensating for the short wings. The very long front teeth intermesh when the jaw closes, making an ideal device for catching fish. As in all rhamphorhynchoids, the tail is long and stiff, the vertebrae lashed together by tendons.

Below: Dorygnathus is similar to a new pterosaur, Cacibupteryx, which was found in Cuba in 2004, extending the range of rhynchosauroids.

Distribution: Germany.
Classification: Pterosauria, Rhamphorhynchoidea.
Meaning of name: Spear jaw.
Named by: Wagner, 1860.
Time: Pliensbachian stage of the early Jurassic.
Size: 1m (3ft) wingspan.
Lifestyle: Fish-hunter.
Species: *D. banthensis.*
D. mistelgauensis.

WING STRUCTURE

The rhamphorhynchoid wing (below top) differed from that of the pterodactyloid, principally in the structure of the supporting limb. The wrist bone was short, and so the weight of the wing was carried mostly on the elongated fourth finger. The bones of the finger were as thick as those of the arm. A rhamphorhynchoid wing was generally longer and narrower than that of a pterodactyloid.

In a pterodactyloid's wing (below bottom) the wrist bones were often as long as the humerus. This put the fingers of the hand further along the leading edge of the wing.

In both there was an extra bone at the base of the wrist, which appears to have supported a membrane that stretched in front of the arm bones between the wrist and the base of the neck. This would have been used to adjust the flow of air over the wing and give some manoeuvrability.

The other main difference between the two pterosaur groups was the presence of a long tail in the rhamphorhynchoids. This consisted of many long vertebrae lashed together into a stiff rod by tendons and bearing a diamond-shaped fin at the end. It would have been used in flight for steering.

Parapsicephalus

This pterosaur is known only from fragmentary skulls from Whitby, England, although some bones from Germany, found in 1970, may represent a more complete skeleton. It was a significant find. This is the first pterosaur skull that was discovered preserved in three dimensions allowing a study of the brain cavity. Since the original find, however, there have been better brain cavities found in specimens of *Rhamphorhynchus* and *Anhanguera*.

Features: *Parapsicephalus* appears to be a typical rhamphorhynchoid pterosaur, so typical that it was originally classed as a species of *Scaphognathus*, and even now is thought by some to be a specimen of *Dorygnathus*. The skull is longer than those usually found in the group. The brain is much smaller than that of a bird, but larger than that of any terrestrial reptile. The angle of the ear structure suggests that the head was held horizontally in flight.

Distribution: North-east England, possibly Germany.
Classification: Pterosauria, Rhamphorhynchoidea.
Meaning of name: Almost an arched head.
Named by: Newton, 1888.
Time: Toarcian stage of the early Jurassic.
Size: 1m (3ft) wingspan.
Lifestyle: Fish hunter.
Species: *P. purdoni,*
P. mistelgauensis.

Left: The cast of the brain shows that the semicircular canals in the ear region are well developed. This implies that Parapsicephalus had a good sense of balance while in flight.

COELOPHYSID MEAT-EATERS

The coelophysids, the most extensive hunters of Triassic times, continued into the Jurassic period and were the main predators of the early part of the period before being replaced by all sorts of new theropods that were beginning to evolve. The coelophysids existed all over the world from North America to Africa and China.

Dilophosaurus

Distribution: Arizona, USA, China.
Classification: Theropoda, Coelophysoidea.
Meaning of name: Lizard with two crests.
Named by: Welles, 1954.
Time: Sinemurian to Pliensbachian stages of the early Jurassic.
Size: 6m (19½ ft).
Lifestyle: Hunter.
Species: *D. wetherilli, D. breedorum, D. sinensis.*

Right: The crests of Dilophosaurus have never been found attached to the skull. However, the standard restoration seems the most convincing.

Familiar from its appearance in the *Jurassic Park* film, unfortunately *Dilophosaurus* was inaccurately portrayed, being much too small and having poison glands and an erectile neck frill, all from the film designer's imagination. Three skeletons were found together, suggesting that it may have hunted in packs. Another species was found in China, showing that the genus was quite widespread.

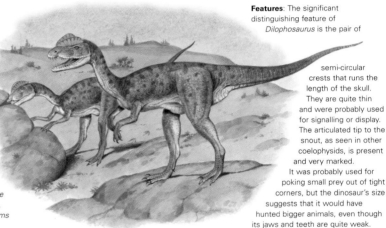

Features: The significant distinguishing feature of *Dilophosaurus* is the pair of semi-circular crests that runs the length of the skull. They are quite thin and were probably used for signalling or display. The articulated tip to the snout, as seen in other coelophysids, is present and very marked. It was probably used for poking small prey out of tight corners, but the dinosaur's size suggests that it would have hunted bigger animals, even though its jaws and teeth are quite weak.

Segisaurus

Lacking hollow, bird-like bones *Segisaurus* was originally thought to have been different from other meat-eating dinosaurs of the time. This later proved not to have been the case, and it is currently classified among the coelophysids. Like the other coelophysids it was built for running and probably jumping, making it an active hunter. It would have been a desert-living animal with the ability to survive sand storms and droughts.

Features: The single, incomplete skeleton of *Segisaurus*, consisting of backbones, ribs, shoulder, pelvic girdles and limb bones, found in desert sandstone, unfortunately does not reveal much about the dinosaur. It seems to have a collar bone, which is unusual in such an early dinosaur, and its hands are long and slender. This suggests that the arms were strong and used for catching prey, probably the small prosauropods that lived in the deserts.

Distribution: Arizona, USA.
Classification: Theropoda, Coelophysoidea.
Meaning of name: Lizard from the Segi Canyon.
Named by: Camp, 1936.
Time: Toarcian stage of the early Jurassic.
Size: 1m (3ft).
Lifestyle: Hunter.
Species: *S. halli.*

Right: The lack of a skull and teeth means that we cannot be sure about the diet of Segisaurus. However, its fast-runner build and clawed hands suggest that it was a meat-eater.

FORMER THEROPOD CLASSIFICATION

Theropods used to be classified in a simple way – coelurosaurs were the small ones, and carnosaurs were the big ones. Under this system the coelophysids were all coelurosaurs, except for *Dilophosaurus* which was too big and was classed as a carnosaur. The coelurosaurs encompassed every small dinosaur from the Triassic period to the end of the Cretaceous. This system of classification, however, did not indicate how or whether the theropods had a common ancestral root, and what the familial relationship was between the different species.

Convenient though this classification might have seemed, it made less sense as new specimens came to light in the 1980s and 1990s. Recognizable groups of related dinosaurs, such as the tyrannosaurs, incorporated the really big examples that were obviously carnosaurs, but also had small, fleet-footed types that were known as coelurosaurs.

The names carnosaur and coelurosaur still exist in dinosaur classification, but now they only apply to certain groups and the details are much more precise and technical.

Below: A carnosaur (top) and a coelurosaur (bottom) according to the old classification.

Podokesaurus

The only remains of this dinosaur were destroyed in a museum fire in 1917, so all recent work has been carried out on casts. The Connecticut valley, USA, where it was found, is famed as the site of extensive dinosaur tracks found in the early nineteenth century. The fossil footprints named *Grallator* may have been formed by *Podokesaurus*.

Features: In its build *Podokesaurus* is very similar to the smaller specimens of the late Triassic *Coelophysis*, from the other side of the North American continent, with its slim body, long hind legs, long tail and long, flexible neck. In fact the two are so similar that many palaeontologists think that they are the same genus. Unfortunately, since the original specimen was destroyed, this can never be proved because the surviving casts are too poor to give much information. The specimen appears to have been a juvenile, incompletely grown, and so the small size quoted could be an underestimation.

Distribution: Massachusetts, USA.
Classification: Theropoda, Coelophysoidea.
Meaning of name: Swift foot lizard.
Named by: Talbot, 1911.
Time: Pliensbachian to Toarcian stages of the early Jurassic.
Size: 1m (3ft).
Lifestyle: Hunter.
Species: *P. holyokensis*.

Right: Podokesaurus is the earliest dinosaur to have been found in the eastern United States. It was discovered in 1910 by local fossil hunter Dr Mignon Talbot, nicknamed the "dinosaur lady".

Syntarsus

More than 30 *Syntarsus* skeletons were found in a bonebed in Zimbabwe, suggesting that a pack was overwhelmed by a disaster such as a flash flood. The cololites, or stomach contents, suggest that they preyed on smaller vertebrates.

Another species of *Syntarsus*, found in Arizona, has a pair of crests on the head, rather like those of *Dilophosaurus* but smaller.

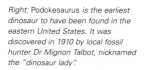

Features: The close similarity between *Syntarsus* and *Coelophysis* has led many to suggest that the two are actually the same genus, with three nimble fingers, a long neck and tail, strong hind legs and a slim body. There seem to have been two sizes of adult – the larger was probably the female and the smaller the male, judging by the

Right: There is a possibility that the name Syntarsus is pre-occupied by an insect. If this proves to be the case the generic name will be changed to Megapnosaurus, and the two known species will be M. rhodesiensis and M. kayentakatae.

size ranges in modern bird flocks. Computer reconstructions of the braincase of *Syntarsus* show it to be quite large compared with that of the earlier herrerasaurids, indicating an increase in intelligence. A bird-like cunning seems to have been evolving at this time.

Distribution: Zimbabwe.
Classification: Theropoda, Coelophysoidea.
Meaning of name: Fused ankle.
Named by: Raath, 1969.
Time: Hettangian to Pliensbachian stages of the early Jurassic.
Size: 2m (6½ft).
Lifestyle: Hunter.
Species: *S. rhodesiensis*, *S. kayentakatae*.

DIVERSIFYING MEAT-EATERS

In early Jurassic times life began to diversify. Different plant-eating animals began to appear in different areas and, as a result, different meat-eating dinosaurs evolved to take advantage of this new variety of food. The coelophysids, who had been the main meat-eaters up to this point, were beginning to be replaced by a variety of other carnivorous hunters.

Sarcosaurus

Early Jurassic Britain consisted of a scattering of low islands across a shallow sea on the northern edge of the supercontinent. The islands were wooded and would have supported plenty of wildlife. *Sarcosaurus* was the big hunter of the time, and its remains are found in the marine lias beds – beds of alternating shale and limestone – having been washed out to sea.

Right: Knowledge of another species, S. andrewsi, also from the early Jurassic period in England, is based on an isolated femur. That now seems to have belonged to a coelophysid.

Features: All we know of this dinosaur is a partial pelvis, a femur and some vertebrae. It is a lightly built, two-footed predator. The state of the bones shows it to have been an adult. Its pelvis is remarkably similar to the later *Ceratosaurus*, and some palaeontologists think it is actually an early species of that genus. Officially, though, it is classed in the neoceratosaurids, a group placed between the coelurosaurids and the ceratosaurids on the evolutionary chart. The bones show similarities with *Ceratosaurus* and also with *Dilophosaurus* and *Liliensternus*.

Distribution: Leicestershire, England.
Classification: Theropoda, Neoceratosauria.
Meaning of name: Flesh-eating lizard.
Named by: Andrews, 1921.
Time: Sinemurian stage of the early Jurassic.
Size: 3.5m (12ft).
Lifestyle: Hunter.
Species: *S. woodi.*

'Saltriosaurus'

Saltriosaurus was found in a limestone quarry in northern Italy, near the Swiss border, in 1996. At the time of writing it has not been scientifically described, and so its name is unofficial. It is important, though, because preliminary studies show it to have been the earliest known member of the Tetanurae, a name that means "stiffened tail", the group to which all the advanced theropods belong.

Right: As well as the way the tail vertebrae are bound together, the tetanurae are characterized by the presence of no more than three fingers, and an extra opening in the skull.

Features: Just one example of *Saltriosaurus* has been found. It consists of about 100 bone fragments, which is about 10 per cent of the skeleton. It seems to have had a long neck and would have weighed more than one tonne. The fact that it is a hunter was proved by the presence of a sharp cutting tooth which is 7cm (2¾in) long. The bones include a wishbone and three-fingered hands, both features of hunting dinosaurs that came in later times. Although it was found in marine limestone, its sheer size suggests that there must have been big landmasses close by.

Distribution: Northern Italy.
Classification: Theropoda, Tetanurae.
Meaning of name: Lizard from Saltrio.
Named by: Dal Sasso, 2000 (but this is unofficial).
Time: Sinemurian stage of the early Jurassic.
Size: 8m (26ft).
Lifestyle: Hunter.
Species: None allocated.

Cryolophosaurus

Distribution: Antarctica.
Classification: Theropoda,
Tetanurae, Carnosauria.
Meaning of name: Frozen
crested lizard.
Named by: Hammer and
Hickerson, 1994.
Time: Pliensbachian stage of
the early Jurassic.
Size: 6m (19½ ft).
Lifestyle: Hunter.
Species: *C. elliotti*.

The name *Cryolophosaurus*, meaning "frozen crested lizard", is a bit of a misnomer. Although its skeleton was found in the Antarctic, this was not a frozen continent in early Jurassic times. The area was much further north, and on the edge of the southern supercontinent of Gondwana – one of the two landmasses formed as the ancient Pangaea split in two. The environment may not have been tropical, but it was certainly much more temperate than today. Other remains in the area consist of prosauropods, pterosaurs and smaller meat-eaters, and there seems to have been quite a breadth of animal life.

The remoteness of the site means that very little field work has been done there, but an expedition in 2004, led by William Hammer of Augusta College, Georgia, USA, who found the original specimen, has yielded even more bones.

Features: *Cryolophosaurus* is a big meat-eating dinosaur with a peculiar crest on its head. In most crested theropods the crest runs fore-and-aft. In *Cryolophosaurus* it runs crossways, turning up in a distinctive quiff that earned the animal the unofficial name "Elvisaurus" before it was formally described. This crest must have been used for display. The skull is tall and narrow, and the lower jaw deep and strong.

DINOSAUR DIVERSIFICATION

The sudden expansion of different types of meat-eating dinosaur in the early Jurassic period must have mirrored the expansive evolution of different types of plant-eating animals. The more types of prey there are, the more kinds of predator evolve to hunt them.

A reason for this expansion may be found in the geography of the time. The single great landmass of Pangaea, which had existed through most of Permian and Triassic times, was beginning to split up. Two main supercontinents were beginning to appear – Laurasia in the north and Gondwana in the south. It would be a long time before they were totally separate, but the cracks were beginning to open.

The rift valleys that formed along the cracks produced arms of the ocean that penetrated deep into the landmasses. The Red Sea, in Egypt, today is an example. At the same time a rising sea level spread shallow seas over much of the continents, spreading moister climates into the interiors, where there had formerly been desert. Lush vegetation began to grow, and this encouraged the development of widespread herbivorous animals, and subsequently carnivorous forms.

The splitting of the landmass was ultimately leading to the formation of new continents, and the spreading of seas on to the continental shelves resulted in the formation of many strings of islands across shallow areas of sea. Isolation is always a spur to evolution, with new species appearing on remote islands, separate from, and different from, living things elsewhere.

THE LATE PROSAUROPODS

Although the sauropods made their presence felt in the early Jurassic period, the more primitive prosauropods continued to survive in many parts of the world. Their remains have been found in early Jurassic rocks of North America and southern Africa, but China seems to be the last bastion of this group, with several prosauropod genera known from the depths of the Asian continent.

Anchisaurus

Though discovered in 1818 *Anchisaurus* was not recognized as a dinosaur until 1885. The two known species may actually represent the male and female versions of the same, with the larger being the female. It was originally thought to have been a Triassic contemporary of *Plateosaurus*, but in the 1970s the New England sandstones in which the fossil was found were proved to be Jurassic in age.

Features: If *Plateosaurus* is the classic example of the biggest medium-size prosauropod, then *Anchisaurus* is the best-known of the smallest medium-size prosauropods. It has the long neck and tail, and a slim body, with long hind legs that allowed it to walk on all fours or with its forelimbs off the ground. The teeth are bigger and the jaw mechanism stronger than those of *Plateosaurus*, suggesting that it ate tougher food.

Left: Between its discovery in 1818 and its proper identification in 1885, this animal lived under a number of names, including Megadactylus and Amphisaurus, both of which were pre-occupied.

Distribution: Connecticut and Massachusetts, USA.
Classification: Sauropodomorpha, Prosauropoda, Anchosauridae.
Meaning of name: Near lizard.
Named by: Marsh, 1885.
Time: Pliensbachian to Toarcian stages of the Jurassic.
Size: 2.5m (8ft).
Lifestyle: Low browser.
Species: *A. major*, *A. polyzelus*.

Jingshanosaurus

This is one of the best-known of the Chinese prosauropods, and is known from a complete skeleton. It is named after the town of Jingshan, or Golden Hill, in Lufeng province, China, close to where the skeleton was found. This town is the site of the Museum of Lufeng Dinosaurs, one of several big dinosaur museums in China.

Features: The complete skeleton shows *Jingshanosaurus* to be a typical large prosauropod, with a heavy body, long neck and tail, and legs that would have allowed a bipedal or quadrupedal mode of travel. It has the characteristic big sauropod claw on the thumb. The skeleton looks like a particularly large version of *Yunnanosaurus*: it may well be the same animal.

Distribution: China.
Classification: Sauropodomorpha, Prosauropoda, Massospondylidae.
Meaning of name: Lizard from Golden Hill.

Named by: Zhang and Young, 1995.
Time: Early Jurassic.
Size: 9.8m (32ft).
Lifestyle: High browser.
Species: *J. dinwaensis*.

Left: Although Jingshanosaurus was a typical prosauropod with the typical prosauropod dentition and presumably vegetarian diet, its describers have suggested that it may have eaten molluscs as well.

Yunnanosaurus

This dinosaur is a well-known Chinese prosauropod. There have been about 20 skeletons found, two of which have skulls. One specimen has more than 60 teeth preserved, giving an insight into the feeding mechanisms of the big prosauropods. However, there is a suggestion that the teeth actually come from a different animal, a sauropod.

Features: The distinctive feature of *Yunnanosaurus* is the teeth. They are very similar to those of one of the later sauropods and, had they been found on their own, would have been identified as part of a sauropod dentition. The wear on them is the same as that of sauropods, and shows that the advanced prosauropods had the same eating strategies as the sauropods. The rest of the skeleton is a typical, medium-size prosauropod.

Distribution: China.
Classification: Sauropodomorpha, Prosauropoda, Massospondylidae.
Meaning of name: Lizard from Yunnan.
Named by: Young, 1942.
Time: Hettangian to Pliensbachian stages of the early Jurassic.
Size: 7m (23ft).
Lifestyle: High browser.
Species: *Y. huangi*.

Left: Yunnanosaurus huangi *is the only species in the genus.* Jingshanosaurus *is possibly a large species of* Yunnanosaurus*. If it is proved to be the same genus it will be renamed* Y. dinwaensis.

LUFENG BASIN

The Lufeng Basin in south-western China consists of continental sediments up to 1,000m (3,280ft) thick. The lower part of it was originally thought to have been Triassic in age, but now the whole sequence is known to be early Jurassic. The sediments were deposited in rivers and lakes, and show a picture of a varied flora and fauna of the interior of the continent in early Jurassic times.

Animals found there consist of small mammals, mammal-like reptiles, crocodiles and dinosaurs, particularly prosauropods. The mammal-like reptiles are unusual since they mostly existed in the Permian period, dying out elsewhere in Triassic times. Everywhere else they were replaced by their descendants, the mammals proper. Their continued existence into the early Jurassic parts of the Lufeng Basin led to the idea that these beds were Triassic in age.

With late surviving prosauropods and mammal-like reptiles, the Lufeng Basin could almost be considered as a kind of 'lost world' with animals surviving into early Jurassic times that had died out elsewhere in the previous Triassic period. Environmental conditions may have remained stable there while they changed in other parts of the world. Such an occurrence is biologically known as a "refugium". A modern example would be mountain-tops that still sustain arctic fauna that has been extinct in the lowlands since the end of the Ice Age.

Massospondylus

The first *Massospondylus* specimen to be found consisted of a few broken vertebrae shipped to Sir Richard Owen, in London, from South Africa in 1854. Since then the skeletons of more than 80 individuals have been found across southern Africa. There has even been a nest of six eggs found that have been attributed to *Massospondylus*. Another possible specimen has been found in Arizona, which may indicate that this was a very widespread animal.

Distribution: Lesotho, Namibia, South Africa and Zimbabwe.
Classification: Sauropodomorpha, Prosauropoda, Massospondylidae.
Meaning of name: Massive vertebra.
Named by: Owen, 1854.
Time: Hettangian to Pleinsbachian stages of the early Jurassic.
Size: 4m (13ft).
Lifestyle: Low browser.
Species: *M. carinatus*, *M. hislopi*.

Features: The teeth are large, some serrated and some flat. It is usually shown as a much more slender animal than other prosauropods of the same size. It has five fingers, but the fourth and fifth are very small. The huge, clawed first finger could be curved over, thumb-wise, across the second and third, making this a versatile hand.

PRIMITIVE SAUROPODS

The primitive sauropods, the big plant-eating dinosaurs with the long necks, originated in Triassic times, but came to prominence during the early Jurassic period. The Victorian palaeontologists named the group "sauropod" because they saw a resemblance in the arrangement of the foot bones to that of a lizard, as the sauropod had five toes compared with three of the contemporary theropods.

"Kunmingosaurus"

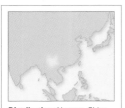

Distribution: Yunnan, China.
Classification:
Sauropodomorpha,
Sauropoda, maybe
Vulcanodontidae.
**Meaning of
name**: Lizard
from Kunming.
Named by: Chao, 1985.
Time: Early Jurassic.
Size: 11m (36ft).
Lifestyle: Herbivore.
Species: "*K. wudingensisu*".

A name that has been given to a fossil animal that has not been formally described in a scientific paper is a *nomen dubium*, a doubtful name. This does not mean that the animal did not exist, but shows that we cannot say too much about it with confidence since other scientists have not had the opportunity to check the original discoverer's research. This is the case with "*Kunmingosaurus*", and the custom is to put the name in inverted commas to show that it is a *nomen dubium*.

Features: "*Kunmingosaurus*" has not been scientifically described. However, there has been an impressive full-body skeleton constructed. All that we can say about the animal is that it is a primitive sauropod and walked on all fours.

Left: The mounted skeleton is assembled from disarticulated bones found in the same quarry in 1954. At first the teeth were thought to have been from Yunnanosaurus *and the jaw from* Lufengosaurus *– both prosauropods.*

Kotasaurus

Known from a single bone bed in Andhra Pradesh in India, the remains of 12 individuals of various sizes of *Kotasaurus* lay in river sandstones. They are possibly the remains of a herd that had been drowned during a river crossing. A mounted skeleton now stands in the Science Centre in Hyderabad.

Features: *Kotasaurus* seems to be intermediate between the prosauropods and the sauropods. Some of the hip bones are prosauropod-like, but others are definitely those of a sauropod, as are the vertebrae and teeth. Studies carried out in 2001 show that its simple vertebrae, its narrow shoulder bone and slim leg bones are all signs of a primitive sauropod. Like other early sauropods, it is a heavy, four-footed animal with a long neck and small head.

Distribution: India.
Classification:
Sauropodomorpha,
Sauropoda, Vulcanodontidae.
Meaning of name: Lizard
from Kota province.
Named by: Yadagiri,
1988.
Time: Toarcian stage of
the early Jurassic.
Size: 9m (30ft).
Lifestyle: Herbivore.
Species: *K. yamanpalliensis*.

Vulcanodon

When it was discovered in 1972, *Vulcanodon* was regarded as a prosauropod. The teeth were definitely those of a meat-eater, which fitted in with the current thinking that the prosauropods were omnivorous. It has now been established that these teeth, for which the animal was named, belonged to an unidentified theropod that had scavenged the carcass. The true teeth of this animal, as well as the skull and neck, are unknown.

Features: The status of *Vulcanodon* as a sauropod was established in 1975 on the basis of the shape of the toe bones. These are broader than they are deep – a sauropod feature – although they seem to be even broader than those of most other sauropods. Another distinguishing feature of *Vulcanodon* is the length of the front legs, which are comparatively long compared to other sauropods. The ichnogenus *Deuterosauropodopus* from Lesotho probably represents the footprints of *Vulcanodon*.

Distribution: Zimbabwe.
Classification: Sauropodomorpha, Sauropoda, Vulcanodontidae.
Meaning of name: Volcano tooth, from the volcanic deposits in which it was found.
Named by: Raath, 1972.
Time: Hettangian stage of the early Jurassic.
Size: 6.5m (21ft).
Lifestyle: Herbivore.
Species: *V. karibaensis*.

Left: The restoration of Vulcanodon *is made with difficulty – the only skeleton found had been eaten by a theropod of some kind.*

Barapasaurus

Barapasaurus seems to have been quite a common early Jurassic sauropod in India. More than 300 specimens representing parts of about six individuals have been found. However, as is common in sauropod dinosaurs, no skulls are known, and neither have the foot bones been found, which poses further difficulties for precise classification.

Features: Although *Barapasaurus* is defined, and has been scientifically described, based on the hip bone, many of the other bones are known as well. The vertebrae in the back have a characteristic deep cleft for the spinal cord, a feature that distinguishes this animal from other sauropods. It appears to have had rather long and slim legs. The teeth are spoon-shaped, like many later sauropods, and used for raking leaves from branches. The narrowness of the hip bones suggest that this dinosaur belongs to the Cetiosauridae, rather than the more primitive Vulcanodontidae.

Distribution: India.
Classification: Sauropodomorpha, Saurischia, Cetiosauridae.
Meaning of name: Big legged lizard.
Named by: Jain, Kutty, Roy-Chowdry and Chatterjee, 1975.
Time: Toarcian stage of the early Jurassic.
Size: 18m (59ft).
Lifestyle: Herbivore.
Species: *B. tagorei*.

Right: Barapasaurus *was the first of the really big sauropods, comparable in length to the giants that appeared later in the Jurassic. A mounted skeleton stands in the Geological Studies Unit of the Indian Statistical Institute, Calcutta, India.*

LITTLE ORNITHOPODS

The early plant-eating ornithopods were small, beaver-sized animals. By the early Jurassic period some had not yet evolved the cheek pouches and the chewing teeth of the later forms, but one group, the heterodontosaurids, had developed a very strange, almost mammal-like arrangement of differently sized teeth. The remains of these animals are known from desert deposits of southern Africa.

Lesothosaurus

The most primitive ornithischians, such as *Lesothosaurus*, had not evolved the complex chewing mechanism that was to characterize the later forms. Instead, they would have crushed their food by a simple up and down chopping action of the jaws. This is quite an unspecialized feeding method, and these animals may well have eaten carrion or insects as well as plants in order to survive.

Features: *Lesothosaurus* is one of the most primitive of the ornithischians, and as such it is difficult to put into a strict classification. It is a small, two-footed plant-eater, built for speed. The head, on the end of a flexible

Right: Lesothosaurus is very similar to the earlier-discovered Fabrosaurus. *However the* Fabrosaurus *material is so poor it is impossible to make direct comparisons. If they are the same genus, then the name* Fabrosaurus *would have to take precedence, being applied first.*

neck, is short, triangular in profile, with big eyes. The teeth are arranged in a simple row and, unlike all other ornithopods, the mouth does not seem to have cheeks. The jaw action is one of simple chopping. The snout ends with a horn-covered, vegetation-cropping beak.

Distribution: Lesotho, South Africa.
Classification: Ornithischia, Fabrosauridae.
Meaning of name: Lizard from Lesotho.
Named by: Galton, 1978.
Time: Hettangian to Sinemurian stages of the early Jurassic.
Size: 1m (3ft).
Lifestyle: Low browser.
Species: *L. diagnosticus*.

Abrictosaurus

The name of the "wide-awake lizard" derives from a dispute between palaeontologists. Tony Thulborn proposed that the heterodontosaurids slept away the hot desert summer as some modern animals do. J. A. Hopson disagreed with this theory, based on the study of the growth of the teeth, and celebrated his notion of a year-round active animal by giving this new dinosaur an appropriate name.

Features: Our knowledge of *Abrictosaurus* is based on two skulls and some fragmentary pieces of the skeleton. The skulls are almost identical to those of *Heterodontosaurus*, with a similar varied arrangement of teeth, except that they lack the prominent tusks. The remains may, in fact, represent female specimens of *Heterodontosaurus*. It is possible that only the males had the tusks and used them for

Right: Only the skull and a few bones of Abrictosaurus *are known, and so the remainder of the restoration is based on the skeleton of* Heterodontosaurus.

display, as with many modern animals. On the other hand, many modern animal groups, such as the pigs, have prominent tusks in some genera but not in others. A wild boar's differ from those of a warthog.

Distribution: Lesotho, South Africa.
Classification: Ornithischia, Heterodontosauridae.
 Meaning of name: Wide-awake lizard.
 Named by: Hopson, 1975.
Time: Hettangian to Sinemurian stages of the early Jurassic.
Size: 1.2m (4ft).
Lifestyle: Herbivore.
Species: *A. consors*.

Lanasaurus

This animal – the 'woolly lizard' – was named in honour of the noted palaeontologist A. W. Crompton, whose nickname was 'Fuzz'. As with most of the heterodontosaurids it is known only from the jaw and teeth. It is possible that it is another specimen of *Lycorhinus*, or one of the other genera of heterodontosaurids that seemed to have abounded in southern Africa in early Jurassic times.

Below: Lanasaurus is closely related – or indeed identical – to Lycorhinus, which is known only from the jaw bone. It was so mammal-like that it was first identified as an early mammal. The name means "wolf snout".

Distribution: South Africa.
Classification: Ornithischia, Heterodontosauridae.
Meaning of name: Woolly lizard.
Named by: Gow, 1975.
Time: Hettangian to Sinemurian stages of the early Jurassic.
Size: 1.2m (4ft).
Lifestyle: Herbivore.
Species: *L. scalpridens*.

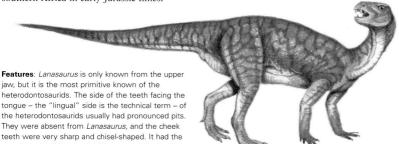

Features: *Lanasaurus* is only known from the upper jaw, but it is the most primitive known of the heterodontosaurids. The side of the teeth facing the tongue – the "lingual" side is the technical term – of the heterodontosaurids usually had pronounced pits. They were absent from *Lanasaurus*, and the cheek teeth were very sharp and chisel-shaped. It had the prominent, tusk-like teeth of the rest of the family.

HETERODONTOSAURID TEETH

So, what was the purpose of the strange dentition of the heterodontosaurids? The chisel-like teeth at the back were obviously for chewing. The muscular cheeks would have held the plant material as it was worked over again and again by the chewing action. The cheeks themselves have not been fossilized, of course, but we surmise that they were there because the tooth rows were set in from the side of the skull, leaving a gap at each side of the mouth that must have been covered over with something. Also, the chopping action of the teeth would have meant that half the food would have escaped the mouth if there had been nothing there to catch it.

The big tusks may have been used for digging up tubers that were part of the desert vegetation. They may also have been used as defensive weapons, protecting the family or herd against the attacks of meat-eating dinosaurs. A third purpose could have been display, being used in ritual combat and competition in the herd hierarchy, as we see in modern warthogs. If this were the case it could explain why only some specimens have been found with the very big tusks. Perhaps only the males sported them.

Below: Heterodontosaurus teeth and skull.

Heterodontosaurus

The original description of *Heterodontosaurus* in 1962 was based on a single skull, but in 1976 it was backed up by the discovery of one of the most complete and well-preserved dinosaur skeletons known – almost caught in an action pose as if it had been running away and frozen in time. It shows a swift-footed animal with long front legs and strong hands bearing five fingers.

Features: The typical heterodontosaurid tooth pattern consists of three pairs of sharp stabbing teeth at the front of the mouth, the rearmost of which are big like canine tusks. It also has a set of closely packed shearing and grinding teeth further back. The rear teeth would have been enclosed in cheeks to hold the food while it was being chewed. The front teeth worked against a horny beak on the lower jaw.

Distribution: South Africa.
Classification: Ornithischia, Heterodontosauridae.
Meaning of name: Differently-toothed lizard.
Named by: Crompton and Charig, 1962.
Time: Hettangian to Sinemurian stages of the early Jurassic.
Size: 1m (3ft).
Lifestyle: Herbivore.
Species: *H. tucki*.

PRIMITIVE ARMOURED DINOSAURS

In early Jurassic times, a new group of dinosaurs began to appear. They were distinguished by the presence of armour on their backs, as bony plates covered in horn, embedded in the skin. Armoured dinosaurs were all plant-eaters and evolved from the ornithischian, or the bird-hipped, line. The evolutionary spur would have been the presence of so many big, contemporary, meat-eating dinosaurs.

Scutellosaurus

This plant-eating dinosaur that would have sought refuge from its enemies in two different ways, either by running away, or by squatting down and letting the armour deter enemies. It had long hind legs and a very long tail for balance that would have enabled it to run. Its strong front legs would have carried its body as it hunkered down in passive defence.

Features: If it were not for the armour plates, this animal would have been regarded as a fabrosaurid, like *Lesothosaurus*. It has the simple dentition, and the long hind legs and tail. The presence of this armour suggests that *Scutellosaurus* was close to the ancestry of the big plated and armoured dinosaurs to come. It would have evolved from purely bipedal ancestors.

Distribution: Arizona, USA.
Classification: Ornithischia, basal Thyreophora.
Meaning of name: Little shield lizard.
Named by: Colbert, 1981.
Time: Hetangian stage of the early Jurassic.
Size: 1.2m (4ft).
Lifestyle: Low browser.
Species: *S. lawleri*.

Left: The armour consists of more than 300 little shields in six different types, ranging from tiny lumps to big stegosaur-like plates.

Scelidosaurus

The first almost complete dinosaur skeleton to be discovered and named was *Scelidosaurus*. As an armoured dinosaur it would have been a slow-moving creature, but the length of its hind legs and weight of its tail seem to suggest that it may have been able to run on two legs for short distances.

Features: This very primitive animal belonged to the group known as the thyreophorans. The group consists of the plated dinosaurs (the stegosaurs) and the armoured dinosaurs (the ankylosaurs). In the past it has been regarded as ancestral to the stegosaurs, but now it is thought to be closer to the ancestors of the ankylosaurs. Its back, sides and tail are covered by an arrangement of armoured scutes with a mosaic of small armoured scales between them.

Distribution: Dorset, England.
Classification: Ornithischia, Thyreophora, Ankylosauromorpha.
Meaning of name: Leg lizard.
Named by: Owen, 1868.
Time: Sinemurian to Pliensbachian stages of the early Jurassic.
Size: 4m (13ft).
Lifestyle: Low browser.
Species: *S. harrisonii*.

Right: Two almost complete skeletons of Scelidosaurus *have been found in southern England, and it is from these skeletons that we can reconstruct the likely appearance of related, less complete, animals.*

Emausaurus

The scientific description of *Emausaurus* is based on an almost complete skull and some skeletal parts found in northern Germany. The skull is about half the size of that of the better-known *Scelidosaurus* but does not seem to have been as heavily armoured, and probably represents an early stage in the evolution of the armoured dinosaurs.

Features: The skull and teeth of *Emausaurus* are very similar to those of *Scelidosaurus*, but the head seems to be wider towards the rear and narrows towards the snout. The jaw joint is very simple, which suggests that the mouth worked in a scissor-like action, but there does not seem to have been any wear on the teeth. Perhaps *Emausaurus* just tore at the plants with its mouth, and swallowed without chewing.

Distribution: Germany.
Classification: Ornithischia, Thyreophora, Ankylosauria.
Meaning of name: Acronym of Ernst-Moritz-Arndt Universität lizard.
Named by: Haubold, 1990.
Time: Toarcian stage of the early Jurassic.
Size: 2m (6ft).
Lifestyle: Low browser.
Species: *E. ernsti*.

Left: Despite its placing with Scelidosaurus *and its relatives here, certain aspects of the skull are stegosaur-like – very close to* Huayangosaurus.

"Lusitanosaurus"

French Jesuit priest and palaeontologist Albert-Felix de Lapparent (1905–75), who hunted dinosaurs all over Europe and North Africa was the finder of "*Lusitanosaurus*". The specimen consisted of a fragment of jawbone a few centimetres (1–2in) long. It appears to be rather like that of *Scelidosaurus* and is about the same age, hence it belongs to the same family.

Features: "*Lusitanosaurus*" is something of a *nomen dubium*. Only parts of the upper jaw and an array of eight teeth are known. These differ from those of *Scelidosaurus* in being taller, and in lacking the scalloped edges of the English specimen. There is no guarantee that the rest of the animal looked anything like *Scelidosaurus*, but it is the best guess until more remains come to light that may confirm or disprove this suggestion.

Distribution: Portugal.
Classification: Ornithischia, Thyreophora, Ankylosauria.
Meaning of name: Lizard from Lusitania (an old name for Portugal).
Named by: de Lappaarent and Zbyszewski, 1957.
Time: Sinemurian stage of the early Jurassic.
Size: 4m (13ft).
Lifestyle: Low browser.
Species: "*L. liasicus*".

Right: Albert de Lapparent, who discovered and named "Lusitanosaurus", made many dinosaur discoveries between the late 1940s and the early 1960s. Most of his work was done in the Sahara Desert, Africa.

MEGALOSAURUS AND ITS CLONES

The big, meat-eating dinosaurs whose fossils have been found in the middle Jurassic rocks, especially those from Europe, have traditionally been regarded as megalosaurs. This classification is undergoing revision. Although these big beasts seem similar to one another, being quite conservative in their build and appearance, they seem to have represented quite a range of animals.

Megalosaurus

The first dinosaur to be studied scientifically and named goes to *Megalosaurus*. It has always been the source of some confusion, however, since for a long time any early or middle Jurassic theropod found in Europe was assigned to the genus *Megalosaurus* without too much study. The scientific term "wastebasket genus" is applied to such a genus.

Features: Although the original *Megalosaurus* remains consist of nothing but a jawbone and teeth, and a few scraps of bone, subsequent discoveries have shown that it gives quite a good impression of a generalized, large meat-eating dinosaur. It has the big head with the long jaws and sharp teeth which are to be expected. It ran on powerful hind legs with the smaller front legs held clear of the ground. The tiny hands have three fingers and the feet have four toes, three of which reach the ground.

Distribution: England, France and perhaps Portugal.
Classification: Theropoda, Tetanurae.
Meaning of name: Big lizard.
Named by: Ritgen, 1826.
Time: Aalenian to Bajocian stages of the middle Jurassic.
Size: 9m (30ft).
Lifestyle: Hunter or shoreline scavenger.
Species: *M. bucklandii,* "*M. dabukaensis*," "*M. phillipsi*", "*M. tibetensis*", although these are *nomen dubia*.

Above: About 50 genera have been misassigned to Megalosaurus, *including Dryptosaurus, Proceratosaurus, Eustreptospondylus, Magnosaurus, Iliosuchus, Metriacanthosaurus, Carcharodontosaurus, Dilophosaurus and even the prosauropod* Plateosaurus.

Eustreptospondylus

This dinosaur was one of those theropods that was casually assigned to *Megalosaurus* until it was found to have been something completely different. The vertebrae are more curved than those of *Megalosaurus*, hence its name. It is known from an almost complete skeleton found in 1871. This skeleton is now mounted and displayed in the University Museum in Oxford, England.

Distribution: England.
Classification: Theropoda, Tetanurae.
Meaning of name: Well-curved vertebrae.
Named by: Walker, 1964.
Time: Callovian stage of the middle Jurassic.
Size: 9m (30ft), but only known from the skeleton of a 6m (19½ft) juvenile.
Lifestyle: Hunter.
Species: *E. oxoniensis*.

Features: *Eustreptospondylus* is a medium to large hunter which seems to have been more closely related to the ancestors of the later *Allosaurus* or of the spinosaurids than to its contemporary *Megalosaurus*. In the one skeleton found, the skull is fragmentary and difficult to reconstruct. The vertebrae are mostly missing their upper parts, suggesting that the skeleton is of a juvenile, and that its bones had not yet had time to knit together properly. Casts made from the disarticulated skull are often used as the basis for the skull of *Megalosaurus* in museum displays.

Above: The single skeleton of Eustreptospondylus *was found in marine clays. It had died on land and had been washed out to sea. Once the decaying body had released all its gases it sank to the bottom and was buried in the fine sediment.*

Poekilopleuron

We cannot say too much about *Poekilopleuron*, since most of its remains were destroyed during World War II when the Museé de la Faculté des Sciences de Caen, France, was bombed. What we do know is based on casts that were subsequently found in the Museum National d'Histoire Naturelle in Paris, France. An almost complete skull and other fragments assigned to *P. valesdunensis* were excavated in Normandy in 1994.

Features: The front limbs – almost all that is known of the original specimen – are very short but remarkably strong, with attachment areas for powerful muscles. From what is known of the rest of the skeleton, this animal, along with *Eustreptospondylus*, seems to belong among the ancestors of the spinosaurid group of theropods, judging from the length of spines on the bones of the middle neck. The skull, from the more recently found material, is quite low and has no ornamentation of any sort.

Distribution: France.
Classification: Theropoda, Tetaneurae.
Meaning of name: Varying side.
Named by: Eudes-Deslongchamps, 1838.
Time: Bathonian stage of the middle Jurassic.
Size: 9m (30ft).
Lifestyle: Hunter.
Species: *P. bucklandii*, *P. valesdunensis*, *"P. schmidtii"* – the *nomen dubium* of some unidentifiable bones from Russia.

A NEW CLASS OF REPTILES

For a long time the quarries in the so-called Stonesfield slate, actually a thinly bedded limestone, yielded fossil remains. The quarries were close to Oxford in England, and were worked for building materials. In about 1815 some bones, including a large jawbone with teeth in place, were discovered and passed to William Buckland, the first Professor of Geology at the University of Oxford. One of his colleagues, James Parkinson, published a drawing of the tooth, which was named *Megalosaurus* in 1822. As there was no concept of a dinosaur at that time, it was assumed that the jawbone and teeth came from some kind of gigantic lizard.

However, other big animals were coming to light, and in 1841 Sir Richard Owen established the classification "dinosauria" to include these discoveries. Even then they were regarded as giant lizards, and were represented as big, dragon-like, four-footed animals.

The most famous of these early restorations are the full-sized statues that were constructed for the Crystal Palace park in south London, in 1854. They are there to this day, erroneous but impressive for the insight they give to the scientific understanding of the time.

Left: The first impressions of Megalosaurus.

Magnosaurus

This is another of those theropods that had originally been assigned to *Megalosaurus*. When it was found to be a different genus it was renamed *Magnosaurus* in view of its similarity to it. The bones seem to have belonged to a juvenile animal. They are now in Oxford University, England, the last resting place of many British megalosaurs.

Features: *Magnosaurus* is not a very well-known animal. The only remains are jaws, teeth and bits of bone. Some of the limb bones are known only from the casts of the internal cavities. The teeth are thicker than those of *Megalosaurus*, and there are fewer in the jaws. It lived some time earlier, and there is still some argument over whether *Magnosaurus* is indeed just an early, small species of *Megalosaurus*.

Distribution: Dorset, England.
Classification: Theropoda.
Meaning of name: Huge lizard.
Named by: von Huene, 1932.
Time: Bajocian stage of the middle Jurassic.
Size: 4m (13ft).
Lifestyle: Hunter.
Species: *M. nethercombensis*, *M. lydekkeri*.

Left: Magnosaurus *was a swift, medium-size hunter. However we have no direct evidence as to its prey – there are hardly any other land-living vertebrate remains found in that particular horizon of the middle Jurassic.*

THEROPODS

The meat-eating dinosaurs ranged from little chicken-sized animals to great dragon-like monsters. There were also many that fell between the two extremes. From China to Australia, South America and Great Britain, the medium-size, meat-eating dinosaurs were widespread in middle Jurassic times. At that time the continents were beginning to split up, but the same kinds of animal still existed everywhere.

Ozraptor

This is the only Jurassic theropod to be found in Australia so far. In middle Jurassic times Australia was at the far reaches of the southern arm of the supercontinent Pangaea. It was still part of the major landmass and so, despite its remoteness, there is no reason why it should not have supported a similar fauna to the rest of the world.

Features: *Ozraptor* is only known from a part of a leg bone, but from it palaeontologists can infer that the animal was a swift-footed carnosaur. The joint at the ankle end is unique among theropods, and seems to have been adapted for fast running. In this respect it resembles that of a basal dromaeosaur, one of the swift hunters that existed in North America in Cretaceous times. There are probably many more Jurassic theropods to be discovered in Australia.

Right: The leg bone of Ozraptor was found in 1966 by college students and was originally thought to have belonged to a turtle. It was not until 30 years later that it was removed from its matrix and seen to be the bone of a dinosaur.

Distribution: Western Australia.
Classification: Theropoda.
Meaning of name: Oz (a colloquial name for Australia) lizard.
Named by: Long and Molnar, 1998.
Time: Bajocian stage of the middle Jurassic.
Size: Maybe 3m (10ft).
Lifestyle: Fast hunter.
Species: *O. subotaii*.

Kaijiangosaurus

This is an obscure theropod from the famous Dashanpu fossil beds of China. As with many potentially interesting dinosaurs it is only known from a partial skeleton from which we can deduce very little. Like its contemporary, *Gasosaurus*, it would have preyed on the vast range of plant-eating sauropod dinosaurs that inhabited this area during middle Jurassic times.

Features: All that is known of *Kaijiangosaurus* is a group of seven neck vertebrae. They are definitely the vertebrae of a carnosaur but are very primitive, lacking the ball-and-socket articulations that we usually find. Their unspecialized nature suggests that *Kaijiangosaurus* may be very close to the ancestors of the carnosaur group. There is even a suggestion that it is the same animal as *Gasosaurus*, which comes from the same time and place.

Above: Kaijiangosaurus and Gasosaurus (inset) would have been the main predators of the sauropods found in the Dashanpu quarries.

Distribution: China.
Classification: Theropoda, Tetanurae.
Meaning of name: Lizard from the Kaijiang River.
Named by: He, 1984.
Time: Bathonian to Callovian stages of the middle Jurassic.
Size: 6m (19½ft).
Lifestyle: Hunter.
Species: *K. lini*.

THE BODY SHAPE OF THE MEAT-EATERS

Plant-eating dinosaurs developed all kinds of different shapes during the 160-million year tenure of the Age of Dinosaurs. There were the big, elephantine, long-necked sauropods, the two-footed ornithopods that were either small enough to sprint away from danger or found safety in herds, the heavy thyreophorans with their decoration of plates or their defence of armour, and the later ceratopsians with their shields and horns.

For all that, the meat-eating dinosaurs seemed to have adopted a single shape and stuck with it for the entire Mesozoic. It was a shape that worked. The simple carnivorous digestive system did not need a big body. The resulting animal was light enough to be carried on two legs. The killing mechanisms – the jaws, teeth and claws – could be carried forward, balanced by the tail behind. The typical theropod was a splendidly designed killing machine.

Enough complete skeletons of theropods are known to tell us that few of them departed from this basic shape. Accordingly when new animals turn up, and they consist only of a few bones and teeth, we can be confident in the prediction that the whole animal would not have differed much from this basic shape.

Right: The shape of the meat-eating dinosaur.

Piatnitzkysaurus

The discovery of this carnosaur in the 1970s was the first indication that relatives of the North American dinosaurs actually existed in South America in middle Jurassic times. The implication is that North and South America were connected at that time, and the seaway between them that was known to exist later had not yet formed, isolating South America as an island continent. It is known from two partial skulls and some skeletal material. It is regarded as having the dinosaur name that is most difficult to spell!

Distribution: Argentina.
Classification: Theropoda, Tetanurae, Carnosauria.
Meaning of name: Piatnitzky's lizard, after a friend of the discoverer.
Named by: Bonaparte, 1986.
Time: Callovian to Oxfordian stages of the middle to late Jurassic.
Size: 4.3m (14ft).
Lifestyle: Hunter.
Species: *P. floresi*.

Features: This early ancestor of the *Allosaurus* group differed from its later relative in the more primitive set of hip bones, a longer arm bone and more powerful shoulders. It has a very similar build to *Allosaurus*, and presumably pursued the same hunting lifestyle. As it is much smaller than the local sauropods, it may have concentrated on hunting juveniles or the weak and ageing members of the plant-eater herds.

Proceratosaurus

Distribution: Gloucestershire, England.
Classification: Theropoda, Tetanurae, Carnosauria.
Meaning of name: Before Ceratosaurus.
Named by: von Huene, 1926.
Time: Bathonian stage of the middle Jurassic.
Size: 2m (6½ft).
Lifestyle: Hunter.
Species: *P. bradleyi*, *P. divesensis*.

When first studied, by noted British palaeontologist Sir Arthur Smith Woodward in 1910, this dinosaur was thought to have been yet another *Megalosaurus*. German dinosaur expert Friedrich von Huene just as erroneously attributed it to the *Ceratosaurus* line when he studied the skull 15 years later. Another specimen of what could be the same genus was found in France in 1923.

Features: *Proceratosaurus* is known only from a fragmentary skull, which shows the base of a horn on the nose. This led von Huene to surmise that it was an ancestor of the later horned *Ceratosaurus*. Although there is not much in the way of skull material to study, it certainly seems to be a carnosaur rather than a ceratosaur, which it would be if it really were related to *Ceratosaurus*.

Right: Despite the paucity of remains, it seems as if Proceratosaurus can be placed in the carnosaur family. That makes it the earliest well-known member of the group.

THE SAUROPODS OF THE DASHANPU QUARRY, CHINA

The Dashanpu quarry, near Zigong in Sichuan Province, China, is the world's most famous middle Jurassic dinosaur site. In recent decades more than 40 tonnes of fossils, or more than 8,000 dinosaur bones, have been excavated here. In 1987 the Zigong Dinosaur Museum opened, celebrating this richness.

Omeisaurus

The first skeleton of this dinosaur was unearthed in 1939 by Chinese palaeontologist Young Chung Chien (or more correctly, Yang Zhongjian) and American Charles L. Camp. As with *Mamenchisaurus*, the length of the neck was not obvious at first. This fact only came to light with the discovery of a more complete skeleton in the 1980s. The long necks may have been used to reach food high in the trees, or over large areas on the ground. Remains of tail clubs found at the same site have been attributed to *Omeisaurus*. However, it is more likely that they belong to large specimens of *Shunosaurus*.

Features:
Omeisaurus, with its extremely long neck, short body and stocky limbs, must have looked rather like the slightly earlier *Mamenchisaurus*. Indeed, some specimens of *Omeisaurus* have been mis-identified as *Mamenchisaurus*. The difference lies in the shape of the vertebrae in the back – the spines are divided in *Mamenchisaurus*, but in *Omeisaurus* they are not. Small features like these help palaeontologists to distinguish one genus from another.

Left: Like most sauropods the hips of Omeisaurus were higher than the shoulders. The nostrils were well forward on the skull – an unusual feature.

Distribution: China.
Classification: Sauropoda.
Meaning of name: Lizard from Mount O-mei.
Named by: Young, 1939.
Time: Kimmeridgian to Tithonian stages of the late Jurassic.

Size: 15m (50ft).
Lifestyle: Browser.
Species: *O. junghsiensis*, *O. tianfuensis*, *O. luoquanensis*.

Shunosaurus

Sometimes known as *Shuosaurus*, this is the best-known of the Chinese dinosaurs. There have been about 20 individual skeletons found, some of them complete. Unusually for a dinosaur, every single bone of *Shunosaurus* is known. Palaeontologists regard this as a very common dinosaur of middle Jurassic times, and probably the most abundant in the eastern part of the Laurasian continent.

Features: The remarkable feature of *Shunosaurus* is the spiked club on the tail, evidently used as a defensive weapon. When first found, in 1979, it was thought to have been an abnormal growth at the site of an injury. However, several fully articulated skeletons have been excavated with the club in place. The skull is relatively long and low with the nostrils pointing sideways. In sauropods the teeth are usually spoon-shaped or pencil-shaped; in *Shunosaurus* they seem to be between the two.

Distribution: China.
Classification: Sauropoda.
Meaning of name: Shu (the old local name for Sichuan Province) lizard.
Named by: Dong, Zhou and Chang, 1983.
Time: Bathonian to Callovian stages of the middle Jurassic.
Size: 9m (30ft).
Lifestyle: Browser.
Species: *S. lii*, *S. ziliujingensis*.

Mamenchisaurus

When it was discovered in 1952, the nature of the *Mamenchisaurus constructus* vertebrae was unclear. The vertebrae were so delicate and poorly preserved that they could not be excavated undamaged. Early restorations of this animal showed only a moderate length of neck. It was with the discovery *M. hochuanensis*, in 1957 that the fantastic length of the neck was appreciated.

Features: *Mamenchisaurus* has the longest neck of any known dinosaur. It consists of 19 vertebrae – the greatest number so far found – and is about 14m (46ft) long, taking up about two-thirds of the length of the entire animal. The vertebrae are very thin and lightweight, made up of fine struts and sheets, rather like the later diplodocids. However, the short, deep skull shows that it belongs to the more primitive euhelopid group.

Distribution: China, Mongolia.
Classification: Sauropoda, Euhelopodidae.
Meaning of name: Lizard from Mamen Brook.
Named by: Young, 1954.
Time: Tithonian stage.
Size: 21m (69ft).
Lifestyle: Browser.
Species: *M. sinocanadorum*, *M. hochuanensis*, *M. youngi*, *M. jingyaninsis*, *M. anyuensis*, *M. constructus*.

Left: The extreme length of its neck may have enabled Mamenchisaurus *to reach in between closely-spaced trees to eat undergrowth in dense woodland.*

Datousaurus

This dinosaur is known from two incomplete skeletons. The skull that has been attributed to *Datousaurus* was actually found some distance away from the skeletons, and so there is some uncertainty as to whether it actually belongs to this animal: the Dashanpu quarry is full of all sorts of sauropods. Hence there is some confusion over its classification.

Features: *Datousaurus* has a neck that is longer than that of most early sauropods, but not as long as those to come. There are 13 vertebrae in the neck. The skull is large and heavy for a sauropod, with the nostrils at the front. The jaws have spoon-shaped teeth. The skeleton is *Diplodocus*-like, but the skull, if it is indeed the skull of the same animal, suggests that *Datousaurus* belongs to a different sauropod line altogether.

Right: Datousaurus is regarded as one of the least abundant of the sauropods that existed in central China in the middle Jurassic.

Distribution: China.
Classification: Sauropoda, Cetiosauridae or possibly Euhelopidae.
Meaning of name: Chieftain lizard.
Named by: Dong and Tang, 1984.
Time: Bathonian to Callovian stages of the middle Jurassic.
Size: 15m (50ft).
Lifestyle: Browser.
Species: *D. bashanensis*.

CETIOSAURS

The cetiosaurs were the most primitive of the sauropod dinosaurs. They were discovered in Europe and were among the first dinosaurs to be recognized – but now they are also known from fossils in Australia, Africa, North America and, significantly, South America. As a rule, their front and hind legs were more or less the same length, unlike most sauropods in which the front legs tend to be shorter.

DEVELOPMENT OF THE EVOLUTIONARY LINE

The sauropods were the natural successors to the prosauropods of the Triassic and early Jurassic periods. They had the same big bodies with the huge plant-food processing gut, and the long necks that enabled the little heads to reach high into trees or in a wide arc on the ground. In fact, many of the more advanced prosauropods looked very much like primitive sauropods, and there is often confusion in the classification about where the line should be drawn between the two.

South America had a wide variety of prosauropods in early times. In the later Cretaceous period there was a large number of sauropods, especially in the advanced titanosaurid group. The late sauropods flourished in South America towards the end of the age of dinosaurs, even though they were on the wane in the rest of the world at the time. However, in between, there is little evidence of what was evolvng. The few middle Jurassic sauropods that we know seem to be closely related to the cetiosaurids that were more typical of Europe at that time.

Cetiosaurus

This was the first of the sauropods to be discovered – in 1825 – and described. As such it suffered the same fate as its contemporary theropod, *Megalosaurus*: for a long time any sauropod remains found in Europe were attributed to it.

Sir Richard Owen, who named it in 1842, the same year that he invented the name "dinosaur", did not recognize it as one of this group of animals. He thought that he was studying the remains of a gigantic crocodile. He saw similarities in the backbones to those of a whale, hence the name he gave it, and assumed that it was an aquatic animal. It was dinosaur pioneer Gideon Mantell who recognized these bones as dinosaur bones in 1854.

Cetiosaurus remains have mostly been found in marine deposits, suggesting that this dinosaur lived close to the sea, or at least close to rivers where the remains could be washed out to sea. The best *Cetiosaurus* skeleton was found in 1968 by workmen digging in a clay pit in the English Midlands. This skeleton is now on display in the Leicester Museum and Art Gallery.

Distribution: England, Portugal and perhaps Morocco.
Classification: Sauropoda, Cetiosauridae.
Meaning of name: Whale lizard.
Named by: Owen, 1842.
Time: Bajocian to Bathonian stages of the middle Jurassic.
Size: 14m (46ft).
Lifestyle: Browser.
Species: *C. mogrebiensis, C. medius, C. conybearei, C. oxoniensis.*

Features: The *Cetiosaurus* skeleton is that of a typical sauropod, with a small head, long neck and tail, and a heavy body supported by elephantine legs. It is quite primitive because the vertebrae are not hollowed out as a weight-saving measure as they are in more advanced members of the group. Instead they are spongy and coarse – the whale-like feature that was noted by Owen.

Amygdalodon

This is the earliest-known South American sauropod. It predates, by tens of millions of years, the varied titanosaur sauropods that were to dominate that continent in Cretaceous times. It is possible that *Amygdalodon*'s ancestors are to be found among the many prosauropods that existed in South America, or it may show that the primitive cetiosaurid group had spread worldwide from its European origin. The name derives from the shape of the teeth, which are almond-like. This is the most primitive sauropod known from South America.

Features: It is known from the teeth, parts of the vertebrae, some ribs and hip bones and a fragment of limb from two individuals. As usual, when only part of a skeleton has been found and identified, we have to assume that the whole animal followed the long-necked sauropod form. Its teeth suggest that it is a member of the cetiosaurid group, mostly known from Europe. If this is the case it indicates how widespread this early line of the sauropods was in middle Jurassic times, when all the continents were still joined as a single landmass.

Distribution: Argentina.
Classification: Sauropoda, Cetiosauridae.
Meaning of name: Almond-shaped tooth.
Named by: Cabrera, 1947.
Time: Bajocian stage of the middle Jurassic.
Size: 13m (43ft).
Lifestyle: Browser.
Species: *A. patagonicus.*

Right: Although we presume
Amygdalodon to have been long-necked
and long-tailed, it seems likely that the
neck and tail were not quite as long as
those in later sauropods.

Patagosaurus

In the late 1970s the skeletons of about a dozen *Patagosaurus* were found together, five of them with skulls, suggesting that they moved about in herds. This discovery makes *Patagosaurus* the best-known of the early South American sauropods. Their presence, along with that of theropods such as *Piatnitzkysaurus*, shows that South America was connected to the other landmasses at that time.

Features: The teeth are similar to those of the earlier *Amygdalodon*. The rest of the skeleton, however, is more advanced. The skeleton resembles that of *Cetiosaurus* because the neck vertebrae have the same kind of undivided spines, and the back vertebrae have shallow cavities rather than the deep hollows of later sauropods. These simple backbones are largely what define the cetiosauridae. However, the hip bones and the tail are different from its European relative. A fine mounted skeleton of *Patagosaurus* stands in the Argentine Museum of Natural Sciences in Buenos Aires, and is dedicated to the pioneer Argentinian scientist Bernardino Rivadavia.

Distribution: Argentina.
Classification: Sauropoda, Cetiosauridae.
Meaning of name: Lizard from Patagonia.
Named by: Bonaparte, 1979.
Time: Callovian stage of the middle Jurassic.
Size: 18m (60ft).
Lifestyle: Browser.
Species: *P. fariasi.*

Right: Patagosaurus moved about in
groups. The finding of two adults and
three juveniles together supports this.

ORNITHOPODS

The ornithopods were plant-eating dinosaurs that existed in middle Jurassic times, living at the same time as the sauropods. Their remains have been found in locations from the prolific quarries of Dashanpu, in China, to the coast of Portugal. However, they were not as abundant or as varied as their lizard-footed cousins. Their time was yet to come.

Yandusaurus

Known from two almost complete skeletons including the skulls, we have a good idea of what *Yandusaurus hongeensis* looked like. Another species of *Yandusaurus*, *Y. multidens*, is probably really *Agilisaurus*. It is one of the dinosaurs from the Dashanpu quarries, which are famed for their sauropod remains.

Features:
Yandusaurus is a typical hypsilophodont. It is a small, two-footed plant-eater with very long legs, and was built for speed. It has a small head and chopping teeth. The food was held in cheek pouches while it was being chewed. Hypsilophodonts were more common in Cretaceous times but their ancestry can be traced back to the middle Jurassic period. Palaeontologists distinguish *Yandusaurus* from other hypsilophodonts by the ridges on the teeth.

Above: The short forearms of Yandusaurus had the full complement of five fingers.

Distribution: China.
Classification: Ornithopoda, Hypsilophodontidae.
Meaning of name: Lizard from Yandu.
Named by: He, 1979.
Time: Bathonian to Callovian stages of the middle Jurassic.
Size: 1.5m (about 6ft).
Lifestyle: Low browser.
Species: *Y. hongheensis*, *Y. multidens* (possibly).

Agilisaurus

Despite the fact that *Agilisaurus* is known from an almost complete skeleton, its actual position in the dinosaur evolutionary hierarchy is still unclear. Paul Barrett, of the Natural History Museum, London, England, suggests that it is too primitive to be a conventional ornithopod. Perhaps it is a fabrosaurid, one of the very primitive ornithopods that had not yet evolved the cheek pouches, or perhaps it is a new family altogether.

Features: The skull of *Agilisaurus* is small, with big eyes, and its teeth are leaf-shaped. The teeth are larger and pointed at the front. The hind limbs are long, much longer than the front limbs. The thigh bone is particularly short compared with the rest of the leg bones – a sign of a fast-running animal with a lightweight foot and all the muscle concentrated near the hip. The tail is long and used for balance while running.

Distribution: China.
Classification: Ornithischia. Fabrosauridae (unproven).
Meaning of name: Fast lizard.
Named by: Peng, 1992.
Time: Bathonian to Callovian stages of the middle Jurassic.
Size: 1.2m (4ft).
Lifestyle: Low browser.
Species: *A. louderbecki*, *Yandusaurus multidens* may well be a species of *Agilisaurus*.

Left: Despite the similarities between the skeletons of Agilisaurus and Yandusaurus (above), these may be quite different animals.

"Xiaosaurus"

This animal is known only from the teeth and jawbone, and a few bones from the rest of the skeleton including the hind leg. So little is known about this small dinosaur that *"Xiaosaurus"* is really a *nomen dubium*. It may even be a species of *Agilisaurus*. It does show, however, that there was quite a range of basal ornithopods in middle Jurassic China.

Right: Even if "Xiaosaurus" is found to be ancestral to the horned dinosaurs, it would still have the general appearance shown here.

Features: The remains, found in the Dashanpu quarries, are so fragmentary that very little can be deduced about this animal. However, an odd thing about the femur is the arrangement of leg muscle attachments. They seem to be very similar to those of the primitive horned dinosaurs, and this has led to speculation that *Xiaosaurus* may be close to the ancestry of such dinosaurs as *Triceratops* that did not flourish until late Cretaceous times.

Distribution: China.
Classification: Ornithopoda.
Meaning of name: Small lizard.
Named by: Dong, Tang, 1983.
Time: Bathonian stage of the middle Jurassic.
Size: 1m (3ft).
Lifestyle: Low browser.
Species: "*X. dasanpensis*".

ESSENTIAL DIFFERENCES

From a distance a small ornithopod would have looked very similar to a small theropod – both would have been standing on hind legs and balanced by a heavy tail. If we visited Mesozoic times we would have to know the difference, as one would be dangerous.

The most obvious difference between the two types is the size of the body. The more complex, plant-eating digestive system of the ornithopod would have meant that its body would be bigger. Then there is the shape of the head. The theropod had long jaws and sharp teeth, and probably eyes pointing forward. The ornithopod would have a smaller head, big eyes on the side, usually cheeks at the side of the mouth and a beak at the front.

The hands would be different, too. The theropod would have three, or even two fingers, and be sporting big, curved, claws. The ornithopod would have four or five fingers, with blunt claws. Finally, the colour. This is something we know nothing about, but it would seem likely that theropods would be brightly coloured, like birds or tigers, while ornithopods would be camouflaged, with greens and browns.

Below: Theropod (left) and ornithopod (right) for comparison.

"Alocodon"

All we know about this dinosaur, unfortunately, are a few distinctive teeth. This is a usual state of affairs in vertebrate palaeontology. Sometimes the teeth can be used to reconstruct the whole animal, but more often they pose more questions than they answer. These teeth were found near Pedrógão, Portugal, one of the few dinosaur sites in that country.

Features: The teeth of *Alocodon* have vertical grooves in them, hence the name. The only other dinosaur teeth that they resemble are those of the hypsilophodont *Othnielia*, which came later in the evolutionary line. Some palaeontologists regard this animal as being intermediate between the lesothosaurids and the hypsilophodonts. A set of sharper teeth furnished the front of the upper jaw. There has even been a suggestion that these teeth come from a primitive thyreophoran, one of the plated or armoured dinosaurs. Whatever the classification, "Alocodon" is featured as a primitive ornithopod here.

Distribution: Portugal.
Classification: Ornithopoda.
Meaning of name: Furrowed tooth.
Named by: Thulborn, 1973.
Time: Middle or late Jurassic.
Size: 1m (3ft).
Lifestyle: Low browser.
Species: "*A. kuehni*".

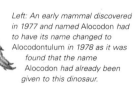

Left: An early mammal discovered in 1977 and named Alocodon had to have its name changed to Alocodontulum in 1978 as it was found that the name Alocodon had already been given to this dinosaur.

EARLY THYREOPHORANS

Heavy, four-footed, bird-hipped dinosaurs, with ornamentation and armour on their backs and tails,
began to spread in middle Jurassic times. These were the thyreophorans. The earliest to come to
prominence, in the late Jurassic period, were the stegosaurs, but they were replaced in Cretaceous times
by the ankylosaurs. The early forms of both were present in the middle Jurassic period.

Huayangosaurus

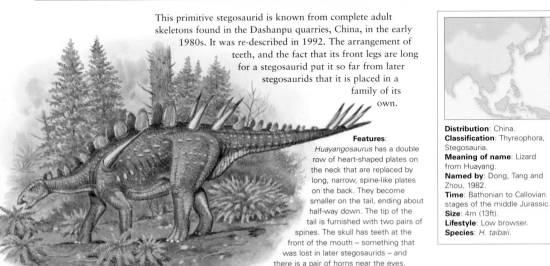

This primitive stegosaurid is known from complete adult
skeletons found in the Dashanpu quarries, China, in the early
1980s. It was re-described in 1992. The arrangement of
teeth, and the fact that its front legs are long
for a stegosaurid put it so far from later
stegosaurids that it is placed in a
family of its
own.

Features:
Huayangosaurus has a double
row of heart-shaped plates on
the neck that are replaced by
long, narrow, spine-like plates
on the back. They become
smaller on the tail, ending about
half-way down. The tip of the
tail is furnished with two pairs of
spines. The skull has teeth at the
front of the mouth – something that
was lost in later stegosaurids – and
there is a pair of horns near the eyes.

Distribution: China.
Classification: Thyreophora,
Stegosauria.
Meaning of name: Lizard
from Huayang.
Named by: Dong, Tang and
Zhou, 1982.
Time: Bathonian to Callovian
stages of the middle Jurassic.
Size: 4m (13ft).
Lifestyle: Low browser.
Species: *H. taibaii*.

Lexovisaurus

An early stegosaur that roamed the islands that dotted
the shallow sea covering Europe in middle Jurassic times
was *Lexovisaurus*. Since it was discovered in the 1880s,
several different names have been given to it, including
Omosaurus and *Dacenturus*. At times it was regarded as
a species of *Stegosaurus* and *Kentrosaurus*, which it
resembled. The definitive study was
done by
R. Hoffstetter
in 1957.

Features: The plates of *Lexovisaurus*
are narrow and short on the neck and
back, and there are several pairs of
long spines on the tail. There is
also a pair of long spines jutting
sideways from the shoulders. Old
restorations show these spines
jutting from the hips.

Distribution: England, France.
Classification: Thyreophora,
Stegosauria.
Meaning of name:
Lizard of the Lexovi tribe
(a tribe of ancient Gaul).
Named by: Hulke, 1887.
Time: Callovian to
Kimmeridgian stages of the
middle and late Jurassic.
Size: 5m (16½ft).
Lifestyle: Low browser.
Species: *L. vetustus*,
L. duobrivensis.

Right: According to contemporary theory, it
makes sense for the shoulder spines to be
in a position where they could damage an
enemy, using a thrusting motion, rather than
at the hip.

Tianchisaurus

When discovered *Tianchisaurus* was named *Jurassosaurus* because it was so unusual for an ankylosaur to be found in rocks as early as the Jurassic period. Its cumbersome species name is derived from the initials of the cast of *Jurassic Park* – Neill, Dern, Goldblum, Attenborough, Peck, Ferrero, Richards and Mazello. Steven Spielberg, the film's director, named it because he financed the research.

Features: This small primitive ankylosaur, found in 1974, has well-developed shoulder armour and its back is covered in scutes. The head is quite heavy and there may or may not be a club on the end of its tail. The jawbone is more stegosaur-like than ankylosaur-like. In the future, it may prove to be a nodosaurid rather than an ankylosaurid. It is certainly the earliest

known of the ankylosaur group, most of which lived in late Cretaceous times.

Distribution: China.
Classification: Thyreophora, Ankylosauria, Ankylosauridae.
Meaning of name: Tian Chi (Heavenly Pool) lizard.
Named by: Dong, 1993.
Time: Bathonian stage of the middle Jurassic.
Size: 3m (10ft).
Lifestyle: Low browser.
Species: *T. nedegoapeferima*.

ARMOURED DINOSAUR CLASSIFICATION

The first breakdown of Sir Richard Owen's 1842 classification of the Dinosauria came in 1870 when Thomas Henry Huxley separated off all forms of dinosaur with armour into the Scelidosauridae family grouping. This was based on the well-preserved skeleton of *Scelidosaurus*. When Harry Govier Seely divided the Dinosauria into the lizard-hipped Saurischia and the bird-hipped Ornithischia, a classification still used today, the armoured dinosaurs fell into the latter group. O. C. Marsh proposed the Stegosauria in 1896 as the group to which the armoured dinosaurs belonged, based on the shared arrangement and shape of the teeth.

The stegosauria group was split into two in 1927 by Alfred Sherwood Romer who was the first to separate the plated dinosaurs (the Stegosauria) from the armoured dinosaurs (the Ankylosauria). *Scelidosaurus* was grouped with the Stegosauria in Romer's classification.

Franz Baron Nopcsa in 1915 had proposed grouping them all together with the horned dinosaurs in a group called the Thyreophora. This never really caught on, but the concept was resurrected in the 1980s by Paul Sereno who used this name (but without the horned dinosaurs) to combine both the plated and the armoured dinosaurs once more. Under the current scheme *Scelidosaurus* is more closely related to the ankylosaurs than to the stegosaurs. This is how the classification stands today.

"Sarcolestes"

Sarcolestes is known only from the left half of the lower jaw. The animal is so obscure that it was first thought to have been a flesh-eating dinosaur, hence the meaning of the name. For almost a century it was regarded as the oldest known ankylosaur, but that was until the discovery of *Tianchisaurus* in China. It is often regarded as a *nomen dubium*.

Features: The jaw is very similar to that of the nodosaurid *Sauropelta*. Its ankylosaur features are the small teeth, extending all the way along the lower jaw bone with the little denticles along the edge, and an armour plate welded to the jawbone. The area that joins the two lower jaws together at the front is also typically ankylosaurid. It was probably a heavily built animal, armoured at the sides and shoulders and lacking a tail club.

Distribution: Cambridgeshire, England.
Classification: Ankylosauria, Nodosauridae.
Meaning of name: Flesh thief.
Named by: Lydekker, 1893.
Time: Callovian stage of the middle Jurassic.
Size: 3m (10ft).
Lifestyle: Low browser.
Species: "*S. leedsi*".

Left: The name of this animal, meaning "flesh thief," could not be more inappropriate. Far from being carnivorous, it was a slow-moving plant-eating animal.

PLESIOSAURS

By the middle and end of the Jurassic the plesiosaurs had really become established. The difference between the long-necked plesiosauroids and the big headed pliosauroids had become quite marked. The shallow shelf seas of the time provided plenty of different ecological niches, and different food sources to exploit, and could support a wide variety of marine animals.

Muraenosaurus

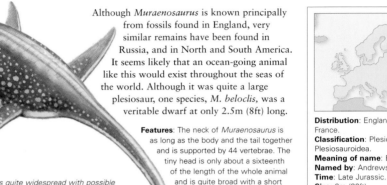

Although *Muraenosaurus* is known principally from fossils found in England, very similar remains have been found in Russia, and in North and South America. It seems likely that an ocean-going animal like this would exist throughout the seas of the world. Although it was quite a large plesiosaur, one species, *M. beloclis*, was a veritable dwarf at only 2.5m (8ft) long.

Features: The neck of *Muraenosaurus* is as long as the body and the tail together and is supported by 44 vertebrae. The tiny head is only about a sixteenth of the length of the whole animal and is quite broad with a short snout. The teeth, 19–22 pairs of them in each jaw, become gradually larger towards the front of the mouth.

Above: Muraenosaurus was quite widespread with possible specimens known as far apart as Wyoming, USA, South America and Russia as well as Europe. Andrews had reconstructed the skull as rather flat and snake-like. However, a new reconstruction by Mark Evans of Leicester Museum, England shows it to have been high and domed.

Distribution: England and France.
Classification: Plesiosauria, Plesiosauroidea.
Meaning of name: Eel lizard.
Named by: Andrews, 1910.
Time: Late Jurassic.
Size: 6m (20ft).
Lifestyle: Fish-hunter.
Species: M. leedsii, M. beloclis, M. purbecki, M. elasmosauroides.

Cryptoclidus

This was the typical long-necked plesiosauroid of the middle and late Jurassic seas. Numerous fossils have been found that bear the tooth marks of something big like *Liopleurodon*, indicating that it was not only the hunter of the seas but also the prey. Sometimes the fossils of isolated paddles are discovered, as if the owner had been crippled by an attack before being eaten.

Features: The typical solid body and long pointed paddles are here, as are the small head and the long neck. However, the neck is not quite as long as in some of the other plesiosauroids and it does not seem to be particularly flexible. The teeth are all about the same length and resemble thin curved needles. The eyes are on the top of the skull and point upwards.

Distribution: England and France.
Classification: Plesiosauria, Plesiosauroidea, Cryptocleidoidea.
Meaning of name: Hidden collar bone.
Named by: Phillips, 1871.
Time: Late Jurassic.
Size: 4m (13ft).
Lifestyle: Fish-hunter.
Species: C. eurymerus, C. richardsoni.

Liopleurodon

This was the top predator of the Jurassic seas. It was the middle and late Jurassic equivalent of the modern great white shark, a veritable aquatic *Tyrannosaurus*. The fossils of half-chewed *Ichthyosaurus* skeletons have been found, as have plesiosaur limb bones with tooth marks corresponding to the bite of *Liopleurodon*, all attesting to the diet of this gigantic animal. The huge specimens that are being found suggest a revised length of about 25m (80ft) and make it the biggest flesh-eating vertebrate known.

Features: *Liopleurodon* has the typical pliosaurid shape, with the huge head, the short neck and the compact streamlined body. The teeth are 20cm (8in) long, a good three quarters of which consists of root embedded deeply into the jaws giving a very strong bite. The strongest are arranged in a rosette at the front of the mouth. The paddles are large and worked on the underwater flight principle found in all plesiosaurs, using the front flippers as wings in the manner of modern sea lions.

Distribution: England, France, Germany, Eastern Europe and possibly Mexico, USA.
Classification: Plesiosauria, Pliosauroidea.
Meaning of name: Smooth-sided tooth.
Named by: Sauvage, 1873.
Time: Late Jurassic.
Size: 25m (80ft).
Lifestyle: Ocean hunter.
Species: *L. macromerius, L. ferox, L. pachydeirus, L. grossouveri, L. rossicus.*

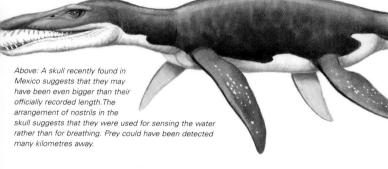

Above: A skull recently found in Mexico suggests that they may have been even bigger than their officially recorded length. The arrangement of nostrils in the skull suggests that they were used for sensing the water rather than for breathing. Prey could have been detected many kilometres away.

PLESIOSAUR MANOEUVRABILITY

There has always been speculation about the flexibility of the neck of plesiosaurs. The classic restoration shows the animal swimming close to the surface of the ocean, with the head held high, on top of a neck held in a graceful swan-like S-shape.

Study of the bones of the neck suggests that this may not have been possible. Both the long-necked genera featured here, *Muraenosaurus* and *Cryptoclidus*, may have been unable to lift the main part of their necks above the horizontal, with the head end flexible enough just to peek upwards. They may, however, have been able to lower their necks to about 45 degrees below the horizontal. This suggests that they swam close to the surface and hunted for fish below them. Their necks seem to have been more flexible sideways than in the vertical plane.

Below: Cryptoclidus, *showing the lowered neck.*

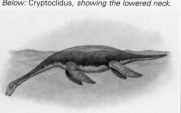

Peloneustes

With its short, streamlined body and long skull, *Peloneustes* must have looked like a giant diving bird, like a penguin or a gannet. The long jaws of *Peloneustes* made up for its short neck in catching fast-moving prey. Stomach contents analysis has revealed a preponderance of horny hooks from the tentacles of squid and other cephalopods.

Features: The long snout and the relatively few teeth of this pliosaur indicate a specialized diet. The big hind paddles suggest a fast and manoeuvrable swimming action, allowing it to pursue the swift soft-bodied squid and belemnites that flourished in the shallow seas of late Jurassic Europe. Head and neck, with 20 vertebrae, are approximately the same length. The underside is solid, formed by a fusion of the shoulder and hip girdles and the belly ribs, giving a completely inflexible body.

Distribution: Europe.
Classification: Plesiosauria, Pliosauroidea.
Meaning of name: Mud swimmer.
Named by: Seeley, 1869.
Time: Late Jurassic.
Size: 3m (10ft).
Lifestyle: Swimming hunter.
Species: *P. philarchus.*

Right: Early restorations showed Peloneustes *with quite a thin and flexible neck. It now seems more likely that the neck was stiff and made a streamlined shape with the head and body.*

RHAMPHORHYNCHOID PTEROSAURS

The late Jurassic was a rich time for pterosaurs. The two main types existed together: the short-tailed pterodactyloids, abundant at the end of the era, were establishing themselves, while the more primitive rhamphorhynchoids (illustrated here) were still abundant and diverse. None of the rhamphorhynchoid line survived into the Cretaceous. They were warm-blooded and covered in fur.

Rhamphorhynchus

Distribution: Germany, England and Tanzania.
Classification: Pterosauria, Rhamphorhynchoidea.
Meaning of name: Beak snout.
Named by: von Meyer, 1847.
Time: Oxfordian to Kimmeridgian stages of the late Jurassic.
Size: 1.75m (5¾ft) wingspan.
Lifestyle: Fish-hunter.
Species: *R. intermedius, R. gemmingi, R. jessoni, R. longicaudus, R. longiceps, R. muensteri. R. tendagurensis.*

This is the animal that gives its name to the whole group. It is the most common genus found in the Solnhofen deposits in Germany. *Rhamphorhynchus* is what most people think of as the typical pterosaur, with its leathery wings and its long tail ending in a vertical diamond paddle.

Features: The jaws of *Rhamphorhynchus* combine a set of very long fangs and a pointed beak, hence the name. The upper jaw has ten pairs of these teeth while the lower jaw has seven, and they project forwards and outwards. The breast bone is broad and strong and carries a forward-pointing crest, giving a wide attachment for strong wing-muscles. The neck is short and compact, holding the head straight out and not at an angle as with birds. The wings are stiffened with fine struts of gristle that radiated from the arm bones in the same pattern as the flight feathers of a bird.

Below: Rhamphorhynchus probably hunted fish by skimming close to the surface of the water. It is possible that it had a throat pouch, like a pelican, for holding its prey.

Batrachognathus

Most rhamphorhynchoids were adapted to hunting fish. *Batrachognathus*, however, had a completely different diet. Its blunt teeth were ideal for crushing insect carapaces, and the broad jaws formed a kind of scoop. Its tiny body would have made it very manoeuvrable in flight. It must have flown over lake surfaces catching insects in flight, as modern swallows do.

Features: *Batrachognathus* has a broad frog-like mouth with little peg-like teeth. It has a high short skull, unlike that of other rhamphorhynchoids. The tail is short. The tail vertebrae are fused together, quite unlike what would be expected from a rhamphorhynchoid and more like that of a pterodactyloid. The wing structure with its short wrists is, however, very much that of its long-tailed relatives.

Left: The range of pterosaur types must reflect a range of feeding strategies, as with the range of types of modern birds.

Distribution: Kazakhstan.
Classification: Pterosauria, Rhamphorhynchoidea.
Meaning of name: Frog face.
Named by: Riabinin, 1948.
Time: Late Jurassic.
Size: 0.5m (1½ft) wingspan.
Lifestyle: Insectivore.
Species: *B. volans.*

Jeholopterus

The extreme specialization of this rhamphorhynchoid has led to active speculation as to its lifestyle. One suggestion by independent palaeontologist David Peters is that the big claws allowed it to cling to the sides of big dinosaurs, while the wide-opening mouth and the pair of strong buttressed fangs at the front could be driven into thick skin to reach blood vessels. *Jeholopterus* was a vampire!

Above: Jeholopterus is known from a complete articulated skeleton, found in fine lake deposits. The deposits are so delicately preserved that they show the hairy covering and also the wing membrane of this pterosaur.

Features: The skull is broad and flattened at the front, giving it a rather cat-like appearance. The jaws are articulated to open wider than those of any other pterosaur. All the teeth are reduced except for a pair in the upper jaws at the front, and these are particularly strong and protruding, deeply rooted in a very strong palate. The claws are large and sharper than those of other pterosaurs, while the tail is short for a rhamphorhynchoid. It is closely related to other short-tailed rhamphorhynchoids like *Batrachognathus*.

Distribution: Northeastern China.
Classification: Pterosauria, Rhamphorhynchoidea.
Meaning of name: Wing from the Jehol geological formation.
Named by: Wang, Zhou, Zhang and Xu, 2002.
Time: Late Jurassic or early Cretaceous.
Size: 1m (3ft) wingspan.
Lifestyle: Insectivore or perhaps blood-sucker.
Species: *J. ninchengensis.*

OTHER RHAMPHORHYNCHOIDS

Scaphognathus has a shorter head than *Rhamphorhynchus* and has long teeth that are set upright rather than pointing forward. It also lacks the beak at the front of the jaws.

Anurognathus is similar to *Batrachognathus*, but with a wingspan of 50cm (1½ft) it is the smallest known pterosaur. Like *Batrachognathus* it has a short head and no tail to speak of, and may have caught insects on the wing.

OTHER PTERODACTYLOIDS

The pterodactloids below were located in the Solnhofen lagoons of Germany.

Germanodactylus has a long, straight, narrow beak with widely-spaced teeth, and a long crest running down the midline of the head.

Diopecephalus is a large pterodactyloid, with a wingspan of about 1.45m (5ft) and a particularly long neck and head.

Across the world, in the famous Morrison Formation of late Jurassic North America, rhamphorhynchoids and pterodactyloids also lived together. The dinosaur beds of Colorado and Utah yielded **Comodactylus**, a rhamphorhynchoid, and **Mesadactylus**, a pterosauroid. These are only known from indivdual bones, but they show that the two groups were cosmopolitan.

Sordes

For a long time palaeontologists debated about whether or not the pterosaurs were warm-blooded, like birds or bats. The argument for being warm-blooded was very strong, as a warm-blooded metabolism would have been needed for active aerial hunting. It was the discovery of *Sordes*, with its coat of hair, a sure sign of warm-bloodedness, which finally settled the argument.

Features: *Sordes* is a small rhamphorhynchoid, similar to the general form presented by *Rhamphorhynchus* itself. The first fossil of *Sordes* found is so well-preserved that the fur-like covering is visible. Short hairs about 6mm (¼in) long cover the whole body but leave the tail naked. There is even hair on the wing membrane and between the toes although it is much sparser here. There seems to be no diamond fin on the tail, but the tail itself is flattened and paddle-like at the end.

Distribution: Kazakhstan.
Classification: Pterosauria, Rhamphorhynchoidea.
Meaning of name: Filth.
Named by: Sharov, 1971.
Time: Late Jurassic.
Size: 60cm (2ft) wingspan.
Lifestyle: Insectivore or fish-hunter.
Species: *S. pilosus.*

Above: The hairy body covering would have helped to regulate the body temperature during active flying, but it would also have helped to reduce the noise of flying, useful for an aerial hunter.

PTERODACTYLOID PTEROSAURS

The pterodactyloids (illustrated here) became the dominant pterosaur group from the late Jurassic onwards. They had short stubby tails and longer necks than the more primitive rhamphorhynchoids, and generally had broader, more manoeuvrable wings. They carried their heads at more of a bird-like angle. Like the rhamphorhynchoids they were warm-blooded and covered in fur.

Ctenochasma

With its array of long fine teeth, *Ctenochamsa* could have been nothing but a filter feeder. It would have rested on all fours in shallow ponds and skimmed tiny animals, such as crustaceans or invertebrate larvae, from the surface, rather like the modern flamingo does. Recent studies, by Stephane Jouve, from the Museum National d'Histoire Naturelle, Paris, have suggested that some juvenile pterosaurs, once regarded as *Pterodactylus*, were actually *Ctenochasma*.

Left: The many teeth of Ctenochasma were not present throughout its life. The hatchling possessed about 60 of them and they developed to the full complement as the pterosaur grew towards maturity.

Distribution: Germany and France.
Classification: Pterosauria, Pterodactyloidea.
Meaning of name: Comb mouth.
Named by: von Meyer, 1852.
Time: Late Jurassic.
Size: 1.2m (3.9ft) wingspan.
Lifestyle: Plankton feeder.
Species: *C. porocristata, C. gracilis.*

Features: The long jaws contain more than 250 fine needle-like teeth that fan outwards at the tip. The two species of the *Ctenochasma* genus are distinguished from each other by the presence of a crest on the skull of *C. porocristata*, which is absent in *C. gracilis*. This crest probably forms the base of a horny display structure. The crest is very lightweight, consisting of very porous bone, to present as little impediment to flight as possible.

Gnathosaurus

The masses of needle-like teeth in the long jaws of this pterosaur were so reminiscent of the jaws of some aquatic-feeding crocodiles that it was classed as a crocodile when first discovered, and remained so until a second specimen was found in 1951. Like *Ctenochasma* it would have been a filter-feeder living off small invertebrates in shallow waters. *Gnathosaurus* was found in the lithographic limestone deposits of southern Germany, where *Ctenochasma* also sieved for food. There must have been a big enough range of food to support at least two different filter feeders.

Features: The skull and lower jaw of *Gnathosaurus* are very similar to those of *Ctenochasma*, but there are fewer teeth – about 130 in all – and those that it does have are thicker. Unlike *Ctenochasma* the snout was expanded and spoon-shaped at the end. A crest ran down the mid-line of the skull for about three quarters of its length. Only the skull and jaw have been found in Germany; we know nothing of the rest of the skeleton.

Distribution: Germany.
Classification: Pterosauria, Pterodactyloidea.
Meaning of name: Jaw lizard.
Named by: von Meyer, 1833.
Time: Late Jurassic.
Size: 1.7m (5½ft) wingspan.
Lifestyle: Filter feeder.
Species: *G. sublatus.*

Right: Only the jaws and a scrap of bone are known from Gnathosaurus. However, the body shape of the pterodactyloids is so conservative that we can restore the appearance of the whole animal with confidence.

Pterodactylus

The word pterodactyl is widely used as a popular term for any of the pterosaur group. The word, originally written as 'pterodactyle', was made up by Cuvier when the fossils were first studied. At first this creature was though to have been a mammal like a bat, and there were even suggestions that it was a swimming creature like a penguin, with the elongated finger supporting a paddle.

Features: *Pterodactylus* is the typical pterodactyloid pterosaur, with a short tail, long wrist bones, a longer neck than that of a typical rhamphorhynchoid, formed by elongation of the individual neck vertebrae, and a head held at an angle, like that of a bird, rather than in line with the neck as in a rhamphorhynchoid. The wing membrane is strengthened by fibres that radiate away from the arm and wrist, rather like the orientation of the flight feathers of a bird.

Right: Many Pterodactylus species have been found, and reassigned to other genera.

Distribution: Germany, France, England, Tanzania.
Classification: Pterosauria, Pterodactyloidea.
Meaning of name: Wing finger.
Named by: Soemmering, 1812 (and Cuvier in 1809).
Time: Kimmeridgian to Tithonian stages of the late Jurassic.
Size: Up to 2.5m (8ft) wingspan.
Lifestyle: Fish hunter.
Species:
P. antiquus,
P. arningi,
P. grandis,
P. manseli,
P. maximus,
P. pleydelli,
P. micronyx,
P. kochi,
P. cerinensis,
P. grandipelvis,
P. suprajurensis.

Gallodactylus

This pterosaur was described from a specimen found in Var in southern France in 1974, but since then a species of *Pterodactylus* found in Germany 120 years earlier has been assigned to the genus. It is also known from the site of Solnhofen, Germany where, along with *Pterodactylus*, it is one of the largest of the Jurassic pterosaurs.

Features: *Gallodactylus* shows another in the range of head shapes that appeared in the pterodactyloids as soon as they evolved. The long jaws are almost toothless except for the tips where fairly slender teeth are bunched together. This is another of the fish-traps evolved in the pterosaur line. The skull is extended back into a crest, presumably used as a display structure. The rest of the body is similar to that of all other pterodactyloids.

Distribution: France and Germany.
Classification: Pterosauria, Pterodactyloidea.
Meaning of name: Finger from Gaul (the Roman name for France).
Named by: Fabre, 1974.
Time: Late Jurassic.
Size: 1.35m (4½ft) wingspan.
Lifestyle: Fish eater.
Species: G. canjuersensis.

Above: Gallodactylus differed from Pterodactylus in the arrangement of teeth in the jaws and the presence of the crest at the rear of the skull. Apart from these features it was very similar to members of that genus.

MORRISON FORMATION MEAT-EATERS

In the second half of the nineteenth century the pioneering palaeontologists made most of their dinosaur discoveries in the vast Morrison Formation of the American Mid-west. The Jurassic rocks here are widespread. Even today new finds are being made, both in the well-worked quarries and in new areas where no-one has looked before. Meat-eating theropods form a spectacular part of the Morrison fauna.

MEAT-EATING THEROPODS

These animals, and many of those in the pages to follow, were found in the famous Morrison Formation of North America, named after a small town to the west of Denver where the formation was first identified.

At the end of the Jurassic period, a shallow seaway spread southwards across the middle of the North American continent. To the west the ancestral Rocky Mountains were rising. The rock fragments that washed down this crumbling mountain range were brought down to the sea and were deposited to form broad deltas and well-watered plains. The sand, silt, mud and pebbles that formed the soil of this plain are now found today as a vast swathe of sandstone and mudstone that stretches from New Mexico far north into Canada. These rocks form what is known as the Morrison Formation.

The dinosaurs that lived on the plains stayed mostly by the water. That is where the thickest vegetation grew. Occasionally they would fall into the rivers or lakes and become buried by the sediment. Sometimes many dead dinosaurs would be washed down together, and the result would be one or another of the famous dinosaur quarries of the Mid-west. The modern climate in this area is quite arid, and with only sparse vegetation masking the rocks, eroding dinosaur bones can be quite easily seen by prospectors. Most of our early knowledge of late Jurassic dinosaurs was obtained in the 1880s by the pioneering, and often rival, palaeontologists who excavated the Morrison Formation.

Ceratosaurus

One of the last of the wiggly-tailed ceratosaurids was *Ceratosaurus* itself, and what a spectacular animal it was! It was not the biggest meat-eating dinosaur of the time, but it was certainly a beast with character. If any dinosaur looked like a medieval dragon, this was it. With horns on the head, a jagged crest right down its back and teeth that seemed too big for its mouth, it must have presented a fearsome aspect to the plant-eaters of the late Jurassic period. It was one of the most common hunters of the time, and may have hunted down its prey in packs or in family groups.

Features: The most obvious feature of this dinosaur is the big horn on the nose and the two smaller ones above the eyes, probably used for display. It has four fingers on the hands – a rather primitive condition – and the arms, although short, are quite strong. One species, *C. ingens*, known only from the teeth, is much larger than the others and may have been one of the biggest theropods.

Distribution: Colorado, Utah, USA, and perhaps Tanzania.
Classification: Theropoda, Neoceratosauria.
Meaning of name: Horned lizard.
Named by: Marsh, 1884.
Time: Tithonian stage of the late Jurassic.
Size: 6m (17½ ft).
Lifestyle: Hunter.
Species: *C. nasicornis, C. dentisulcatus, C. ingens, C. magnicornis, C. meriani, C. roechlingi.*

Torvosaurus

Although all that has been found officially of this dinosaur is the forelimb, other bones, such as the jawbone, part of the skull, the neck bones and parts of a hip have been attributed to it. Put together these give the picture of a very big, meat-eating animal. It was found in the same quarry, the Dry Mesa Quarry in the mountains of Colorado, as the biggest of the late Jurassic North American sauropods.

Features: This dinosaur is actually regarded by some palaeontologists as a species of *Megalosaurus*. It is as big as *Allosaurus* – the best-known of the late Jurassic meat-eating dinosaurs – but it seems to be more massive. The short arms are very powerful and carry big claws. Some palaeontologists surmise that it may have been too heavy to hunt and was merely a scavenger, eating the bodies of dinosaurs that had already been killed.

Distribution: Colorado, USA, and possibly Portugal.
Classification: Theropoda, Tetanurae.
Meaning of name: Savage lizard.
Named by: Galton and Jensen, 1979.

Time: Tithonian stage of the late Jurassic.
Size: 10m (33ft).
Lifestyle: Hunter or scavenger.
Species: *T. tanneri*.

Left: Dry Mesa quarry, as it appears today – a ledge high on an arid hillside overlooking a deep valley. The fossils formed from a build-up of corpses in a Jurassic river.

Left: A leg bone found in 1998 in Portugal may belong to Torvosaurus. *However, it has also been suggested that it belongs to* Allosaurus, *or even the distinctly Portuguese* Lourinhanosaurus.

Edmarka

Known only from parts of a skull, some ribs and a shoulder bone from three individuals, one of which is a juvenile, *Edmarka* is not the best-known of the Jurassic theropods. In fact, it probably represents a species of *Torvosaurus* (or *Megalosaurus*, if its taxonomy can be correctly established). It was found at the famous dinosaur site of Como Bluff in Wyoming, USA, where it would have preyed upon the big sauropods of the time, or scavenged from their dead bodies.

Distribution: Wyoming, USA.
Classification: Theropoda, Tetanurae.
Meaning of name: Named after Dr William Edmark, the

Features: *Edmarka* may have been the heaviest meat-eating dinosaur of the Jurassic period, weighing maybe two tonnes. The presence of *Edmarka* in the late Jurassic rocks of Wyoming seems to uphold Dr Bob Bakker's idea that each area and time period produced one particularly large meat-eating dinosaur that was the dominant predator, and the presence of this one prevented the evolution of any other.

Left: As with all other big heavy meat-eating dinosaurs there is much debate as to whether Edmarka *was an active hunter or a slow-moving scavenger, or even both.*

Colorado State University scientist.
Named by: Bakker, Krails, Siegwath and Filla, 1992.
Time: Kimmeridgian stage of the late Jurassic.
Size: 11m (36ft).
Lifestyle: Hunter.
Species: *E. rex*.

THEROPODS OF CLEVELAND-LLOYD

One of the most prolific of the quarries in the Morrison Formation is Cleveland-Lloyd in Utah. More than 12,000 bones have been recovered here, representing 11 species of animal. It seems likely that complete skeletons are present here, but they are all disarticulated and jumbled. Most of the specimens that have been identified are of meat-eating dinosaurs.

Marshosaurus

This theropod was discovered in Dry Mesa quarry in Colorado, and the Cleveland-Lloyd quarry in Utah. The latter is famous for the large number of skeletons of the bigger theropod *Allosaurus*. Whatever lifestyle *Marshosaurus* pursued, it must have been in the presence of these enormous *Allosaurus* meat-eaters. Perhaps *Allosaurus* scavenged from the kills of a more active *Marshosaurus*.

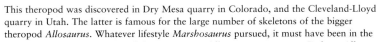

Features: *Marshosaurus* must have been an active hunter. It has sharp, curved teeth with serrated edges. What we know of the skull suggests that the head was quite long. The short and massive upper arm bone shows that it had small but powerful arms. At this time it is difficult to say what kind of theropod it was – some palaeontologists suggest it was a carnosaur, like *Sinraptor*, while others say that it may be an early dromeosaur.

Left: The species name, M. bicentisimus, commemorates the fact that the United States of America was celebrating 200 years of its foundation in the year that this dinosaur was discovered.

Distribution: Utah and Colorado, USA.
Classification: Theropoda, Tetanurae.
Meaning of name: Marsh's lizard (after dinosaur pioneer Othniel Charles Marsh).
Named by: Madsen, 1976.
Time: Tithonian stage of the late Jurassic.
Size: 5m (16½ft).
Lifestyle: Hunter.
Species: *M. bicentisimus*.

Stokesosaurus

Another of the theropods found in the Cleveland-Lloyd quarry, *Stokesosaurus* is based on very few scattered remains – mostly a hip bone, jaw bone and braincase – that are very different from the other theropod remains found there. Some of the bones seem to resemble those of late Cretaceous theropods, and this has led some scientists to surmise that it may be a primitive tyrannosaurid.

Distribution: Utah, USA.
Classification: Theropoda, Tetanurae, Tyrannosauridae (unproven).

Meaning of name: Lee Stokes's (an American palaeontologist) reptile.
Named by: Madsen, 1974.
Time: Tithonian stage of the late Jurassic.
Size: 4m (13ft).
Lifestyle: Hunter.
Species: *S. clevelandi*. *Iliosuchus incognitus* may be the same genus.

Features: *Stokesosaurus* seems to have been very similar to another dinosaur called *Iliosuchus*, found in England in the 1930s, and these are now regarded as the same animal. The braincase that has been found has many tyrannosaurid features, but there is still some uncertainty as to whether or not this actually belonged to *Stokesosaurus* – the bones are so scattered. Verification would have an effect on the classification of this animal. If it is a tyrannosaurid, it is the earliest that has been found.

Right: A jawbone and a braincase found in South Dakota may also have belonged to Stokesosaurus.

Allosaurus

The most well-known of the late Jurassic theropods must be *Allosaurus*. It was unearthed by Othniel Charles Marsh during the "bone wars", and since then many specimens have been found. The Cleveland-Lloyd quarry alone has yielded more than 44 individuals. Species attributed to *Allosaurus* have been found as far away as Tanzania and Portugal. Sauropod bones have been found bearing marks gouged by *Allosaurus* teeth.

Features: *Allosaurus* is a familiar animal, with massive hind legs, strong S-shaped neck, a huge head with jaws that could bulge sideways to bolt down great chunks of meat; sharp, serrated, steak-knife teeth with 5cm- (2in-) long crowns; and short, heavy arms with three-fingered hands

Distribution: USA, possibly Portugal and Tanzania.
Classification: Theropoda, Tetanurae, Carnosauria.
Meaning of name: Different lizard.
Named by: Marsh, 1877.
Time: Tithonian to Kimmeridgian stages of the late Jurassic.
Size: 12m (39ft).
Lifestyle: Hunter or scavenger.
Species: *A. tendagurensis, A. fragilis, A. amplexus, A. trihedrodon, A. whitei* and *A. (Saurophaganax) maximus?*

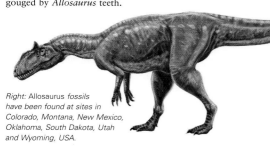

Right: Allosaurus *fossils have been found at sites in Colorado, Montana, New Mexico, Oklahoma, South Dakota, Utah and Wyoming, USA.*

bearing ripping claws that were up to 15cm (6in) long. This enormous carnivore would have hunted the biggest plant-eaters of the time including the massive sauropods, the remains of which were found in the Cleveland-Lloyd quarry.

CLEVELAND-LLOYD QUARRY

The Cleveland-Lloyd quarry lies 38km (30 miles) south of Price, Utah, close to the community of Cleveland. Dinosaur excavations started on the site in 1929 and over the next 10 or 12 years they continued thanks to financing by Malcolm Lloyd, a Philadelphia lawyer.

The University of Utah began a thorough excavation in 1960, headed by William Lee Stokes, and another in 2001 with Utah state geologist James H. Madsen in charge. The site was designated a National Natural Landmark in 1966. The excavations are open to view by the public, and skeletons from the site are on view in more than 60 museums worldwide.

The skeletons found at the site include sauropods and stegosaurs, but most appear to be the bones of meat-eaters. It is possible that the site represents a watering hole in an abandoned river meander on an arid Jurassic plain. Such a site would have attracted plant-eating animals from all over, and the meat-eaters would have converged, finding easy pickings among the weak and dehydrated sauropods and stegosaurs. The mud of the water hole would have hampered any escape. The skeletons appear to have been torn apart on the spot, and the bones show signs of having been trampled underfoot. Subsequent floods would have buried all this and set the fossilization process in motion.

Saurophaganax

The few bones known of *Saurophaganax* were excavated in the 1930s, but were not studied seriously until the 1990s. It turned out to be very similar to *Allosaurus* but a great deal bigger. After Don Chure's naming of it in 1995, David K. Smith re-analysed it in 1998 and came to the conclusion that it represented a particularly big species of *Allosaurus*.

Features: The description of *Allosaurus* can just as well apply to *Saurophaganax*, so close are the two animals. The differences lie in the sheer size of *Saurophaganax*, and in the shape of the neck and tail vertebrae. A complete mounted skeleton of *Saurophaganax* is on display in the Sam Noble Museum, in Oklahoma City, but most of it is made up of sculpted bones scaled up from those of the *Allosaurus* from the Cleveland-Lloyd quarry.

Distribution: Oklahoma, USA.
Classification: Theropoda, Tetanurae, Carnosauria.
Meaning of name: The greatest reptile-eater.
Named by: Chure, 1995.
Time: Kimmeridgian stage of the late Jurassic.
Size: 12m (39ft).
Lifestyle: Hunter or scavenger.
Species: *S. maximus.*

SMALLER THEROPODS

*We often think of theropods as being big, fierce animals. However, some of them were quite small.
In modern faunas we have the big carnivores like lions and wolves, but we also find smaller types
such as weasels and raccoons. It was the same in the dinosaur age – various sizes of carnivore
evolved to hunt different-sized prey.*

Compsognathus

One of the first complete dinosaur skeletons ever found was also one of the smallest. *Compsognathus* was found perfectly preserved in the fine lithographic limestones of Bavaria, Germany. In its build and appearance it is very similar to the first bird, *Archaeopteryx*, which was also found here, giving an early indication of the close relationships between the birds and the dinosaurs.

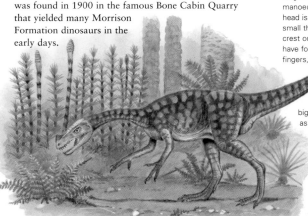

*Right: Compsognathus
is usually shown with two
fingers on each hand.
However, some of
the finger
bones may
have been
missing from the original
specimen, and it may
have had three.*

Features: Although the length of *Compsognathus* is given as 1m (3ft), this gives the wrong impression. Most of that is tail and neck. The body is about as big as a chicken. It was evidently a meat-eater, as the German specimen has the remains of a lizard in its stomach cavity, its last meal before death. A recent discovery, by Peter Griffiths, was the presence of unhatched eggs surrounding the fossil. This specimen had been a female and the eggs had been ejected from the body cavity by the trauma of her death. The French specimen is of a slightly larger species.

Distribution: Germany and France.
Classification: Theropoda, Tetanurae, Coelosauria.
Meaning of name: Pretty jaw.
Named by: Wagner, 1859.
Time: Tithonian stage of the late Jurassic.
Size: about 1m (3ft).
Lifestyle: Hunter.
Species: *C. longipes, C. corallestris.*

Ornitholestes

The name of this dinosaur suggests that it hunted birds, and many early restorations show it doing so – invariably chasing *Archaeopteryx*, a bird that actually lived on the other side of the world. It is more likely to have hunted small reptiles and ground-living mammals. Its skeleton was found in 1900 in the famous Bone Cabin Quarry that yielded many Morrison Formation dinosaurs in the early days.

Features: This fox-sized animal, with a long neck and tail, hunted on two legs. It is lightly built with hollow bones, and its long tail – making up more than half the length of the animal – gave it manoeuvrability while hunting. Its head is deeper than that of other small theropods, and has a bony crest on the snout. Its hands have four fingers, one of which is tiny and almost invisible, while another is big and could be used as a thumb.

Distribution: Utah and Wyoming, USA.
Classification: Theropoda, Tetanurae, Coelosauria.
Meaning of name: Bird thief.
Named by: Lambe 1904.
Time: Tithonian stage of the late Jurassic.
Size: 2m (6½ft).
Lifestyle: Hunter.
Species: *O. hermanni.*

*Left: The teeth were not
built for snatching but
for cutting. Ornitholestes
probably chased its prey
and grasped it with its
versatile hands, using
both teeth and claws
to kill it.*

Coelurus

For a long time *Coelurus* was thought to have been another specimen of *Ornitholestes*. However, studies by John Ostrom in 1976 and Jacques Gauthier in 1986 show that the hands are like those of the maniraptorans. In contrast, the neck is nothing like that of a maniraptoran and it is unclear where this animal fits into the dinosaur family tree.

Distribution: Wyoming, USA.
Classification: Theropoda, Tetanurae, Coelurosauria.
Meaning of name: Hollow tail.
Named by: Marsh, 1879.
Time: Tithonian stage of the late Jurassic.
Size: 2m (6½ft).
Lifestyle: Hunter.
Species: *C. fragilis*.

Right: Specimens of Coelurus *found in four locations in the same quarry may have come from the one individual. That individual may not even have been fully grown, and so the size estimate here may be on the small side.*

Features: This is another of the small, hunting dinosaurs. It has a strangely down-curved jaw with sharp, curved teeth. The hands are long but not particularly strong, with a wrist joint similar to that of a bird, and very flexible fingers. The "hollow tail" part of the name refers to the deep excavations in the vertebrae of the back and tail, something like those found as a weight-saving measure in sauropods.

THEROPOD DIET

What was the prey of the small theropods? Sometimes we have direct evidence, as in the lizard bones found in *Compsognathus's* stomach. For the rest we have to make inferences from the other animals that lived in the area.

We know that small reptiles and amphibians abounded in late Jurassic times. The Morrison Formation has fossils of snakes and lizards, and frogs and salamanders. There were also small mammals, none of which would have been bigger than a rat, although the only physical fossils we have of them are the teeth. All of these animals would have been prey for the small, active theropods. Although doubt has been thrown on the many early illustrations of *Ornitholestes* chasing *Archaeopteryx*, it is likely that there were birds around at that time. Active small dinosaurs may have ambushed them on the ground, as do cats today. The abundant pterosaurs may also have been part of the small theropod diet.

Then there are the insects and other invertebrates. We know the remains of many types of insects from Jurassic rocks, including dragonflies and beetles. The small lightweight dinosaurs would have fed on some of them.

Below: Compsognathus skeleton with a lizard skeleton in the stomach region.

Elaphrosaurus

This theropod is one of the fossils found in the late Jurassic site at Tendaguru, in what is now Tanzania, by the famous German expeditions in the 1920s. Palaeontologists are unsure about the relationships of this animal, but it seems likely to have been a ceratosaurid. It would have been too small to tackle the big sauropods of the area, and probably hunted the small ornithopods.

Features: It looks rather like one of the later ostrich-mimics – and was once classified as an early member of this group – but the legs are shorter in proportion and the body very long and shallow-chested. We have a good idea of what the body is like, but the skull is missing. Small teeth like those of *Coelophysis* were found near the skeleton and seem to have belonged to *Elaphrosaurus*, suggesting that the head also was *Coelophysis*-like. It is the shortest theropod dinosaur in stature, going by the height of the hips compared with the overall length of the animal.

Distribution: Tanzania.
Classification: Theropoda.
Meaning of name: Light lizard.
Named by: Janensch, 1920.
Time: Tithonian to Cenomanian stages of the late Jurassic and early Cretaceous.
Size: 6m (19½ft).
Lifestyle: Hunter.
Species: *E. bambergi, E. gautieri, E. iguidensis.*

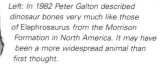

Left: In 1982 Peter Galton described dinosaur bones very much like those of Elaphrosaurus *from the Morrison Formation in North America. It may have been a more widespread animal than first thought.*

BRACHIOSAURIDS AND CAMARASAURIDS

The sauropods were the big plant-eaters of late Jurassic times. The plains of North America and the forests of Europe would have been alive with herds of these vast animals roaming to and fro, from one feeding place to another. There were several different groups of sauropods at that time, one of which, the Macronaria, consisted of the brachiosaurids and camarasaurids.

Giraffatitan

Distribution: Tanzania.
Classification: Sauropoda, Macronaria.
Meaning of name: Giraffe giant.
Named by: Olshevsky, 1991 (after Janensch, 1914).
Time: Kimmeridgian to Tithonian stages of the late Jurassic.
Size: 22m (72ft).
Lifestyle: High browser.
Species: *G. brancai*.

Although the name may be unfamiliar, *Giraffatitan* is one of the best-known sauropods in the world. Under the name *Brachiosaurus brancai* it has stood for 80 years as an exhibit in the Humboldt Museum of Berlin, Germany, as the biggest mounted dinosaur skeleton in existence. It was one of the many dinosaurs collected early in the twentieth century at Tendaguru, in what was then German East Africa.

Features: What was originally regarded as *Brachiosaurus* was given a new name after studies by Greg Paul in 1988 and George Olshevsky in 1991 showed that it was different from the original *Brachiosaurus*. The main difference is in the vertebrae of the back which, in *Giraffatitan*, produces a ridge between the shoulders like a horse's withers. To be fair, not all palaeontologists agree with this, and many books will continue to describe this animal as a species of *Brachiosaurus*.

Left: The famous mounted skeleton in the Humboldt Museum, Germany, is a composite of at least five individual skeletons found at the same site. In total 34 individuals were found at Tendaguru.

Lusotitan

It has long been known that brachiosaurids were not restricted to North America and Africa, where the most famous finds were made. Remains have also been found in late Jurassic rocks of Europe. In 1975 a partial skeleton was found in Portugal and named *Brachiosaurus atalaiensis* by Lapparent and Zybszewski. It is now thought to be a separate genus. Specimens of *Lusotitan* consisting of isolated vertebrae have been found in five different localities in Portugal.

Distribution: Portugal.
Classification: Sauropoda, Macronaria.
Meaning of name: Giant from Portugal.
Named by: Teles, Antunes and Mateus, 2003.
Time: Tithonian stage of the late Jurassic.
Size: 25m (82ft).
Lifestyle: High browser.
Species: *L. atalaiensis*.

Features: *Lusotitan* is known to be a brachiosaurid because of the low spines on the vertebrae and the muscle attachment of the long arm bone. As with all brachiosaurids, *Lusotitan* has very long forelimbs and high shoulders, allowing it to reach up into trees for food. No skull is known, but it would be the typical brachiosaurid pattern, with the very high and open nostrils, and the spoon-shaped teeth.

Left: The spikes on the backs of some restorations are conjectural. There is some evidence to suggest that spines existed on the backs of the other major sauropod group, the diplodocids, and some palaeontologists believe that the macronaria had them as well.

MACRONARIA

The macronaria are the sauropods with the boxy heads. The distinctive head shape was due to the fact that the holes in the skull representing the nostrils were much bigger than the eye sockets. And since the nostrils were on top of the head, it is easy to see why the rather strange-looking skull has been the subject of a great deal of scientific speculation.

It was once thought that the high nostril meant that the animal could remain submerged in a lake and breathe without any of its body showing. That idea has been dismissed. Some scientists drew attention to the fact that the nostrils on an elephant's skull are extremely large as well, and that maybe these sauropods had an elephant-like trunk. But this does not make much sense, since, with a long neck, the added reach of a trunk would seem to be superfluous. Anyway, a sauropod's skull lacks the broad plates of bone that would be needed to anchor the muscles that make up a trunk.

The most likely hypothesis is that the big nasal cavities would have been filled with moist membranes in life, and would have kept the interior of the skull and the little brain cool under the hot sun that would have beaten down on the late Jurassic plains of North America.

Brachiosaurus

The best-known *Brachiosaurus* skeleton in the world is now thought to be a different genus – *Giraffatitan* (far left). However, an even bigger animal, *Ultrasaurus*, found in the Dry Mesa Quarry in Colorado, is now regarded as a particularly big specimen of *Brachiosaurus*. The original *Brachiosaurus* was discovered as two partial skeletons in the Morrison Formation near Fruita in Utah in 1900 by Elmer G. Riggs.

Distribution: Colorado and Utah, USA.
Classification: Sauropoda, Macronaria.
Meaning of name: Tall-chested arm lizard.
Named by: Riggs, 1903.
Time: Kimmeridgian to Tithonian stages.
Size: 22m (72ft).
Lifestyle: High browser.
Species: *B. altithorax*.

Left: The position of the neck whether it was vertical or horizontal – is an on-going debate among scientists.

Features: About half of the height of *Brachiosaurus* is due to the neck. This, with its long front legs and tall shoulders, meant that it could reach high up into the trees to feed. Even its front feet contributed to its high reach – the fingers are long and pillar-like, and arranged vertically in the hand. Despite its fame, it is one of the rarest of the sauropods from the Morrison Formation.

Camarasaurus

This must have been one of the most abundant of the Morrison Formation sauropods, judging by the number of remains found. It often thought of as a small animal. This is because the best skeleton found is of a juvenile, perfectly articulated, lying in the rock of Dinosaur National Monument in Utah, USA, and mounted in Pittsburgh Museum.

Features: The "chambered lizard" in its name refers to the cavities in the backbone, designed to keep down the weight of the animal. Other cavities are present in the skull, which is merely a framework of bony struts, with enormous nostrils and spoon-shaped teeth. The forelimbs and hind limbs are approximately the same length, making the back of the animal horizontal. It is the Morrison Formation sauropod that is less bulky than *Brachiosaurus*, but not as slim as *Diplodocus*.

Distribution: New Mexico to Montana, USA.
Classification: Sauropoda, Macronaria.
Meaning of name: Chambered lizard.
Named by: Cope, 1877.
Time: Kimmeridgian to Tithonian stages of the late Jurassic.
Size: 20m (66ft).
Lifestyle: Browser.
Species: *C. supremus, C. grandis, C. lentus, C. lewisi.*

DIPLODOCIDS

If the macronarians (the brachiosaurids and camarasaurids) went for height to be impressive, the diplodocids went for length. Among these sauropods we find the longest land animals that ever existed. The slimness of the neck and tail meant that, despite its length, a diplodocid was not a particularly heavy animal. The tail was twice the length of the body and neck together, and ended in a whiplash.

Diplodocus

The familiar long, low sauropod is known as *Diplodocus*. It is well known from the many casts of the graceful skeleton of *D. carnegii*, the second species to be found. The casts, which appear in museums throughout the world, were excavated, reproduced and donated with finance provided by the Scottish-American steel magnate Andrew Carnegie in the early years of the twentieth century.

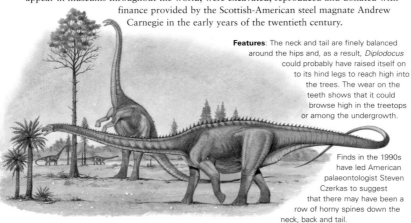

Features: The neck and tail are finely balanced around the hips and, as a result, *Diplodocus* could probably have raised itself on to its hind legs to reach high into the trees. The wear on the teeth shows that it could browse high in the treetops or among the undergrowth.

Finds in the 1990s have led American palaeontologist Steven Czerkas to suggest that there may have been a row of horny spines down the neck, back and tail.

Distribution: Colorado, Utah and Wyoming, USA.
Classification: Sauropoda, Diplodocidae.
Meaning of name: Double beam.
Named by: Marsh, 1878.
Time: Kimmeridgian to Tithonian stages of the late Jurassic.
Size: 27m (89ft).
Lifestyle: Low or high browser.
Species: *D. longus, D. carnegiei, D. hayi.*

Supersaurus

The dinosaur *Supersaurus* was one of the big sauropods found in the Dry Mesa Quarry, Colorado, by dinosaur hunter Jim Jensen. Unfortunately that site is such a jumble – probably representing a log-jam of bones in a Morrison Formation river – that the skeletons are all mixed up. One of Jensen's giants (then called *Ultrasaurus*) actually consisted of a shoulder blade of *Brachiosaurus* and ribs from *Supersaurus*.

Features: The vertebrae and the partial shoulder and hip that have been found of this animal show that it is closely related to *Diplodocus*. It may even be a large species of *Diplodocus* that stood 8m (26ft) high at the shoulder. The tallest vertebrae are as tall as a standing child. There are even bigger bones from Dry Mesa in the basement of Brigham Young University, in Salt Lake City, USA, that are yet to be examined and could alter our interpretation of what this dinosaur looked like.

Distribution: Colorado, USA.
Classification: Sauropoda, Diplodocidae.

Meaning of name: Super lizard.
Named by: Jensen, 1985.
Time: Kimmeridgian to Tithonian stages of the late Jurassic.
Size: 30m (98ft).
Lifestyle: Low or high browser.
Species: *S. vivianae.*

Right: The body proportions are based on those of Diplodocus *to which* Supersaurus *is obviously related. However, the neck may be longer than that of* Diplodocus.

Seismosaurus

Currently holding the record for the longest dinosaur known, *Seismosaurus* has taken 13 years to excavate, largely because of its size. It is only known from a single specimen, consisting of most of the vertebrae, part of the pelvis, the ribs and some stomach stones. Some palaeontologists regard it as a species of *Diplodocus*. The two are certainly closely related.

Features: The amazing length of this animal is entirely due to the length of the neck and tail. The body is not particularly big for a diplodocid, and the legs are quite stubby in comparison. Chevron-shaped protrusions beneath the backbones probably supported the muscles that helped the mobility of the neck and tail. The characteristic whip-like diplodocid tail has a downward curve not far from the hips, suggesting that the distant part of it trailed on the ground.

Distribution: New Mexico, USA.
Classification: Sauropoda, Diplodocidae.
Meaning of name: Earthquake lizard.
Named by: Gillette, 1991.
Time: Kimmeridgian stage of the late Jurassic.
Size: 40m (131ft).
Lifestyle: Low or high browser.
Species: *S. hallorum*.

Right: Seismosaurus *was not the only huge sauropod that inhabited the river plains that produced the Morrison Formation. Others included* Supersaurus *and* Diplodocus *itself. Obviously the range of food was enough to support several related big animals.*

DIPLODOCID TEETH

Above: The skull of Diplodocus.

The teeth of the diplodocids were distinctive. They had pencil-like teeth arranged like the teeth of a rake in front of the long jaws. There were no teeth at the back. The wear on these teeth suggests that the animals could browse from high up in the trees or from the ground. (The term graze nowadays refers to eating grass, but there was no grass in Mesozoic times so the term is used to mean low browsing.) The articulation of the neck bones and the musculature suggests that diplodocid necks were usually held horizontally, which would suggest that low browsing was the normal mode of feeding. The long neck would provide access to a large area of ferns or horsetails without moving the big body too far.

This tooth arrangement meant that these animals could not chew food. They would have spent their time raking and swallowing to obtain enough to feed. To help digestion they swallowed stones. These gathered in a gizzard and ground up the tough plant material. When the stones became smooth, they regurgitated them and swallowed fresh ones. Little piles of smooth stones are known from the Morrison Formation.

Barosaurus

This species is known from five partial skeletons from the Morrison Formation, three of them in Dinosaur National Monument in Utah, USA. The African species, until recently known as *Gigantosaurus*, was part of the Tendaguru fauna, and known from four skeletons. The rearing skeleton of *Barosaurus* in the American Museum of Natural History, at a height of 15m (49ft), is the tallest mounted skeleton in the world, only made possible by modern techniques of producing casts of fossil bones in lightweight materials.

Distribution: South Dakota and Utah, USA, and Tanzania.
Classification: Sauropoda, Diplodocidae.
Meaning of name: Slow heavy lizard.
Named by: Marsh, 1890.
Time: Kimmeridgian to Tithonian stages of the late Jurassic.
Size: 27m (89ft).
Lifestyle: Low or high browser.
Species: *B. lentus*, *B. africanus*.

Left: The way that Barosaurus, *and the other diplodocids, was balanced at the hips suggests that it could rear up on its hind legs for feeding.*

Features: *Barosaurus* is very much like *Diplodocus* – indeed the limb bones are indistinguishable between the two genera – but its tail bones are shorter and its neck bones at least one-third longer, one of which is 1m (3ft) long. The length of the neck has led some palaeontologists to suggest that there were several hearts along its length to enable the blood to reach the brain when it was feeding from high trees.

MORE DIPLODOCIDS

The diplodocids seem to have evolved first in Europe with animals such as Cetiosauriscus *from middle Jurassic times. They reached their heyday in North America in the late Jurassic period; the Morrison Formation is full of them. They continued into the early Cretaceous period and by that time were spreading into the Southern Hemisphere.*

Apatosaurus

One of the most popular dinosaurs, *Apatosaurus*, keeps changing its identity. For almost a century it went by the evocative name *Brontosaurus*, the "thunder lizard". In 1877 Othniel Charles Marsh discovered *A. ajax* and named it. Two years later he found a more complete animal which he named *Brontosaurus excelsus*. It was not until the twentieth century that it was realized that these were actually two species of the same genus. When there is confusion of this kind, with one animal given two names, it is the first name given that is deemed to be the valid one, in this case *Apatosaurus*. The official change took place in 1903.

Features: *Apatosaurus* is a heavily-built diplodocid. The vertebrae have a groove along the top. This held a strong ligament that supported the weight of the neck and the tail, like the cables on a suspension bridge.

The "deceptive lizard" of the name refers to the fact that the chevron bones attached to the vertebrae look confusingly like those of the aquatic reptile *Mosasaurus*. Although the head is about the size of that of a horse, the brain is only as big as that of a cat. The whole skeleton is similar to that of *Diplodocus*, but is much more stocky and massive, going for weight rather than length.

Distribution: Colorado, Oklahoma, Utah and Wyoming, USA.
Classification: Sauropoda, Diplodocidae.
Meaning of name: Deceptive lizard.
Named by: Marsh, 1877.
Time: Kimmeridgian to Tithonian stages of the late Jurassic.
Size: 25m (82ft).
Lifestyle: Low or high browser.
Species: *A. ajax, A. excelsus, A. louisae, A. montanus.*

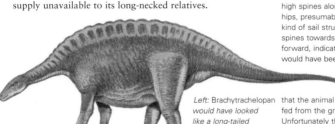

Brachytrachelopan

Found in Patagonia by a farmer out looking for lost sheep, this dinosaur surprised everybody because it was a sauropod – one that actually had a neck that was shorter than its back. Evidently *Brachytrachelopan* evolved its short neck to browse vegetation that grew close to the ground – a food supply unavailable to its long-necked relatives.

Features: The most amazing feature of *Brachytrachelopan* is the extremely short neck – more like that of a stegosaur than a diplodocid. Like other members of the dicraeosaur group it has high spines along the back and hips, presumably supporting a kind of sail structure in life. The spines towards the front curve forward, indicating that the head would have been held low and

Distribution: Argentina.
Classification: Sauropoda, Dicraeosauridae.
Meaning of name: Short-necked shepherd.
Named by: Rauhut, Remes, Fechner, Cladera and Puerta, 2005.
Time: Tithonian stage of the late Jurassic.
Size: 10m (33ft).
Lifestyle: Low browser.
Species: *B. mesai.*

Left: Brachytrachelopan would have looked like a long-tailed stegosaur, but without the armour plates. that the animal fed from the ground. Unfortunately the skull and tail are missing from the single skeleton known.

Eobrontosaurus

Many palaeontologists, including Robert T. Bakker, were unhappy with the renaming of the original *Brontosaurus* genus to *Apatosaurus*, and so when a recently discovered species of *Apatosaurus*, named *A. yahnahpin* by Filla, James and Redman in 1994, was found to be different enough to merit its own genus name, Bakker resurrected a form of the name *Brontosaurus* and named it *Eobrontosaurus*.

Below: Unusually, the single specimen of Eobrontosaurus that has been found contained the gastralia, the belly ribs. These are not usually preserved in a sauropod. Eobrontosaurus seems to be a more primitive animal than any of the accepted species of the Apatosaurus genus.

Features: *Eobrontosaurus* is distinguished from *Apatosaurus* by two specific features. The first is the slightly thicker neck produced by the bigger ribs, which is supported by the neck vertebrae. The second is that *Eobrontosaurus* has a slightly different arrangement of shoulder bones – a little more like those of a macronarian.

Distribution: Wyoming, USA.
Classification: Sauropoda, Diplodocidae.
Meaning of name: Early Brontosaurus.
Named by: Bakker, 1998.
Time: Kimmeridgian to Tithonian stages of the late Jurassic.
Size: 20m (66ft).
Lifestyle: Browser.
Species: *E. yahnahpin*.

APATOSAURUS

Above: The skull of Apatosaurus *is long and narrow.*

The shape of the head also created uncertainty. A restoration of *Apatosaurus* (or *Brontosaurus*) carried out in the early twentieth century shows a dinosaur with a short boxy head like that of *Camarasaurus*. This is because the skull of *Apatosaurus* had not been found at the time, and when the skeleton was mounted in the American Museum of Natural History in New York, USA, the technicians built a *Camarasaurus*-type skull for it.

A long and narrow skull, like that of *Diplodocus*, had actually been found close to one of the skeletons at Dinosaur National Monument, but only the director of the Carnegie Museum in Pittsburgh, W. J. Holland, thought that this was the true *Apatosaurus* skull. He was overruled by the much more influential H. F. Osborne. The correct skull was finally established in the 1970s.

Dicraeosaurus

The only late Jurassic diplodocid found on the southern continents was *Dicraeosaurus*. It was a member of the Tendaguru fauna and, along with the other animals, showed that the same families of dinosaurs existed in North America and Africa at that time. However, *Dicraeosaurus* was so different from the North American forms that it has been given its own family, Dicraeosauridae.

Features: For a diplodocid, *Dicraeosaurus* has a strangely short neck with only 12 vertebrae, far fewer than any of the other late Jurassic diplodocids except for *Brachytrachelopan*. The vertebrae have extremely long spines that are deeply cleft in the neck and form a kind of low sail over the back. These features would have made it look bigger in profile and would have helped to deter predators. The tail has the typical diplodocid whiplash that would have been used as a weapon.

Distribution: Tanzania.
Classification: Sauropoda, Dicraeosauridae.
Meaning of name: Two forked lizard.
Named by: Janensch, 1914.
Time: Kimmeridgian stage of the late Jurassic.
Size: 20m (66ft).
Lifestyle: Low or high browser.
Species: *D. hansemanni, D. sattleri*.

Left: A full skeleton of Dicraeosaurus *is mounted in the Humboldt Museum in Berlin, Germany, beside that of its Tendaguru neighbour Giraffatitan.*

ORNITHOPODS

The two-footed ornithopod dinosaurs were present at the same time as the sauropods of the late Jurassic and were also plant-eaters. The ornithopods, however, were mostly quite small animals, scampering about at the feet of their big lizard-hipped cousins, and presumably exploiting food sources that were not available to bigger animals.

Camptosaurus

The original species of *Camptosaurus* is based on ten partial skeletons, ranging from juveniles to adults. The species is well known from the mounted skeletons of a juvenile and an adult collected by Fred Brown and William H. Reed in the 1880s in Wyoming, USA, and put on display in the Smithsonian Institution in Washington D.C. The English species is a *nomen dubium* and may not even be an ornithopod.

Features: *Camptosaurus* is very similar to its cousin *Iguanodon*, a genus which was a characteristic feature of the landscape of the later early Cretaceous period in Europe. However, its head is longer and lower, and it has four toes on the back foot rather than three. Its heavy body can be carried on four legs or on two, both its front five-fingered feet and its hind feet carrying weight-bearing hooves. Its long mouth contains hundreds of grinding teeth and it has a beak at the front. Its food would be kept in cheek pouches while chewed.

Right: It is possible that the original species of Camptosaurus *is the only true one. The others may be species of* Iguanodon.

Distribution: Colorado, Oklahoma, Utah and Wyoming, USA, and England.
Classification: Ornithopoda, Iguanodontia.
Meaning of name: Flexible lizard.
Named by: Marsh, 1885.
Time: Kimmeridgian to Tithonian stages of the late Jurassic.
Size: 3.5–7m (11½–23ft).
Lifestyle: Low browser.
Species: *C. dispar*, *C. leedsi*, *C. depressus*, *C. prestwichii*.

Othnielia

Othniel Charles Marsh found this little dinosaur in 1877 and gave it the name *Nanosaurus*, meaning "little lizard". It was renamed after its original finder by Peter Galton exactly 100 years later. It is known from the almost complete skeletons of two individuals, but the only parts of the skull known consist of a few pieces of bone and teeth. It is said that the skull of one skeleton was stolen by collectors before it could be excavated.

Features: This small herbivore was a very fast runner, with compact thighs to hold the leg muscles, and easily moved lightweight legs with long shins and toes. It has five-fingered hands and four-toed feet. Its tail is stiff and straight, with the bones lashed together with tendons, to give balance while running. It is similar to the better-known *Hypsilophodon* with its big eyes and beak, but differs from it by having enamel on both sides of its teeth.

Distribution: Colorado, Utah and Wyoming, USA.
Classification: Ornithopoda.
Meaning of name: Othniel's one.
Renamed by: Galton, 1977 (originally named in 1877).
Time: Kimmeridgian to Tithonian stages of the late Jurassic.
Size: 1.4m (4½ft).
Lifestyle: Low browser.
Species: *O. rex*.

Right: The tiny ornithopods like Othnielia *must have scampered about among the feet of the huge Morrison Formation sauropods, eating the ground-hugging plants that were not available to their big neighbours.*

Drinker

If *Othnielia* was named after one of the greatest American dinosaur pioneers of the nineteenth century, Othniel Charles Marsh, then the dinosaur *Drinker* was named after the other, namely Edward Drinker Cope. It is known from skeletons of an adult and a juvenile. Although there are quite a few remains known, not much has been published about this dinosaur.

Features: *Drinker* differs from its close relative *Othnielia* by having a more flexible tail, and its teeth seem to have more complex crowns. The chewing action that they produced would have worked on a ball of food held in the cheeks. Its toes are long and spreading, suggesting that it may have lived in swampy conditions. However, the differences between the two dinosaurs are so slight that instead of two different dinosaur species we may just be looking at two species of *Othnielia*.

Distribution: Wyoming, USA.
Classification: Ornithopoda.
Meaning of name: Charles Drinker Cope's one.
Named by: Bakker, Galton, Siegwarth and Filla, 1990.
Time: Kimmeridgian to Tithonian stages of the late Jurassic.
Size: 2m (6ft).
Lifestyle: Low browser.
Species: *D. nisti*.

LUSITANIAN BASIN

In late Jurassic times the area of Portugal was a broad, shallow plain called, by geologists, the Lusitanian Basin. It had a warm climate, and rivers flowed into the basin from the nearby mountains. The resulting vegetation supported a large number of late Jurassic dinosaurs.

The fossils found at the site represent brachiosaurs, camarasaurs, diplodocids, stegosaurs and meat-eating theropods. There are also fossils of crocodiles, turtles, pterosaurs and mammals. The fauna is very similar to that of the North American Morrison Formation, indicating that the two areas, which are now far apart, were very close geographically at that time.

Studies of the rich fossil fauna from the site began in 1982 when a farmer found a dinosaur bone in his field. The fossil belonged to the carnosaur *Lourinhanosaurus*. The Museum of Lourinhã, Portugal, is now a world-famous institution, housing the dinosaurs found in the local area. In 1993 local scientist Isabel Mateus found fossilized dinosaur eggs in a sea cliff just to the north of the Lusitanian Basin. These eggs were identified by Philippe Taquet of the National Museum of Natural History in Paris, France, as theropod eggs. There is speculation that these were the eggs of *Lourinhanosaurus*. Since then much of the local dinosaur fauna, including the ornithopod *Draconyx*, has been collected and studied by Octavio Mateus, Isabel's son.

Draconyx

The dinosaur *Draconyx* is known from a scattering of bones from a single individual. The remains include teeth, vertebrae, leg bones and finger and toe bones, including the claws, after which it was named. The best material comes from the hind limbs. It is now housed in the Museum of Lourinhã, the principal centre for Portuguese dinosaur study.

Features: *Draconyx* is an early iguanodont of medium size, seemingly closely related to *Camptosaurus*. It is about the same size, and has the same heavy hind legs and short front legs. It differs from *Camptosaurus* by the shape of the thigh bone and the arrangement of toe bones – it has a vestigial first toe whereas *Camptosaurus* does not, and it lacks the fifth toe that *Camptosaurus* possesses. Otherwise the beasts are very similar.

Distribution: Portugal.
Classification: Ornithopoda, Iguanodontia.
Meaning of name: Dragon claw.
Named by: Mateus and Antunes, 2001.
Time: Late Kimmeridgian to early Tithonian stages of the late Jurassic.
Size: 7m (23ft).
Lifestyle: Low browser.
Species: *D. loureiroi*.

Left: Draconyx *is another dinosaur excavated from the prolific Lourinhã dinosaur quarries of Portugal.*

STEGOSAURS

The stegosaurs were the plated dinosaurs of Jurassic and early Cretaceous times. They appear to have evolved in Asia – most of the primitive forms are found in China – and then migrated to North America and to Africa. They probably evolved from the same ancestors that gave rise to the armoured dinosaurs. Stegosaurus is one of the most recognizable of the dinosaurs, and is the state fossil of Colorado, USA.

Stegosaurus

Although *S. armatus* was the first stegosaurus species to be found, *S. stenops*, found by Othniel Charles Marsh, is the more familiar species. The back plates were once thought to have

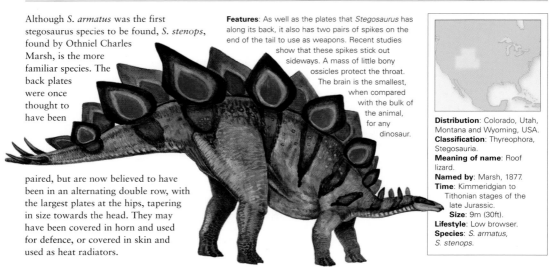

paired, but are now believed to have been in an alternating double row, with the largest plates at the hips, tapering in size towards the head. They may have been covered in horn and used for defence, or covered in skin and used as heat radiators.

Features: As well as the plates that *Stegosaurus* has along its back, it also has two pairs of spikes on the end of the tail to use as weapons. Recent studies show that these spikes stick out sideways. A mass of little bony ossicles protect the throat. The brain is the smallest, when compared with the bulk of the animal, for any dinosaur.

Distribution: Colorado, Utah, Montana and Wyoming, USA.
Classification: Thyreophora, Stegosauria.
Meaning of name: Roof lizard.
Named by: Marsh, 1877.
Time: Kimmeridgian to Tithonian stages of the late Jurassic.
Size: 9m (30ft).
Lifestyle: Low browser.
Species: *S. armatus*, *S. stenops*.

Tuojiangosaurus

The best-known of the many Chinese stegosaurs, and the first to have been discovered is *Tuojiangosaurus*. It is known from two partial skeletons, one of which is 50 per cent complete. Its similarity to *Stegosaurus* shows that the two animals obviously had the same lifestyle and ate the same food – leafy plants growing low on the ground.

Features: *Tuojiangosaurus* has 15 pairs of small, pointed plates running down the neck, back and tail, as well as two pairs of spikes on the tail. As with all stegosaurs, it has a long head, with spoon-shaped teeth and a toothless beak. The teeth are very like those of *Stegosaurus* – small, coarsely serrated and vertically grooved.

Distribution: China.
Classification: Thyreophora, Stegosauria.
Meaning of name: Lizard from the Tuo River.
Named by: Dong, Li, Shou and Zhang, 1973.
Time: Kimmeridgian to Tithonian stages of the late Jurassic.
Size: 7m (23ft).
Lifestyle: Low browser.
Species: *T. multispinus*.

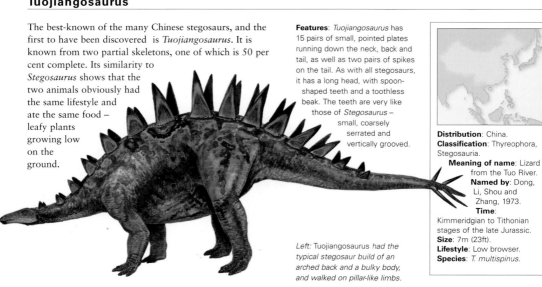

Left: Tuojiangosaurus *had the typical stegosaur build of an arched back and a bulky body, and walked on pillar-like limbs.*

Chungkingosaurus

This stegosaur is known from several partial skeletons. It was one of the smaller, more primitive members of the stegosaur group. On the evolutionary scale it seems to lie somewhere between *Kentrosaurus* and *Stegosaurus*. The incomplete remains suggest that it has more than the usual two pairs of spikes on the end of the tail. The head is deep and narrow, and it has large nostrils.

Features: The distinguishing feature of *Chungkingosaurus* is the shape of the upper arm bone, which seems to be very primitive but with very broad ends, which is unusual. The hip bones are also primitive. Over the back the plates do not have strong bases, and may not have been firmly embedded in the skin. The tail plates appear to have been stronger.

Distribution: China.
Classification: Thyreophora, Stegosauria.
Meaning of name: Lizard from Chunking.
Named by: Dong, Zhou and Zhang, 1983.
Time: Oxfordian stage of the late Jurassic.
Size: 4m (13ft).
Lifestyle: Low browser.
Species: *C. jiangheiensis*.

Right: The plates along the back of Chungkingosaurus *are very thick and narrow, and could almost be regarded as spikes.*

DINOSAUR PLATES

Ever since the first *Stegosaurus* was found, there has been much scientific discussion as to how the plates were arranged. None has ever been found in fossil form attached to the skeleton of a dinosaur.

An early idea of their arrangement was that they lay horizontally over the body, overlapping like the scales of a pangolin, and forming a protective shield. However, it was soon realized that they must have stuck out vertically.

For a long time the perceived image was of the plates being arranged in pairs, forming a symmetrical double row down the neck, back and tail. Now the accepted arrangement is of a double alternating row. Other interpretations abound, however. There is a suggestion that they formed a single row, but overlapped one another.

Scientific understanding of the function of the plates has always been subject to conjecture too. The obvious interpretation is that they were covered in horn, and formed an armour to protect the neck and back. Bob Bakker even suggests that their bases were muscular, and the points could have been turned towards any attacker.

In the 1970s it was suggested that the plates were covered in skin rather than horn, and used as heat exchangers, warming the blood in the morning sun and dissipating the heat at midday. If this worked for broad-plated stegosaurs like *Stegosaurus*, it would not have done so for the narrow-plated stegosaurs like *Chungkingosaurus*, since they would not have provided the surface area needed to make the system work.

Chialingosaurus

The first important stegosaur to have been found in China was *Chialingosaurus*. It is known from only one partial skeleton. Like others of the family it would have fed close to the ground, from the ferns and cycads that abounded at the time. Many stegosaurs were well-balanced at the hips, and seem to have been able to rise to their hind legs to feed from low-growing branches as well.

Features: This is a medium-size stegosaur, with a high, narrow skull. Its plates are small. They appear to be almost disc-shaped over the neck, but tall and spike-like over the hips and tail. They run in two rows from the neck to the tail. Its front legs are long for a stegosaur. The tall extensions on the tail vertebrae suggest muscular hips and perhaps an ability to rear on its hind legs.

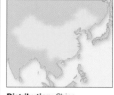

Distribution: China.
Classification: Thyreophora, Stegosauria.
Meaning of name: Jia-ling River lizard.
Named by: Young, 1959.
Time: Middle to late Jurassic.
Size: 4m (13ft).
Lifestyle: Low browser.
Species: *C. kuani*, *C. guangyuanensis*.

MORE STEGOSAURS

The most familiar image we have of the stegosaurs is of Stegosaurus *itself, with the big, broad back plates and the spiked tail. However,* Stegosaurus *represents something of an extreme. Most stegosaurs had much smaller back structures, and many of them had spines rather than plates. The more primitive forms even had spines on their shoulders.*

Kentrosaurus

This stegosaur was excavated between 1909 and 1912 from the Tendaguru site by a team from Germany. Several hundred *Kentrosaurus* bones were found, suggesting that something like 70 individuals died there. The group find suggests that it may have been a herding animal. Two mounted skeletons were prepared for the Humboldt Museum in Berlin, Germany, but one was destroyed by bombing during World War II.

Features: The nine pairs of plates on the neck and back are very much narrower than those of *Stegosaurus,* and the five pairs of spines run in a double row right down the tail. Another pair of spines projects sideways from the shoulders. Unlike more advanced stegosaurs, it seems *Kentrosaurus* did not have ossicles across its body embedded in the skin. The little skull contains a tiny brain with well-developed olfactory bulbs. This suggests *Kentrosaurus* had a very good sense of smell, which would have been useful in food gathering.

Distribution: Tanzania.
Classification: Thyreophora, Stegosauria.
Meaning of name: Pointed lizard.
Named by: Henning, 1915.
Time: Kimmeridgian stage of the late Jurassic.
Size: 5m (16½ ft).
Lifestyle: Low browser.
Species: *K. aethiopicus, K. longispinus.*

Dacentrurus

Remains of *Dacentrurus* are found throughout the late Jurassic period, and they seem to represent individuals ranging in size from 4–10m (13–33ft), making it one of the biggest of the stegosaurs. The youngest remains are common in the Lourinhã site in Portugal. The French specimens were lost when the museum at Le Havre was destroyed during World War II.

Features: This was the first stegosaur to be discovered, by Owen in 1875, and was named *Omosaurus.* The forelimbs are longer than those of other stegosaurs, and the back is lower. The pieces of armour that we know suggest that it may have had spines rather than plates. In build it is a very primitive stegosaur, but the vertebrae and the hips have some quite advanced features. It is possible that the remains found represent more than one species.

Distribution: England, France and Portugal.
Classification: Thyreophora, Stegosauria.
Meaning of name: Very pointy lizard.
Named by: Lucas, 1902.
Time: Oxfordian to Kimmeridgian stages of the late Jurassic.
Size: 10m (33ft).
Lifestyle: Low browser.
Species: *D. armatus, D. phillipsi.*

Right: Dacentrurus lived between 154 and 150 million years ago.

"Yingshanosaurus"

The dinosaur known as *"Yingshanosaurus"* seems to be a *nomen dubium* – a name that has no scientific backing – and there has been little published work on the specimen. However, an almost complete skeleton was excavated from the Sichuan basin in China in the 1980s. It is a stegosaur with a double row of plates, two pairs of tail spines and a huge pair of shoulder spines.

Features: The features that make this animal unique are the wing-like shoulder spines, known scientifically as "parascapular spines". Other stegosaurs possess them, but none are as large. It seems that shoulder spines are only present in the primitive stegosaurs; the more advanced types such as *Stegosaurus* and *Wuerhosaurus* show no evidence of them.

Distribution: China.
Classification: Thyrophora, Stegosauria.
Meaning of name: Lizard from Yingshan.
Named by: Zhou, 1984.
Time: Late Jurassic.
Size: 5m (16½ft).
Lifestyle: Low browser.
Species: 'Y. jichuanensis'

Left: Note the large-scale spines on the shoulders. Stegosaurs used to be restored with these spines on the hips, but it makes more sense from a defensive angle to have them on the shoulders.

Hesperosaurus

The skeleton of this stegosaur was found in 1985, not far above base of the Morrison Formation, during the excavation of a *Stegosaurus* skeleton. It was complete, only missing the limbs, which were probably lost by erosion as it lay near the surface. It is now on display in the Denver Museum of Natural History, USA. *Hesperosaurus* is the oldest known stegosaur from North America.

Features: *Hesperosaurus* is distinguished by its short and wide skull, and deep lower jaw. Its armour consists of at least 10 plates, probably 14, which are oval, and the traditional two pairs of defensive spikes on the end of the tail. Unlike those of *Stegosaurus*, the plates are longer than they are high, and the spikes point in more of a backward direction. Its closest relative seems to have been *Dacentrus* of contemporary Europe.

Distribution: Wyoming, USA.
Classification: Thyreophora, Stegosauria.
Meaning of name: Western lizard.
Named by: Carpenter, Miles and Cloward, 2001.
Time: Kimmeridgian to Tithonian stages of the late Jurassic.
Size: 6m (17½ft).
Lifestyle: Low browser.
Species: *H. mjosi*.

Left: The species name, H. mjosi, honours Ronald G. Mjos who collected and prepared the specimen and mounted a cast of it for display in the Denver Museum of Natural History, USA.

THE
CRETACEOUS
PERIOD

By the Cretaceous period the supercontinent Pangaea had split apart

into individual continents that could be recognizable today. In the

Southern Hemisphere, though, the landmasses of Australia and

Antarctica were still joined – in a supercontinent called Gondwana.

With isolated continents, different animals evolved, so that the

dinosaurs of South America differed from those of North America and

Europe. Tree ferns and conifers were beginning to give way to flowering

plants, with the conifers being pushed to high latitudes and

mountainous areas. However, the inhabitants of these forests were quite

remarkable – for this was the climax of the Age of Dinosaurs.

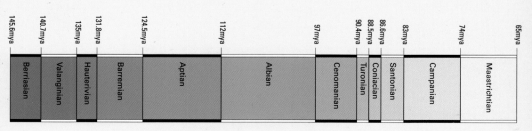

145.6mya	140.7mya	135mya	131.8mya	124.5mya	112mya	97mya	90.4mya	88.5mya	86.6mya	83mya	74mya	65mya
Berriasian	Valanginian	Hauterivian	Barremian	Aptian	Albian	Cenomanian	Turonian	Coniacian	Santonian	Campanian	Maastrichtian	

Top: The world as it would have appeared in Cretaceous times.

*Above: The timeline shows the different chronological stages that make up the
Cretaceous period. (mya = million years ago)*

PTEROSAURS

In the early Cretaceous the pterodactyloid pterosaurs evolved into all kinds of different types. The body and wing arrangement were quite conservative, but there were great differences in the shape of the head, indicating different food types and lifestyles. We have good evidence of pterosaurs that lived by the sea, lakes and rivers. There must have been more from inland and mountain areas that were never fossilized.

Dsungaripterus

Distribution: China.
Classification: Pterosauria, Pterodactyloidea.
Meaning of name: Wing from the Junggar Basin.
Named by: Young, 1964.
Time: Early Cretaceous.
Size: 3m (10ft) wingspan.
Lifestyle: Shellfish-eater.
Species: *D. brancai, D. weii.*

The strange mouth has been interpreted as an adaptation to feeding on shellfish, with the narrow beak used for prizing the shells from their substrate and the bony knobs used for crushing them. Since the specimens found are all a long way from any sea, the shellfish must have been freshwater types, from rivers or lakes.

Features: The remarkable feature of this large pterodactyloid is the head. The jaws are pointed and turned upwards at the tips, like specialist forceps. The sides of the mouth are armed with bony knobs in the place of teeth. The eye socket is surprisingly small, but much of the side of the deep skull is taken up by a cavity that combines the nostril with the other holes usually found there. A thin crest sticks up along the mid-line of the skull, and another juts out behind.

Left: The body of Dsungaripterus is very strong, with the back vertebrae fused together in the shoulder region, giving a structure similar to that of a bird.

Plataleorhynchus

Nicknamed the Purbeck spoonbill after the area in southern England where it was found, this creature must have worked the shallow waters of the swampy lowlands that existed at the time that the limy sediment of the famous Purbeck building stone was deposited. Like its namesake bird, the spoonbill, it would have filtered the water for organisms such as crustaceans and insect larvae.

Features: The only part of this pterosaur known is the tip of the long narrow jaws, in a specimen in which the top side is exposed. This is flat with a spoon-shaped end, and is lined with small pointed teeth that curve outwards. This is superficially similar to the structure in *Gnathosaurus*, but the arrangement of jaw bones is different; the maxilla at the front is very small, suggesting that it evolved independently. This is an example of convergent evolution. The rest of this pterosaur is unknown, but as with the other little-known examples, we can restore the living appearance of the animal by reference to the other pterosaurs in the group.

Below: At the end of the Jurassic and beginning of the Cretaceous southern England was a wide swampy plain, covered in slow rivers and lakes – an ideal environment for animals that fed on shallow-water creatures.

Below: Gnathosaurus for comparison.

Distribution: Southern England.
Classification: Pterosauria, Pterodactyloidea.
Meaning of name: Spoonbill beak.
Named by: Howse and Milner, 1995.
Time: Tithonian stage of the late Jurassic to the Berriasian stage of the early Cretaceous.
Size: 2.5m (8ft).
Lifestyle: Filter feeder.
Species: *P. streptophorodon.*

Criorhynchus

A number of pterosaurs had crests on the tips of their jaws, both the upper and the lower. *Criorhynchus* was the first to be found. The crests were probably used for cleaving the water as the pterosaur skimmed across at wave height, making it easier to catch fish just under the surface.

Features: A semicircular crest at the tip of the upper jaw is matched by a similar one on the lower, giving the impression of a continuous vertical disc when the jaws are closed. The first specimen was so fragmentary that it was not until the discovery of a more complete specimen of the closely related *Tropeognathus* more than a century later, in 1987, that the arrangement of crests on the skull was determined.

Distribution: England.

Classification: Pterosauria, Pterodactyloidea.
Meaning of name: Battering ram snout.
Named by: Owen, 1874.
Time: Cenomanian stage of the middle Cretaceous.
Size: 5m (16ft) wingspan.
Lifestyle: Fish-hunter.
Species: *C. simus*.

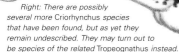

Right: There are possibly several more Criorhynchus *species that have been found, but as yet they remain undescribed. They may turn out to be species of the related* Tropeognathus *instead.*

OTHER PTEROSAUR GENERA

It is impossible to cover all the different genera of pterosaur here, even the ones that we know from good fossils. Other early cretaceous types include:
Phobetor A close relative of *Dsungaripterus*, but only about half the size and with straighter forceps jaws.

Domeykodactylus Another of the *Dsungaripterus* group but from South America – the first to be found outside China.

Doratorhynchus Known from a single neck vertebra which is particularly long, suggesting that it is an early member of the group that covered the giant pterosaurs of the late Cretaceous, such as *Quetzalcoatlus*.

Ornithodesmus A large British pterosaur with a long rounded duck-like beak lined with short teeth. It is quite unlike any other known pterosaur and is classed in a family of its own.

Santanadactylus A Brazilian form whose discovery opened the study of pterosaurs in South America.

Pterodaustro

Filter feeding was taken to an extreme in this pterosaur, the first to be found in South America. It must have rested close to the water and swept the surface with its bristle-combed lower jaw, scooping out small invertebrates, rather like a modern flamingo does. It has been suggested that its fur was pink – a pigment imparted by the crustaceans on which it fed, as in the plumage of modern flamingos.

Features: The jaws are very long and slim, and sweep upwards in a bow-shape. There are no teeth in the lower jaw, instead there is a brush-like arrangement of long elastic bristles, up to 500 of them, into which the upper jaw fits when closed. The upper jaw has many short blunt teeth that comb food from the bristles and into the mouth. The hands are small and the feet big for a pterosaur.

Distribution: Argentina.
Classification: Pterosauria, Pterodactyloidea.
Meaning of name: Southern wing.
Named by: Bonaparte, 1969.
Time: Early Cretaceous.
Size: 1.3m (4¼ft).
Lifestyle: Filter feeder.
Species: *P. guinzaui* and possibly two others, unnamed.

Left: Pterodaustro *was the first pterosaur to have been found in South America. Several others have been discovered since.*

BIG PTEROSAURS

The pterosaurs continued to evolve, reaching their greatest diversity in early Cretaceous times. After this they began to decline, as the birds evolved and took over their niches. This competitive evolutionary pressure encouraged the pterosaurs to adopt more and more bizarre forms as they adapted to very restricted lifestyles.

Tapejara

The big lightweight crest of *Tapejara* is a mystery. The obvious explanation is that it is a display structure, but it is possible that it may have had something to do with aerodynamics, and was used as a windsurfer sail. The short mouth, likewise, is a puzzle. It may have been a fruit-eater, like modern heavy-billed birds such as hornbills or toucans, or it may have lived on fish or carrion.

Distribution: Araripe Plateau, north-eastern Brazil.
Classification: Pterosauria, Pterodactyloidea.
Meaning of name: Old one.
Named by: Kellner, 1989.
Time: Aptian stage of the Cretaceous.
Size: Probably 5m (16ft) wingspan (but only skulls are known).
Lifestyle: Fruit-eater, fish-eater or scavenger.
Species: *T. wellnhoferi, T. imperator, T. navigans.*

Right: If Tapejara *had been a fruit-eater, then it probably served an important ecological purpose, helping to spread seeds over a wide area, just like modern fruit-eating birds.*

Features: The distinctive feature of *Tapejara* is the extraordinary head crest. Two bony struts protrude from the skull, one broad and pointing directly upwards and one narrow and sweeping back. In the species *T. navigans* the back section is almost non-existent, but in *T. imperator* it is long and slim. In life these would have been joined by a flap or a sail, giving a broad display structure. The front of the short snout is turned down as a strong beak.

Tupuxuara

This pterosaur was found in the same area and from the same period as the other spectacularly crested pterosaur, *Tapejara*. The skin-covered bony crest of *Tupuxuara* was well-endowed with blood vessels, which suggests that the crest could change colour, or at least the intensity of the colour, according to the animal's mood and activity. This would have made for an elaborate communication system for threat or mating displays. Perhaps only the males possessed the crest.

Right:
The name Tupuxuara means "familiar spirit" in the language of the Tupi, the original inhabitants of the area of Brazil in which it was found. Likewise the name Tapejara means "old one" in the same language, and refers to a being in local mythology.

Features: The crest sweeps upwards and backwards in a semicircular plate, making the skull longer than the body itself. The jaws are narrow and toothless. Because of the spectacular crest, *Tupuxuara* was thought to have belonged to the same family as *Tapejara*, but detailed study of the skeleton suggests that it is closer to the giants like *Quetzalcoatlus*. Blood vessels in the crest indicate that it was covered by skin. It may even have had a keratinous extension, that would have made it even bigger.

Distribution: Ceará, Brazil.
Classification: Pterosauria, Pterodactyloidea.
Meaning of name: Familiar spirit of local Tupi Indian mythology.
Named by: Kellner and Campos, 1988.
Time: Aptian stage of the Cretaceous.
Size: 5.5m (17ft) wingspan.
Lifestyle: Fish-eater or meat-eater.
Species: *T. leonardii, T. longicristatus.*

Anhanguera

The shape of the brain in one particularly well-preserved specimen of *Anhanguera* shows that it could co-ordinate information from all parts of the body. The wing was used as a sensory organ and the brain was able to control the position and attitude of the body while keeping its eyes focused on its prey. This ability probably applies to the rest of the pterodactyloids.

Features: This large pterosaur has very long wings and crests on both the upper and lower jaws, about halfway along. The legs are quite small, and the hip bones relatively weak. Their articulation indicates that it is impossible to bring the legs vertically under the body like a bird, and so when at rest the legs must have been splayed out at the side.

Distribution: Northeastern Brazil.
Classification: Pterosauria, Pterodactyloidea.
Meaning of name: Old devil.
Named by: Campos and Kellner, 1985.
Time: Aptian stage of the Cretaceous.
Size: 4m (13ft) wingspan.
Lifestyle: Fish-eater.
Species: *A. santanae*, *A. blittersdorffi*.

Left: The skull of Anhanguera *is as long as its body.*

PTEROSAUR EGGS

It had always been assumed that pterosaurs laid eggs. However, when a fossil egg is found it is difficult to tell what kind of animal laid it.

In early 2004 a tiny egg, no more than 5cm (2in) in diameter, was found in the famous late Jurassic and early Cretaceous lake deposits in Liaoning province in China. It was the fossil of a soft egg, like that of a snake or a turtle, rather than one with a hard shell as we find in birds and dinosaurs. So well preserved was this egg that when it was dissected it was found to contain a young pterosaur with long wings tightly folded. The baby would have had a 27cm (11in) wingspan when it hatched.

The long wing bones had solidified in the egg before hatching, which suggested that as soon as the youngster hatched it would have been ready for flight. This means that the youngsters would have had to fend for themselves as soon as they were free from the shell, like crocodiles and turtles, and would have had little in the way of parental attention. This is in keeping with the observation that pterosaurs grew slowly, as all their youthful energy would have gone into finding food rather than growing.

Cearadactylus

This genus is known only from a single damaged skull with the jaws still articulated. The resemblance of the head to that of some fish-eating dinosaurs and crocodiles suggests that *Cearadactylus* hunted fish. The bunch of fine teeth at the end of the jaw would have been ideal for holding on to slippery prey which it would then swallow whole.

Features: *Cearadactylus* has long jaws, with the long needle-like teeth restricted to the tip and forming a kind of a rosette. The teeth of the upper and lower jaw tip interlock when the jaws close. Apart from these, the teeth in the rest of the jaw are quite small and conical. The upper jaw has a notch close to the end. In fact the whole skull is reminiscent of that of the fish-catching dinosaur *Baryonyx*.

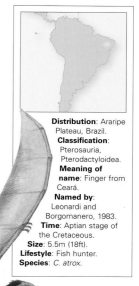

Distribution: Araripe Plateau, Brazil.
Classification: Pterosauria, Pterodactyloidea.
Meaning of name: Finger from Ceará.
Named by: Leonardi and Borgomanero, 1983.
Time: Aptian stage of the Cretaceous.
Size: 5.5m (18ft).
Lifestyle: Fish hunter.
Species: *C. atrox*.

EARLY CRETACEOUS CARNOSAURS

In early Cretaceous times the theropods began to diversify, and all kinds of new meat-eating dinosaurs started to evolve. However, the traditional line of theropods, the carnosaurs, which were the main hunters of the big plant-eaters of the Jurassic period, continued to flourish, and to prey on the plant-eating dinosaurs of the time.

Acrocanthosaurus

The presence of the fin on this animal originally led scientists to think that it must have been related to the spinosaurs. However, the fin seems different in construction to that of these later dinosaurs, and the rest of the animal is definitely carnosaurian. Footprints found in Texas, USA, may be those of *Acrocanthosaurus* and, if so, show that it hunted in packs.

Features: Imagine an *Allosaurus* with a low fin down the middle of its back. That is what the *Acrocanthosaurus* looks like. The fin is supported by spines from the vertebrae, 50cm (20in) over the back and progressively lower down the neck and tail. This fin was probably thicker and fleshier than that of the spinosaurs. It would have been bright, and used for identification and signalling. Its 68 thin, sharp, serrated teeth show that it was a hunter. The feet are quite small and adapted to walking on firm dry ground. The hand has three grasping fingers.

Distribution: Oklahoma, Texas, Utah and perhaps Maryland, USA.
Classification: Theropoda, Tetanurae, Carnosauria.
Meaning of name: High-spined lizard.
Named by: Stovall and Langston, 1950.
Time: Aptian to Albian stages of the Cretaceous.
Size: 8m (26ft).
Lifestyle: Hunter.
Species: *A. atokensis*.

Above: Acrocanthosaurus hunted in packs.

Left: Allosaurus.

Neovenator

Judging by the number of remains found close to one another on the Isle of Wight, *Neovenator* may have been a pack hunter. As an agile and fast hunter, it would have been the main predator of the herds of *Iguanodon* and *Hypsilophodon* that grazed the horsetail-choked marshes that existed in the area.

Features: About 70 per cent of the skeleton of *Neovenator* has been found. It is a very sleek and streamlined animal. The skull is narrow and has a particularly large nostril cavity. The arrangement of five teeth on the premaxilla, the front bone of the lower jaw, shows that it is related to *Allosaurus*. Its skeletal features suggest that it may have been on the ancestral line to the big theropods, such as *Carcharodontosaurus* and *Giganotosaurus*, that were to come. *Neovenator* is the first allosaurid to have been found in Europe, and apart from *Baryonyx* is the biggest meat-eater from the area and time.

Below: The one species of Neovenator found is named after the family Salero, on whose land the fossil was found.

Distribution: Isle of Wight.
Classification: Theropoda, Tetanurae, Carnosauria.
Meaning of name: New hunter.
Named by: Hutt, Martill and Barker, 1996.
Time: Barremian to Aptian stages of the Cretaceous.
Size: 6–10m (19½–33ft).
Lifestyle: Hunter.
Species: *N. salerii*.

Afrovenator

By the 1990s this was the only almost complete skeleton found of an African theropod. It was uncovered by a team from the University of Chicago, USA, led by Paul Sereno. The only parts missing were the lower jaw, some ribs and vertebrae, and the toe bones. Its similarity to the North American *Allosaurus* suggests that the two continents were still united at that time.

Above: The species name is from Abaka in Niger, where it was found.

Features: The strong hind legs show that *Afrovenator* was built for active hunting, and the strong arms, which are longer than those of its relative, *Allosaurus*, with the big curved claws, were perfectly designed for catching and holding prey. The skeleton is quite lightweight for the size of animal, and the tail is stiffened by overlapping bony struts, all features of a fast-moving beast. The skull is low, and did not have much in the way of crests.

Distribution: Niger.

Classification: Theropoda, Tetanurae.
Meaning of name: African hunter.
Named by: Sereno, J. A. Wilson, Larsson, Dutheil and Suess, 1994.
Time: Hauterivian to Barremian stages of the late Cretaceous.
Size: 8–9m (26–30ft).
Lifestyle: Hunter.
Species: A. abakensis.

CHARACTERISTICS OF THE TETANURAE

The Tetanurae, the group to which the carnosaurs and all the other advanced theropods belong, is characterized by the presence of three fingers or fewer, an extra opening in the side of the skull behind the nostril, and the stiffened tail.

The stiffening of the tail was brought about by a series of overlapping, bony struts, formed by bony material built up along the tendons that lashed the vertebrae of the tail together. The purpose was to hold the tail stiffly behind, while balancing the teeth and the jaws at the front. In later Tetanurans, such as the maniraptorans, this was taken to an extreme. The bony struts could envelop up to ten vertebrae behind the one from which they protruded. This made the tail stiff and inflexible like a tightrope walker's balancing pole.

An even greater extreme is found in the Tetanuran's present-day representatives. Our modern birds do not appear to have a bony tail at all. The original long dinosaurian tail has now evolved into a bony lump called the pygostyle, in which all the vertebrae are fused together. This is used as a base from which the lightweight feather tail sprouts, and anchors the muscles that control the feathered fan in flight.

Fukuiraptor

The "raptor" part of the name demonstrates that when this dinosaur was originally named it was thought to have been one of the maniraptorans, with the big killing claw on the second toe of the hind foot. Instead this claw was found to be a claw from the hand, and the animal was reclassified as a primitive carnosaur. Other Japanese theropods, *Kitadanisaurus* and *Katsuyamasaurus*, are now thought to have been *Fukuiraptor*.

Features: The big claws on the hand are distinctive, and led to the original mis-identification of the family. The jawbones into which the teeth are set are also similar to those of maniraptorans, and added to the initial confusion. *Fukuiraptor* seems to have been related both to *Sinraptor* from China, and the allosaurs found in Australia.

Distribution: Fukui Prefecture, Japan.
Classification: Theropoda, Tetanurae, Carnosauria.
Meaning of name: Plunderer from Fukui.
Named by: Azuma and Currie, 2000.
Time: Albian stage of the Cretaceous.
Size: 4.2m (14ft).
Lifestyle: Hunter.
Species: F. kitadaniensis.

Above: The Fukuiraptor skeleton that was uncovered is of an immature individual, and the adult would have been bigger than the 4.2m (14ft) length given here.

MEDIUM-SIZE THEROPODS

At the beginning of the Cretaceous period, new types of plant-eating dinosaur began to evolve. The late Jurassic period, which preceded the Cretaceous, was the heyday of the plant-eating sauropods, but now the plant-eating ornithopods were coming into their own. Many were much smaller animals, and as a result smaller theropods, described here, evolved to prey on them.

Huaxiagnathus

For more than 150 years, the smallest theropods were assumed to be the compsognathids. *Compsognathus* and *Sinosauropteryx*, the only two compsognathids hitherto known, were no bigger than chickens. So it was a surprise when the complete articulated skeleton of a compsognathid as big as a cassowary came to light in the bone-littered lake beds of Liaoning, China, in 2004.

Features: *Huaxiagnathus* has long arms, strong and slender hind legs and a flexible neck. The skull is very lightweight, formed merely of an assemblage of bony struts, and the teeth are very small but curved and sharp. Although it was later than *Compsognathus* in the evolutionary line, *Huaxiangnathus* appears more primitive, lacking some of the adaptations of the forelimbs and with a very non-specialized body plan. The details of its skeleton indicate that the compsognathids may have been ancestral to the maniraptorans.

Distribution: Liaoning, China.
Classification: Theropoda, Tetanurae, Compsognathidae.
Meaning of name: Jaw from eastern China.
Named by: Hwang, Norell, Ji and Gao, 2004.
Time: Early Cretaceous.
Size: 1.5m (5ft).
Lifestyle: Hunter.
Species: *H. orientalis*.

Right: Huaxiagnathus is the biggest compsognathid so far known, and probably hunted small reptiles.

Dilong

The earliest indisputable tyrannosaurid was found in 2004 as a semi-articulated skeleton, along with scattered bones of three other specimens in the lake deposits of the Yixian Formation, in Liaoning, China. The imprints of the feathers were clearly seen, and this has led to renewed speculation that even the biggest of the Cretaceous meat-eating dinosaurs were feathered, at least in their small juvenile stages.

Features: Although the skeleton is similar to that of primitive coelurosaurs, the skull is definitely that of a tyrannosaurid, with its strongly bonded front bones and the front teeth with the characteristic D-shaped cross-section. The arms are longer than those of later tyrannosaurids, and it has three fingers. The entombing rocks where *Dilong* was found are so fine-grained that the covering of primitive feathers was preserved, the first indisputable proof we have that the early tyrannosaurids were covered with feathers.

Distribution: Liaoning, China.
Classification: Theropoda, Tyrannosauria.
Meaning of name: Emperor dragon from Chinese mythology.
Named by: Xu, Norell, Kuang, Wang, Shao and Jia, 2004.
Time: Early Cretaceous.
Size: 1.5m (5ft).
Lifestyle: Hunter.
Species: *D. paradoxus*.

Above: The feathers, or more accurately protofeathers, of Dilong consisted of branched structures about 3cm (1¼in) long. They appear to have covered the entire animal and would have been used for insulation.

Nqwebasaurus

The earliest coelurosaurids to be found on any of the southern continents was *Nqwebasaurus*. The find suggests that the family spread throughout the world before the southern supercontinent of Gondwana split from northern Laurasia. It is known from a skeleton that is 70 per cent complete, and was given the nickname 'Kirky', from the Kirkwood Formation in which it was found.

Features: The body of *Nqwebasaurus* is similar to that of a conventional coelurosaurid. It has a small body, a long flexible neck and long running legs. It has a very long first finger, partially opposable like a thumb, that bore a particularly large claw. Stomach stones were found with the skeleton. They usually indicate a vegetarian diet, but *Nqwebasaurus* was definitely a meat-eater. The stones may have come from the stomach of a plant-eating animal which it had killed and eaten.

Distribution: Eastern Cape Province, South Africa.
Classification: Tetanurae, Coelurosauria.
Meaning of name: Lizard from Nqweba, the local name for the Kirkwood District of South Africa.
Named by: de Klerk, Forster, Sampson, Chinsamy and Ross, 2000.
Time: Early Cretaceous.
Size: 0.8m (2½ft).
Lifestyle: Hunter.
Species: *N. thwazi*.

THE EVOLUTIONARY LINE OF BIRDS

It is now accepted that birds are descended from the Tetanuran theropod dinosaurs. In fact, some scientists insist that birds must be referred to as dinosaurs, and the animals that are more traditionally referred to as dinosaurs should be called "non-avian dinosaurs".

The anatomy of birds and dinosaurs is so similar that there is hardly a doubt about the connection. However, one anomaly persists. The Tetanurans had a maximum of three fingers on the hand, with some, such as the later tyrannosaurids, having only two, and the alvarezsaurids only one. Birds' wings are likewise made up of three fingers, but they are a different three fingers. In Tetanurans the fingers are the first, second and third, and they have lost the equivalent of the human little finger and ring finger. Studies of bird embryos show that a bird's three fingers are the second, third and fourth, missing the thumb and little finger. If these animals had common ancestors, then we would expect the same fingers to be involved. This matter has yet to be resolved.

Eotyrannus

Amateur fossil collector Gavin Leng discovered *Eotyrannus* preserved in hard mudstone high on the cliffs of the coast of the Isle of Wight, in 1996. The skeleton was about 40 per cent complete, and there was enough of it for it to be identified as one of the earliest tyrannosaurids known. It would have preyed on some of the smaller herbivores of the time.

Features: *Eotyrannus* is evidently an early tyrannosaurid judging by the heavy skull, the D-shaped front teeth and the arrangement of the shoulder and limb bones. As in other early tyrannosaurids the hands are long compared with those of later members of the group, and in fact the second finger is almost as long as the forearm. The only skeleton found is a juvenile, and so the adult may have been longer than the 4.5m (15ft) stated.

Distribution: Isle of Wight, England.
Classification: Theropoda, Tetanurae, Tyrannosauroidea.
Meaning of name: Dawn tyrant.
Named by: Hutt, Naish, Martill, Barker and Newbery, 2001.

Time: Barremian stage of the early Cretaceous.
Size: 4.5m (15ft).
Lifestyle: Hunter.
Species: *E. lengi*.

Left: Eotyrannus *is very similar to* Dilong *from the other side of the world, and it has been suggested that they are actually the same genus.*

LITTLE FEATHERED DINOSAURS

The fine-grained lake deposits from the early Cretaceous period of Liaoning, in China, have long been famous for their exquisite fossils. However, it was not until the 1990s that they became well-known to Western scientists. Since this date a steady stream of beautifully preserved dinosaur material has been removed from the site, much of it seeming to show transitional stages between dinosaurs and birds.

Microraptor

The feathers of this little dinosaur show that it was a glider, and it probably represented an intermediate stage between ground- or tree-dwelling dinosaurs and birds with a flapping flight. It could spread out its arms and legs, and form a gliding surface enabling it to fly from tree to tree.

Features: The remarkable feature of this dinosaur is the distribution of feathers. Like a bird, it has flight feathers along the arms, but unlike a bird it also has them along the legs. Other adaptations to flight include a very short body with few vertebrae in the back, making the body stiff and strong. It is possibly a dromaeosaurid, but has similarities to the troodontids and also the birds.

Left: Microraptor had long feathers on both the arms and the legs. When spread, these would have provided an effective gliding surface. The tail also was feathered, presumably forming a steering organ.

Distribution: Liaoning, China.
Classification: Theropoda, Tetanurae, Coelurosauria.
Meaning of name: Small plunderer.
Named by: Xu, Zhou, Wang, Kuang, Zhang and Du, 2003.
Time: Barremian stage of the early Cretaceous.
Size: 40–60cm (16–24in).
Lifestyle: Insectivore.
Species: *M. zhaoianus*, *M. gui*.

Sinosauropteryx

This was the first of the Liaoning dinosaurs discovered with a covering of feathers or feather-like structures. The presence of feathers gives strength to the theory that all small meat-eating dinosaurs had feathers, or were at least insulated in some manner. This would support an active lifestyle for a warm-blooded predator. The bones of a mammal were found in the stomach of one specimen.

Features: Apart from the feathers, this is a typical small meat-eating dinosaur, similar to *Compsognathus*, with the short arms and long tail. The feathers are not like the branched feathers of modern birds, but would have formed more of a fuzzy or downy covering. Each filament is up to 3cm (1⅛in) long. *Sinosauropteryx*

Distribution: Liaoning, China.

Above: There is no evidence as to the colour or pattern of the plumage of feathered dinosaurs, but it seems likely that they were as varied as modern birds.

has a very long tail, with 64 vertebrae, unlike the stumpy tail of a modern bird. In the first skeleton found, the feathers were only obvious along the back. Some scientists, unable to accept the idea that dinosaurs had feathers, interpreted this as a continuous crest of skin. Skeletons with clearer feather impressions have since been found.

Classification: Theropoda, Tetanurae, Coelurosauria.
Meaning of name: Chinese bird with feathers.
Named by: Ji Q. and Ji S., 1996.
Time: Barremian stage of the early Cretaceous.
Size: 1.3m (4ft).
Lifestyle: Hunter.
Species: *S. prima*.

ARCHAEORAPTOR

In 1999 an interesting fossil came to the West from the prolific fossil site at Liaoning, in China. It appeared to be a feathered bird but with the tail of a dinosaur.

It came to light through a devious route, the finder having circumvented the legal process of exporting fossils from China. It was sold to an American museum for $80,000. The purchaser, dinosaur researcher and museum-owner Stephen Czerkas, began to prepare a scientific paper on it, but potential co-authors were unwilling to help because of its murky history. To legitimize the procedure, Chinese palaeontologist Xu Xing was sent to America to work on the specimen, in anticipation of its being returned to China.

The resulting paper did not meet the stringent requirements for scientific publication, but nevertheless the popular magazine *National Geographic* publicly announced the name of this new specimen as *Archaeoraptor*. In fact the specimen was discovered to be a fake. A slab of stone containing the body fossil of *Microraptor zhaoianus* had been glued to a fossil containing the tail of a dromaeosaurid, before it had reached the dealer in China. The finder thought that he could make the fossil look more interesting, and hoodwinked the palaeontologists of America and China for months.

Caudipteryx

Known from several almost complete skeletons that include the feathers. The feathers of *Caudipteryx* show that it was a close relative of birds, but not the ancestor – many types of fully formed birds existed at that time in the Liaoning area. It may have waded at the edge of a lake on its long legs, or even perched on branches.

Distribution: Liaoning, China.
Classification: Theropoda, Tetanurae, Coelurosauria.
Meaning of name: Feathered tail.
Named by: Ji Q., Currie, Norell and Ji S., 1998.
Time: Barremian stage of the early Cretaceous.
Size: 70–90cm (27½–35in).
Lifestyle: Hunter.
Species: C. zoui, C. dongi.

Features: The distribution of feathers on this animal is quite distinctive. As well as a covering of fine insulating feathers, it also has long branching feathers on the arms and the end of the tail. Since these are asymmetrical they are not flight feathers, and were probably used for display. Dark bands in the fossil feathers are probably remains of the colour pattern. It has a short face with big eyes and long, sharp front teeth.

Left: The species C. zoui is named after Sou Jiahua, vice-president of China at the time of its discovery .

Protarchaeopteryx

This dinosaur is known from two specimens found that show it to have been a small, feathered theropod. It is less well known than the other feathered dinosaurs of the area, but like its contemporary, *Caudipteryx*, it has long, asymmetrical feathers on the arms and the tip of the tail which were used for display rather than flight. It would have chased small prey along the ground on its long legs.

Below: The lakeside environments of early Cretaceous China must have been very colourful, with so many plumed and feathered animals, presumably all displaying to one another to attract mates or warn off rivals, as modern birds do.

Features: Long legs and a long neck, as well as long, clawed fingers on its hands, show that this is a running animal that grabbed at swift animals on the ground. The wrists are jointed so that the long hands could shoot out forwards to grab prey. The feathers on the arms, and on the short tail, would have been used for display, or even as aerodynamic structures to help steer at speed. It probably fed on insects and small vertebrates that it would be able to run down and snatch.

Distribution: Liaoning, China.
Classification: Theropoda, Tetanurae, Coelurosauria.
Meaning of name: Before Archaeopteryx.
Named by: Ji Q. and Ji S., 1997.
Time: Barremian stage of the early Cretaceous.
Size: 70cm (27½in).
Lifestyle: Hunter.
Species: P. robusta.

FAST HUNTERS

The early Cretaceous period saw the development of several lines of very active theropods. They included the maniraptorans, the so-called "raptors" with the killer claws on the hind foot, and the ostrich-mimics – omnivorous, fleet-footed theropods that probably lived like modern ground-living birds. They would all have been warm-blooded and covered in feathers.

Deinonychus

Known from more than nine skeletons, this is the animal over which the debate about whether dinosaurs were warm- or cold-blooded began. One remarkable deposit has several *Deinonychus* skeletons scattered around the remains of an ornithopod, *Tenontosaurus*, indicating that it was a pack hunter. It was the prototype for the "raptors" in *Jurassic Park*, although modern representations have them covered in feathers.

Below: The tail of Deinonychus was held stiff and straight. Each vertebra had bony tendons growing from it that clasped several of the vertebrae behind, solidifying the whole structure into an inflexible pole with only limited movement at the base for balance.

Distribution: Montana, Oklahoma, Wyoming, Utah and Maryland, USA.
Classification: Theropoda, Tetanurae, Coelurosauria, Deinonychosauria.
Meaning of name: Terrible claw.
Named by: Ostrom, 1969.
Time: Aptian to Albian stages of the Cretaceous.
Size: 4m (13ft).
Lifestyle: Pack hunter.
Species: *D. antirrhopus*.

Features: This medium-sized member of the theropods has a killer claw on the hind foot. Its light, bird-like body is balanced by a stiff, straight tail, and the brainpower is enough to keep the animal balanced while it slashed away with the killer claw on the second toe. The long, heavily clawed hands are angled so that the palms face inwards, enabling it to clutch firmly at its prey.

Utahraptor

This dinosaur is only known from a single specimen consisting of parts of the skull, the claws of the hand and feet, and some tail vertebrae. It is the biggest of the known deinonychids. It is also the earliest; the individuals of the group seem to have become smaller as time went on. *Utahraptor* would have preyed on the big sauropods of the time.

Below: The species name U. ostrommaysorum (illustrated) honours John Ostrom (who first defined the group) and Chris Mays (from the Dinamation International Corporation that financed the excavation).

Distribution: Utah.
Classification: Theropoda, Tetanurae, Coelurosauria, Deinonychosauria.
Meaning of name: Hunter from Utah
Named by: Kirkland, Gaston and Burge, 1993.
Time: Barremian stage of the early Cretaceous.
Size: 6m (19½ft).
Lifestyle: Hunter.
Species: *U. spielbergi, U. ostrommaysorum*.

Features: This dinosaur is essentially a scaled-up version of *Deinonychus*. The killing claw is about 35cm (14in) long, and would have been held up clear of the ground when the animal was stalking. The hands are proportionally larger than those of *Deinonychus*, and have much more blade-like claws which are just as important as the toe claw when hunting. The leg bones are twice as thick as those of the much larger *Allosaurus*, suggesting that they were built for power not speed.

Pelecanimimus

An important group of theropods of the late Cretaceous period were the ornithomimids, the bird-mimics. These were generally built like ostriches and had toothless beaks. *Pelecanimimus* was an early toothed form with a long, shallow snout. Its partial skeleton was found in Spain in lake deposits that were fine enough to show skin features. It may have been a fish-eating animal.

Distribution: Spain.
Classification: Theropoda, Tetanurae, Coelurosauria, Ornithomimosauria.
Meaning of name: Pelican-mimic.
Named by: Perez-Moreno, Sanz, Buscalloni, Moratalla, Ortega and Rasskin-Gutman, 1994.
Time: Hauterivian to Barremian stages of the early Cretaceous.
Size: 2m (6½ft).
Lifestyle: Generalized hunter.
Species: P. polydon.

Features: The 220 tiny teeth in the jaws of this little dinosaur represent the biggest number of teeth in any known theropod. The fine lake deposits in which the specimen was found have preserved the impression of a pouch of skin beneath the jaw, hence the name, and also a soft, fleshy crest at the back of the head. The skin impression shows a wrinkled surface devoid of hair or feathers, unlike other small theropods.

Left: Pelecanimimus *was probably similar to the ancestral form from which the later ostrich-mimic dinosaurs evolved. However, the advanced hand shows that it could not have been a direct ancestor.*

ORNITHOMIMID TEETH

The ornithomimids were the toothless, ostrich-like running dinosaurs of late Cretaceous times. However, their ancestry is to be found in the early Cretaceous period. Because of their lack of teeth and their general build, it has always been assumed that the ornithomimids were omnivorous, feeding on a mixture of insects and small reptiles, and fruit and seeds when available. The often repeated statement that the theropods were the meat-eating dinosaurs of the Mesozoic period may be an over-simplification.

There has been a long debate about the cause of the toothlessness. Some palaeontologists thought that the teeth of a more conventional theropod just became increasingly fewer, and finally disappeared altogether. Others maintained that the teeth became smaller and smaller, so that the jaw became more and more saw-like until, finally, the teeth became so small that they faded away. The discovery of the early ornithomimid, *Pelecanimimus*, with its jaws full of tiny teeth, suggests the latter is correct.

Scipionyx

This baby dinosaur was discovered in the late 1980s by amateur collector Giovanni Todesco, but it wasn't until he had seen the film *Jurassic Park* that he took it to professional palaeontologists for examination. They recognized it to be the most perfectly preserved theropod ever discovered, with traces of the intestines, windpipe, liver and even muscle fibres.

Features: The extreme youth of this single, articulated specimen makes it difficult to classify. It appears to be a maniraptoran and resembles *Composognathus*, but the hand is different. Because it is a hatchling, the skull is much bigger in relation to its body than would be the case in the adult. The position of the preserved liver suggests that *Scipionyx* breathed like a crocodile instead of a bird.

Distribution: Benevento Province, Italy.
Classification: Theropoda, Tetanurae, Coelurosauria.
Meaning of name: Claw of Scipio (a Roman general) and also Scipione Breislak (a geologist who first studied the formation in which it was found).
Named by: dal Sasso and Signore, 1998.
Time: Aptian stage of the early Cretaceous.
Size: 24cm (9in).
Lifestyle: not known.
Species: S. samniticus (after Samnium, the ancient name of the local region).

SPINOSAURIDS

One of the last major dinosaur groups to come to the public's attention was the spinosaurids. They were theropod meat-eaters from the early Cretaceous period, with long, narrow snouts armed with many straight, sharp teeth, and they had a distinctive big claw on the thumb. With these adaptations it seems that these dinosaurs were probably specialized fish-eaters.

Baryonyx

Amateur fossil collector William Walker found a huge fossil claw bone in a clay pit in southern England in 1983, and subsequently a team from the (then) British Museum (Natural History), now the Natural History Museum, London, led by Angela Milner and Alan Charig, excavated the almost complete skeleton. This is the first spinosaurid skeleton found, although individual bones were already known.

Features: *Baryonyx* is a large theropod with distinctive narrow jaws and small teeth. The teeth are straighter than those of other meat-eaters, and there is a peculiar rosette of longer teeth at the tip. The forelimb is very strong and armed with a heavy claw. The neck makes less of an S-shape than in other theropods, and the neck and skull form an almost continuous line. The animal's diet is suggested by fish scales and *Iguanodon* bones found in the stomach.

Distribution: Southern England.
Classification: Theropoda, Spinosauridae.
Meaning of name: Heavy claw.
Named by: Charig and Milner, 1986.
Time: Barremian stage of the early Cretaceous.
Size: 10m (33ft).
Lifestyle: Fish-hunter or scavenger.
Species: *B. walkeri*; see also below.

Right: Baryonyx *is the biggest meat-eater found in Europe from the early Cretaceous period.*

Suchomimus

Suchomimus lived in east Africa a little later than *Baryonyx* lived in Europe. It was big enough to stand in up to 2m (6½ft) of water, chasing fish. However, the east African beds in which it was found contained very few other meat-eating dinosaurs, probably making it the main hunter of the area as well.

Features: When *Suchomimus* was uncovered in the Tenere Desert, Niger, in the late 1990s, scientists were amazed at how similar it looked to *Baryonyx*, with its big claws and long jaws. The only difference is the tall spines along the backbone, which would have supported a low fin in life. However, recent studies of *Baryonyx* suggest that it too might have supported such a fin. It may well be that *Suchomimus* is really a new species of *Baryonyx*.

Distribution: Niger.
Classification: Theropoda, Spinosauridae.
Meaning of name: Crocodile-mimic.
Named by: Sereno, Beck, Dutheil, Gado, Larsson, Lyon, Marcot, Rauhut, Sadleir, Sidor, Varriccio, G. P. Wilson and J. A. Wilson, 1998.
Time: Aptian stage of the early Cretaceous.
Size: 11m (36ft).
Lifestyle: Fish hunter, predator or scavenger.
Species: *S. tenerensis*; see also above.

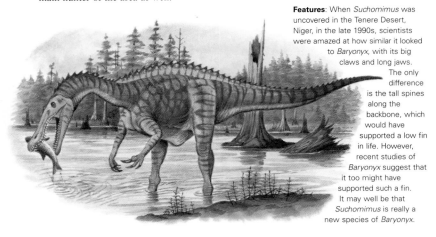

Spinosaurus

The first remains of this remarkable animal were found in Egypt by a German expedition in 1911, and were then lost when the Alte Akademie museum in Munich, Germany, in which it was stored, was destroyed by bombing in 1944. In 1996 Canadian palaeontologist Dale Russell found more remains in Morocco. As the villain of the film *Jurassic Park III*, *Spinosaurus* caught the public's imagination in 2001.

Right:
In early books
Spinosaurus is restored with a short deep head, like that of a carnosaur. That was before the discovery of skull material in Morocco, and the realization of how closely related Spinosaurus was to the others of the group whose skulls were well known.

Features: The most significant feature of the skeleton of *Spinosaurus* is the array of spines sticking up from the backbone, reaching heights of almost 2m (6½ft). In life this would have been covered by skin to form a fin or a sail. It may have acted as a heat regulation device, absorbing warmth from the sun or shedding excess body heat into the wind. It may also have been brightly coloured and used for signalling.

Distribution: Egypt and Morocco.
Classification: Theropoda, Spinosauridae.
Meaning of name: Spined lizard.
Named by: Stromer, 1915.
Time: Albian to Cenomanian stages of the Cretaceous.
Size: Maybe up to 17m (56ft).
Lifestyle: Fish hunter, predator or scavenger.
Species: *S. aegyptiacus*, *S. maroccansus*.

SPINOSAURID JAWS

The jaws and teeth of the spinosaurids are very different from those of any other meat-eating dinosaur. The snout is extremely narrow and very long. The teeth are much straighter than those of other meat-eaters, and on the lower jaw

Top: Inside jaw view.
Above: The spinosaurid upper jaw with teeth.

they are very numerous and small. The tip of the upper jaw carries a separate rosette of teeth that corresponds with a hooked structure at the tip of the lower jaw. The nostrils are placed well back on the snout.

These adaptations seem to have been well suited for catching fish. The narrow snout would cleave the water, the small sharp teeth would seize small, slippery prey, and the nostrils would be clear of the surface. In the modern world we see adaptations such as these in the gavial of the Far East. They are also present in river dolphins.

The spinosaurids probably did not rely on a fish diet, and it would be unreasonable for a *Spinosaurus*, as big as a *Tyrannosaurus*, to feed exclusively on such small prey. Stomach contents, and the presence of spinosaurid teeth in the bones of other animals, suggest that they may have fed on land-living animals too. The big claw may have been a killing weapon, but the specialized teeth seem to have been unsuited for hunting. The spinosaurids probably filled out their fish diet with the carrion of dead animals.

Irritator

The name of this dinosaur is derived from the frustration felt by British palaeontologist Dave Martill when faced with the skull. When it was obtained from Brazil it had been doctored by the finder to try to make it look more spectacular and marketable. After it had been prepared properly, it was seen to be the skull of a spinosaurid.

Features: Only the skull of *Irritator* is known, but it obviously came from a spinosaurid. It is the only one known from South America, although another, *Angaturama*, has been described. *Angaturama*, however, is generally regarded as another name for *Irritator*. In 2004, Eric Buffetaut found the tooth of a spinosaurid, probably *Irritator*, embedded in the backbone of a Brazilian pterosaur, suggesting that the diet of these animals was not confined to fish.

Distribution: Brazil.
Classification: Theropoda, Spinosauridae.
Meaning of name: Irritator.
Named by: Martill, Cruikshank, Frey, Small and Clarke, 1996.
Time: Albian stage of the early Cretaceous.
Size: 8m (26ft).
Lifestyle: Fish hunter, predator or scavenger.
Species: *I. challengeri* (also *Angaturama* which is not valid).

Left: The specific name
I. challengeri (illustrated) refers to Professor Challenger, the hero of Sir Arthur Conan Doyle's novel The Lost World in which live dinosaurs are found existing in South America.

LATE DIPLODOCIDS AND BRACHIOSAURIDS

By Cretaceous times the diplodocids and the brachiosaurids, the main types of sauropods from the Jurassic period, were beginning to die away, but the sauropods themselves were by no means finished. There were still some interesting members, although their line was being taken up by an, until now, lesser known sauropod group.

Amargasaurus

In late Jurassic Argentina *Amargasaurus* was closely related to *Dicraeosaurus*. It too had tall spines jutting up from its backbone. In all diplodocids the spines were split in two, but in *Amargasaurus* this splitting was carried to an extreme. The single, associated skeleton found lacks the tail, and so the total length is uncertain.

Features: Like *Dicraeosaurus*, this diplodocid has a very short neck when compared with the group. The spines sticking up from the backbone between the shoulders and the hips undoubtedly carried a sail, probably used for signalling or for temperature regulation. Those on the neck are paired. They did not carry a sail, as this would have hampered the movement of the neck. They were more likely to have been covered in horn and were possibly used as defensive weapons.

Distribution: Argentina.
Classification: Sauropoda, Dicraeosauridae.
Meaning of name: Lizard

Below: Dicraeosaurus.

from Amarga Canyon.
Named by: Salgado and Bonaparte, 1991.
Time: Hauterivian stage of the early Cretaceous.
Size: 12m (39ft).
Lifestyle: Browser.
Species: *A. cazaui, A. groeben.*

Nigersaurus

This dinosaur was found during a very productive field season in Africa by a team from the University of Chicago, USA, in 1997. As well as parts of several adult specimens of *Nigersaurus*, the expedition found the remains of hatchlings. The strange jaw probably signified a change in the vegetation, with low-growing herbs and flowering plants appearing in the landscape for the first time.

Features: The "Mesozoic lawnmower" is the nickname given to this diplodocid on account of the broad, straight front edge to its mouth that projects to each side of its skull, and which is packed with hundreds of needle-like teeth that form a rake-like cutting edge in a single, straight line. The skull is short for a diplodocid. The rest of the body was typical for a diplodocid, although rather smaller than its earlier cousins.

Distribution: Niger.
Classification: Sauropoda, Diplodocimorpha.
Meaning of name: Lizard from Niger.
Named by: Sereno, Beck, Dutheil, Larsson, Lyon, Mousse, Sadler, Sidor, Varricchio, G. P. Wilson and J. A. Wilson, 1999.
Time: Early Cretaceous.
Size: 15m (49ft).
Lifestyle: Low browser.
Species: *N. taqueti.*

Right: The straight front to the mouth was similar to that of the titanosaurid Bonitasaura. Presumably the two had similar feeding styles.

OTHER GENERA

Other late North American brachiopods include the following:

Sonorasaurus – about one-third of the size of *Brachiosaurus*, and found in Albian rocks in Arizona in 1995.

Venenosaurus – known from the partial skeleton of a juvenile specimen from the Barremian stage of the Cretaceous, in Utah, USA, found in 2001. Some palaeontologists think it was a titanosaurid.

"Pleurocoelus" – specimens attributed to this genus have been found as far apart as Maryland, Texas and Utah, USA. It was first described by Marsh in 1888. This wastebasket taxon is poorly understood because of the number of sauropods that have been wrongly assigned to it. There is even a set of footprints that are supposed to have been made by '*Pleurocoelus*'. Even the original specimens described by Marsh have been re-interpreted as a titanosaur.

Left: The brachiopods listed above had the same general appearance (shown here).

Cedarosaurus

The details about early Cretaceous sauropods, particularly the brachiosaurids, of North America are not well-known, and have usually been placed in the wastebasket taxon '*Pleurocoelus*'. However, more distinct animals, such as *Cedarosaurus*, have been discovered in recent years. Much of one side of the skeleton of a single *Cedarosaurus* individual is known, except for the neck and head.

Features: As with other brachiosaurids, *Cedarosaurus* has front legs that are longer than the hind legs, with long finger bones held vertically, and the humerus and femur the same length. The neck and head are unknown, but they were probably held high as in others of the group. The tail is quite short for a brachiosaurid. Despite the fact that the build is definitely brachiosaurid, the bone structure has certain similarities to that of the later titanosaurids.

Distribution: Utah, USA.
Classification: Sauropoda, Macronaria.
Meaning of name: Lizard from Cedar Mountain.
Named by: Tidwell, Carpenter and Brooks, 1999.
Time: Barremian stage of the early Cretaceous.
Size: 14m (46ft).
Lifestyle: High browser.
Species: *C. weiskopfae*.

Above: With its high shoulders and its presumably long neck, Cedarosaurus would have browsed from high treetops.

Sauroposeidon

An articulated series of four neck vertebrae found in a prison yard in Oklahoma, USA, in 1994 are attributed to *Sauroposeidon*. The vertebrae were so big they were first thought to be fossilized tree trunks. They are very similar to those of *Brachiosaurus*, but about 15–25 per cent bigger, and it must have looked very similar.

Features: The structure of the neck bones suggest that *Sauroposeidon* is proportionally slimmer than *Brachiosaurus*. The neck, when raised, reaches nearly 20m (65½ft) above the ground. As well as being sculpted into thin plates and fine struts, as in other brachiosaurids, the bones have tiny air cells inside them to keep the weight down. It is probably the last of the brachiosaurids of North America, although there is a vertebra known from Mexico that is dated from the Campanian stage of the late Cretaceous.

Distribution: Oklahoma, USA.
Classification: Sauropoda, Macronaria.
Meaning of name: Lizard of Poseidon.
Named by: Wedel, Cifelli and Sanders vide Franklin, 2000.
Time: Albian stage of the early Cretaceous.
Size: 30m (98ft).
Lifestyle: High browser.
Species: *S. proteles*.

Above: Brachiosaurus.

Right: Sauroposeidon is named after Poseidon, the ancient Greek god of earthquakes.

TITANOSAURIDS - THE NEW GIANTS

The titanosaurids were the last group of sauropods to evolve. They may have been related to the macronarians, or even the diplodocids, but their peg-like teeth are the only obvious similarity to the latter. They tended to have broader hips than the other sauropods, resulting in wider trackways and so the group can be identified by their footprints. The genus Titanosaurus *itself is now a wastebasket genus.*

Malawisaurus

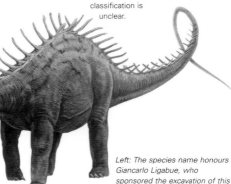

This is the earliest known titanosaurid from Africa, found as an 80 per cent complete, associated skeleton. A prepared and mounted skeleton is now on display in the Cultural Museum Centre in Karonga, Malawi. Its remains were originally classed in the wastebasket taxon *Gigantosaurus* – a name that has caused confusion in recent years after the finding of the giant theropod *Giganotosaurus*.

Features: The skulls of all sauropods are flimsy, and rarely preserved. That of *Malawisaurus* is the first known of the titanosaurids, although it is incomplete, consisting of lower and upper jaws and some teeth. The jaw and teeth show that it had a steep face, sloping down from the eyes to the snout, which has become the general model for the restoration of titanosaurid heads. Another general feature of many titanosaurids is the presence of armour on the back, but this is not present in all examples. The hips are more sturdy than those of other sauropods, being fused to six vertebrae rather than the more usual five.

Above: Originally Malawisaurus *was thought to have lacked bony armour, but mineral nodules found around the skeleton seem to have been fossils of scutes similar to those of other titanosaurids.*

Distribution: Mwakashunguti in the Zambezi valley, Malawi.
Classification: Sauropoda, Titanosauria.
Meaning of name: Lizard from Malawi.
Named by: Jacobs, Winkler, Downs and Gomani, 1993 (but originally named as *Gigantosaurus* by Haughton, 1928).
Time: Aptian stage of the early Cretaceous.
Size: 9m (30ft).
Lifestyle: Browser.
Species: *M. dixeyi*.

Agustinia

In the late 1990s an astonishing sauropod was found in Patagonia. If the armour had been found separately it would have been assumed to have come from a stegosaur or an ankylosaur because no other known sauropod has armour like it. It was originally named *Augustia*, but that name was found to have been given already to another animal.

Features: The most astounding feature of this medium-size sauropod is the arrangement of plates on the back. They are like the plates of a stegosaur, but turned sideways. They are rectangular with some drawn out into sideways-pointing spikes. As for the rest of the body, some features define it as a diplodocid, while others are definitely titanosaurid. At the moment the classification is unclear.

Distribution: Argentina.
Classification: Sauropoda, Titanosauria (unproven).
Meaning of name: Agustin Martinelli's thing (after the discoverer).
Named by: Bonaparte, 1998.
Time: Aptian stage of the early Cretaceous.
Size: 15m (49ft).
Lifestyle: Browser.
Species: *A. ligabuei*.

Left: The species name honours Giancarlo Ligabue, who sponsored the excavation of this remarkable animal.

Phuwiangosaurus

Until the discovery of the partly articulated skeleton of *Phuwiangosaurus* in 1992, all known Asian sauropods had been of a very primitive type. Its presence, and the presence of other unique dinosaurs, suggests that this part of South-east Asia was separated from the main Asian landmass of the time and supported a quite different fauna.

Features: *Phuwiangosaurus* differs from other known Asian sauropods because its teeth are narrow rather than spoon-shaped, and its neck vertebrae are broad and flattened from top to bottom, rather than from side to side. The vertebrae have Y-shaped spines. Skull pieces subsequently discovered indicate a skull shape similar to that of the later nemegtosaurids, indicating that they, too, may be part of the great titanosaurid family.

Distribution: Thailand.
Classification: Sauropoda, Titanosauria.
Meaning of name: Phu Wiang county lizard.
Named by: Martin, Buffetaut and Suteethorn, 1994.
Time: Early Cretaceous.
Size: 20m (65½ft).
Lifestyle: Browser.
Species: *P. sirindhornae*.

JOBARIA

Above: Jobaria, a titanosaurid that is a mix of different animals.

A chimera was a monster in ancient Greek mythology, and a mix of several animals. It had the body of a lion, the tail of a dragon and two heads, a lion's and a goat's. In palaeontological parlance a chimera is also a mix of animals, and throughout the literature of palaeontology we find examples, such as the supposed titanosaurid *Jobaria*.

It was found in Tanzania by the German expeditions of the early twentieth century, and named after the great German palaeontologist Werner Janensch. It has been classified as a titanosaurid because of the shape of the legs and feet, but some of the rest of the skeleton appears to come from some sort of diplodocid. It seems that two different animals died and were fossilized close to one another, giving rise to the confusion.

Chubutisaurus

When it was discovered, *Chubutisaurus* was placed in its own family and thought to have been related to the brachiosaurids. However, it did not have the long front legs associated with the brachiosaurids and it is now regarded as a primitive titanosaurid, although this is far from certain. It is known from two partial skeletons but, as usual, the skulls are missing.

Features: This huge sauropod has big air spaces in its vertebrae, the tail is short and the front legs shorter then the hind. Its brachiosaurid features include the articulation of the vertebrae and the great length of the finger bones. It has few resemblances to the titanosaurids, and its classification here is based on the fact that nearly all South American sauropods are members of the titanosaurid group. It is regarded as a very early and unspecialized form.

Distribution: Argentina.
Classification: Sauropoda, Titanosauria.
Meaning of name: Lizard from the Chubut Province.
Named by: del Corro, 1974.
Time: Albian stage of the early Cretaceous.
Size: 23m (75½ft).
Lifestyle: Browser.
Species: *C. insignis*.

Left: A number of South American sauropods known only from fragmentary material have been classed as brachiosaurids. It seems more likely that, like Chubutisaurus, they were actually very primitive titanosaurids.

HYPSILOPHODONTS AND IGUANODONTS

After their appearance in the Triassic period and their establishment in the Jurassic, albeit in a rather minor role compared to that of the sauropods, the ornithopods really flourished in the Cretaceous period. Soon they were to become the most diverse and abundant of the plant-eating dinosaurs, and by early Cretaceous times they had already diversified into their major evolutionary lines.

Hypsilophodon

Several skeletons of this dinosaur have been found in the last 150 years, and the original image of this was of a smaller version of *Iguanodon* that could climb trees. Its build and size, similar to the modern tree kangaroo, along with a mistaken observation on the toe bones, led to this idea. It is now known to have been a fast-running ground-dweller.

Below: The common image of Hypsilophodon *is of a small animal. However, all specimens so far found are of juveniles, and so it is difficult to estimate the adult size.*

Features: *Hypsilophodon* is often regarded as the gazelle of the dinosaur world. Its legs are long and lightweight, with the muscles concentrated around the hips and the thigh bone, a sure sign of a running animal. The toe bones are not evolved for perching as was originally thought, but for speed. The deep skull contains several front teeth behind the beak, as well as chewing teeth at the back.

Distribution: Isle of Wight, England, and Spain.
Classification: Ornithopoda, Hypsilophodontia.
Meaning of name: High ridged tooth.
Named by: Huxley, 1869.
Time: Barremian to Aptian stages of the Cretaceous.
Size: 2.3m (7½ ft) estimated as an adult size.
Lifestyle: Browser.
Species: *H. foxii, h. wielandi.*

Leaellynasaura

When it was discovered high up in a sea cliff on the south coast of Australia, this little ornithopod was a surprise. In early Cretaceous times this area of Australia was well within the Antarctic Circle. The discovery was the first indication that dinosaurs could cope with the extreme cold and long periods of darkness in high latitudes close to the poles.

Features: The skull of *Leaellynasaura* is distinctive because of the particularly large eye sockets. They, with a big brain cavity, indicate big eyes and possibly the ability to see in the dark. This, in turn, suggests that this dinosaur had a metabolism that allowed it to survive in Antarctic conditions. Classification is a bit uncertain – it would be regarded as a hypsilophodont but for a difference in the shape of the thigh bone and the ridges of the teeth.

Distribution: Victoria, Australia.
Classification: Ornithopoda.
Meaning of name: Leaellyn's (daughter of the discoverers) female lizard.
Named: T. Rich and P. Rich (Leaellyn's parents), 1989.
Time: Aptian to Albian stages of the Cretaceous.
Size: 2m (6½ ft).
Lifestyle: Browser.
Species: *L. amicagraphica.*

Left: The species name L. amicagraphica *honours Friends of the Museum of Victoria, and the National Geographical Society, which funded the research.*

Tenontosaurus

Judging by the number of remains that have been found, including 25 skeletons and scattered bones and teeth, *Tenontosaurus* must have been one of the most abundant herbivores in early Cretaceous North America. It was certainly attractive to meat-eaters – one skeleton has been found surrounded by the bodies of several *Deinonychus* that had been killed while attacking it.

Features: *Tenontosaurus* is like a hypsilophodontid but lacks the teeth on the front part of the jaw. Otherwise it is like an iguanodontid, but the classification is still not clear. Its distinctive feature is its very long tail – longer than the rest of the body – and the network of tendons that supports the spine. Its long forelimbs and strong finger bones suggest that it walked on all fours for most of the time.

Distribution: Western North America.
Classification: Ornithopoda, Iguanodontia.
Meaning of name: Tendon lizard.
Named by: Ostrom, 1970.
Time: Aptian to Albian stages of the Cretaceous.
Size: 6.5m (21ft).
Lifestyle: Low browser.
Species: *T. tillettorum*, *T. dossi*.

Below: Tenontosaurus *was the prey of the North American plains.*

Left: Deinonychus.

EXTREME COLD

The area that is now Victoria in southern Australia was, in early Cretaceous times, deep within the Antarctic Circle. This meant that anything living there would have been subjected to deep cold and a long, dark winter. The vegetation consisted of conifers with thick-skinned needles that were adapted to cold and dryness. There were also ferns, which suggest that the climates were not dry all the time. The mean annual temperature measured by radioactive isotopes in the rocks formed at the time, and by comparing the fossil plants with modern counterparts, was somewhere between 0° and 10°C, like the modern Hudson Bay area, Canada.

The landscape inhabited by the Victoria dinosaurs consisted of a deep rift valley, formed as the continent of Australia was beginning to rip away from that of Antarctica. It could be that such a valley gave shelter from the winter conditions. In any case, it was a hostile world for dinosaurs.
Nevertheless several types inhabited it. As well as the remains of *Leaellynasaura* and other small ornithopods, such as *Atlascopcosaurus*, found in the so-called Dinosaur Cove, were theropods resembling allosaurids and oviraptorosaurids, and even a possible early ceratopsian.

Jinzhousaurus

The Yixian Formation, in Liaoning, China, has yielded little half-bird, half-dinosaur animals. The discovery of *Jinzhousaurus* shows that largish dinosaurs also existed by the Yixian lake. The dinosaur discovery was also important in determining the age of the formation. The presence of such an obviously Cretaceous animal helped to establish that the beds had not been laid down in the earlier Jurassic period, as had previously been surmised.

Features: To look at, *Jinzhousaurus* resembles a small *Iguanodon*. However, its skull shows some advanced features that put the animal somewhere on the evolutionary track towards the later hadrosaurids – the duckbills. It is possible that the iguanodontid line split in three at this time, one producing the iguanodontids proper, the second producing the hadrosaurids, and the third producing something between, of which *Jinzhousaurus* is the only example so far found.

Distribution: Liaoning, China.
Classification: Ornithopoda, Iguanodontia.
Meaning of name: Jinzhou lizard.
Named by: Wang and Xu, 2001.
Time: Barremian stage of the early Cretaceous.
Size: 7m (23ft).
Lifestyle: Low browser.
Species: *J. yangi*.

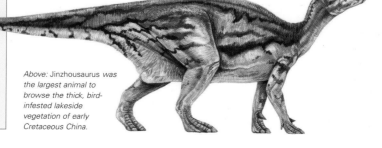

Above: Jinzhousaurus *was the largest animal to browse the thick, bird-infested lakeside vegetation of early Cretaceous China.*

IGUANODONTS

Iguanodontids tended to be bigger than hypsilophodontids, and this used to serve as the difference between the two groups. However, there are a number of differences, especially in the teeth, the forelimbs and the hips. The size of the typical iguanodontids meant that they were too heavy to spend much of their time on hind legs, and so they were basically four-footed animals.

Iguanodon

Famed as being one of the first dinosaurs to be scientifically recognized, *Iguanodon* became something of a wastebasket taxon over the years. It was thought to have been a four-footed, rhinoceros-like animal until complete skeletons were found in a mine, in Belgium, in the 1880s. Thereafter, it was restored in a kangaroo-like pose. Now it is largely regarded as a four-footed animal once more.

Features: *Iguanodon* is the archetypal ornithopod. Its head is narrow and beaked, with tough, grinding teeth. Its hands consist of three weight-bearing fingers with hooves. It has a massive spike on the first finger used for defence or gathering food, and a prehensile fifth finger that works like a thumb. The hind legs are heavy and the three toes are weight-bearing. The long, deep tail balanced the animal as it walked.

Distribution: England, Belgium, Germany and Spain.
Classification: Ornithopods, Iguanodontia.
Meaning of name: Iguana tooth.
Named by: Boulenger and van Beneden, 1881.
Time: Barremian and Valanginian stages of the early Cretaceous.
Size: 6–10m (19½–33ft).
Lifestyle: Browser.
Species: *I. bernissartensis, I. anglicus, I. atherfieldensis, I. dawsoni, I. fittoni, I. hoggi, I. lakotaensis, I. ottingeri.*

Left: Although Iguanodon *was found and named by Mantell in 1825, the description was based only on teeth. In 2000 the International Commission on Zoological Nomenclature ruled the type species to be* I. bernissartensis *described in 1881, based on complete skeletons from Belgium.*

Altirhinus

Once regarded as a species of *Iguanodon* and called *I. orientalis*, there are enough differences to place *Altirhinus* in a genus of its own. It is known from five partial skeletons and two skulls, which are preserved in enough detail to show the distinctive features. It may represent an intermediate stage between the iguanodontids and the hadrosaurids.

Below: Altirhinus *looked just like* Iguanodon *except for the tall nasal region on the head.*

Features: *Altirhinus*, as its name suggests, has a very high nasal region on the skull. This may have been an adaptation to an enhanced sense of smell. It had a greater number of teeth than *Iguanodon*, which provided it with a more efficient food-gathering technique. The beak is wider and flatter than that of *Iguanodon*, rather more like that of one of the later hadrosaurids. It retains the thumb spike, so distinctive of *Iguanodon* and its relatives.

Distribution: East Gobi Province, Mongolia.
Classification: Ornithopoda, Iguanodontia, Hadrosauridae.
Meaning of name: High nose.
Named by: Norman, 1998.
Time: Aptian and Albian stages of the Cretaceous.
Size: 8m (26ft).
Lifestyle: Browser.
Species: *A. kurzanovi.*

Ouranosaurus

This African genus had the hands of a typical iguanodontid, but the skull had a broad, flat beak like a hadrosaurid. It lived at the same time and in roughly the same area as the sail-backed meat-eater, *Spinosaurus*. Its back structures may have been adaptations to life in the hot dry environment found at the time.

Features: The most obvious feature of *Ouranosaurus* is the huge array of spines jutting up and forming a picket fence along the backbone. The back bone is always shown as supporting a sail, and being used for heat control or signalling, but it is just as likely to have been the basis of a fatty hump that would store nourishment and energy for lean times. Modern hump-backed animals, such as the buffalo and camel, have humps supported by similar skeletal structures. The arid climates of early Cretaceous North Africa may have called for the evolution of specialist food-storage devices.

Distribution: Niger.
Classification: Ornithopoda, Iguanodontia, Hadrosauridoidea.
Meaning of name: Brave monitor lizard.
Named by: Taquet, 1976.
Time: Aptian stage of the early Cretaceous.
Size: 7m (23ft).
Lifestyle: Browser.
Species: *O. nigerensis*.

IGUANODON DISCOVERY

Iguanodon was discovered in the 1820s in Sussex, England, by a local country doctor, Gideon Mantell, and his wife Mary. Over several years they unearthed teeth and several bones. There was much discussion among the scientific establishment of the day about what kind of animal the remains, especially the teeth, came from. A fish and a hippopotamus were suggested by the foremost biologists in London and Paris. Eventually Mantell noted the similarity between the teeth and those of a modern iguana lizard – hence the Latin name.

With no other living comparison, Mantell first restored *Iguanodon* as a gigantic lizard, walking on all fours and menaced by a similarly dragon-like, four-footed *Megalosaurus* that had also been found at that time. As such they were restored as full-size statues in the grounds of the Crystal Palace in Sydenham, south London, England, where they stand to this day.

It was only with the discovery of about 40 *Iguanodon* skeletons in a coal mine at Bernissart in Belgium, in the 1880s, that it was obvious what kind of animal *Iguanodon* was. The coal mine was closed for two years while the fossils were excavated. This famous find was studied over the next 40 years by Louis Dollo from the Royal Museum of Natural History, in Brussels.

Muttaburrasaurus

The most complete dinosaur found in Australia so far is *Muttaburrasaurus*, and it is known from two skeletons. The first was found in 1963 by rancher Doug Langdon. The one found at Lightning Ridge in New South Wales had its bones replaced by opal. When alive, this animal may have lived in herds in open woodlands, and fed on ferns, cycads and conifers.

Features: This iguanodontid has a hollow, bony bump on its snout in front of its eyes. This may have had something to do with a sense of smell or an ability to make a noise. The teeth are evolved for slicing rather than for grinding as in the other members of the group. The hands have not been found so we do not know if it had the typical iguanodontid arrangement of fingers with the middle three strong and weight-bearing.

Distribution: Central Queensland, and New South Wales, Australia.
Classification: Ornithopoda, Iguanodontia.
Meaning of name: Lizard from Muttaburra Station.
Named by: Bartholomai and Molnar, 1981.
Time: Albian stage of the early Cretaceous.
Size: 7m (23ft).
Lifestyle: Browser.
Species: *M. langdoni*.

Left: The first skeleton was kicked to bits by grazing cattle as it lay exposed, and some bones were taken home as souvenirs by locals. When its importance was known, most of the skeleton was subsequently recovered.

EARLY HORNHEADS

As with many of the groups of Cretaceous dinosaurs, the ceratopsians, or the horned dinosaurs, seem to have evolved in central Asia and migrated across to North America, where they later flourished. They appear to have evolved from typical, two-footed plant-eaters from the ornithopod line – the basic, primitive ceratopsian body, showing many similarities to that of the generalized two-footed plant-eater.

Psittacosaurus

Once regarded as an ornithopod, albeit one with a peculiar head, *Psittacosaurus* is now regarded as a transitional form between the primitive ornithopods and the horned dinosaurs – the ceratopsians. *Psittacosaurus* has more species than any other dinosaur, and in times to come it may be split into several genera. Recent studies reveal a series of spines on the tail.

Features: The skull is deep and narrow, and carries a heavy beak. The upper part of the beak is supported by a bone, the rostral, that is only found in the ceratopsians. The back of the skull carries a ridge of bone, probably an anchor for the heavy jaw muscles. This gives the head a square profile which, along with the heavy beak, is rather like that of a parrot. There are no teeth at the front of the mouth, and those at the back are built for chopping. Cheek pouches would have held the food as it was chewed.

Distribution: Thailand, China, Mongolia.
Classification: Marginocephalia, Ceratopsia.
Meaning of name: Parrot lizard.
Named by: Osborn, 1923.
Time: Aptian stage of the early Cretaceous.
Size: 2m (6½ft).
Lifestyle: Low browser.
Species:
P. mongoliensis,
P. mazongshanensis,
  P. meileyingensis,
  P. meimongoliensis,
  P. ordosensis,
  P. sattayaraki,
P. sinensis, P. zinjiangensis.

Yaverlandia

A skull fragment, all that we know of this animal, suggests that it is an early pachycephalosaur. However, this classification is open to dispute. The top of the head is quite flat, and suggests that the heavily domed head that typifies the group evolved very gradually. The other possible pachycephalosaur from Europe is represented by a tooth found in late Cretaceous Portugeuese rocks.

Below: If Yaverlandia *proves to be a pachycephalosaur, then the body shape would have been as shown. However, this restoration must be regarded as speculative.*

Features: The roof of the skull shows a thickening – the only feature that suggests that this animal is a pachycephalosaur. Yet the traces of the brain shape that are visible in the specimen, particularly the areas that deal with the sense of smell, are quite different from those of known pachycephalosaurs. The way the skull bones knit together is also dissimilar, and so it may not be a pachycephalosaur at all. See the similar confusion that surrounds *Majungatholus* (right).

Distribution: Isle of Wight, England.
Classification: Marginocephalia, Pachycephalosauria.
Meaning of name: From Yaverland Point.
Named by: Galton, 1971.
Time: Barremian stage of the early Cretaceous.
Size: 2m (6½ft).
Lifestyle: Low browser.
Species: Y. bitholus.

Archaeoceratops

This dinosaur is known from two skeletons, one almost complete but lacking the forelimbs, found during the Sino-Japanese Silk Road Dinosaur Expedition in 1992–3. Its discovery seems to suggest that the ceratopsians evolved first in Asia and later evolved into two lines, one of which migrated to North America, where the dinosaurs later flourished and became the great-horned dinosaurs of the late Cretaceous.

Features:
Archaeoceratops is a small, lightweight animal that had the ability to walk on all fours or run on its hind legs. It is one of the most primitive ceratopsians known, with a barely developed neck frill. The head is quite large for the size of the body, and it still retains the three or four teeth in the front of the mouth that are so distinctive of its ornithopod ancestors. There is no sign of any horns.

Left: Archaeoceratops *was a small, fleet-footed rabbit-sized animal, quite unlike its lumbering descendants.*

Distribution: Gansu Province, China.
Classification: Marginocephalia, Ceratopsia.
Meaning of name: Ancient horned face.
Named by: Dong and Azuma, 1997.
Time: Early Cretaceous.
Size: 80cm (31in).
Lifestyle: Low browser.
Species: *A. oshimai*.

THE EARLY CERATOPSIANS

The ceratopsians were the horned dinosaurs. A typical image of such a beast is of a huge rhinoceros-like animal with a solid shield of bone around its neck, and a set of wicked horns ready to inflict deadly damage in defence or offence.

This was true of the later ceratopsians, but the ancestral forms were quite different. The neck shield seems to have evolved before the horns. It probably originated as a supporting shelf that held the powerful jaw muscles that the primitive ceratopsians needed for processing their main food. They seem to have been well adapted to feeding on cycads and cycad relatives. The strong, sharp beak would have been ideal for selecting the most nutritious part of the plant and ripping it out. The slicing teeth would have chopped up the tough leaves while holding the pulp in the cheek pouches. Very powerful jaw muscles would have been needed for this action which, recent research suggests, involved a forward-and-back as well as an up-and-down action.

As this ridge became bigger it would have functioned as a display structure as well, probably brightly coloured and used in attracting mates or scaring off rivals. The final purpose, that of defence, would have evolved later, once the animals had evolved into big types that would have been too heavy to run away or hide from big meat-eaters.

Liaoceratops

The early Cretaceous lake beds in Liaoning, China, not only produced the fabulous half-bird, half-dinosaur animals that originally made them famous, but also early members of the ankylosaurs, and of the ceratopsians. This fox-sized animal is the earliest-known of the horned dinosaur line. It would have used its shield and horns for display rather than for defence, and probably defended itself by running away.

Features: The large head has a pair of horns, pointing sideways, one under each eye. A frill is present and seems to have acted as an attachment for the jaw muscles, judging by the pitted texture that indicates muscle attachment. Its teeth are adapted for slicing rather than grinding. As with other lightweight dinosaurs, it is designed for running on hind legs as well as walking on all fours. It may belong to a line that gave rise to both the psittacosaurids and the ceratopsians proper.

Distribution: China.
Classification: Marginocephalia, Ceratopsia.
Meaning of name: After the Chinese province and village where it was found.
Named by: X. Xu, P. J. Makovicky, X. L. Wang, M. A. Norell, and H. L. You, 2002.
Time: Barremian stage of the early Cretaceous.
Size: 1m (3ft).
Lifestyle: Low browser.
Species: *L. yanzigouensis*.

Left: For all its distinctive ceratopsian features, Liaoceratops *is in some ways even more primitive than* Psittacosaurus, *belonging to the group from which the ceratopsians evolved. Ceratopsian evolution was more complex than first imagined.*

POLACANTHIDS AND OTHER EARLY ANKYLOSAURS

The ankylosaurs had armour that consisted of bony plates that were set into the skin and covered in horn. Their main feature was the armoured pavement that covered the back. Defence consisted of sideways pointing spikes, sharp plates on the tail, or a mace at the tail's end.

Polacanthus

This dinosaur is known from an almost complete skeleton, lacking the skull, and all sorts of isolated bits. Like *Acanthopholis* it was one of the first dinosaurs to be discovered and studied. Another early Cretaceous ankylosaur, *Hylaeosaurus*, also from southern England, was once thought to be the same animal but it differs in the arrangement of the bones in the shoulder.

Features: The armour of *Polacanthus* and its relatives is distinctive. Over the neck, shoulders and back there is a series of spikes that stick up and sideways, with the tallest over the shoulder region. Over the hips is a "buckler" consisting of a compressed mass of bony scutes. Along each side of the tail are thin, sharp plates pointing outwards. We do not know its skull, but it was probably broad at the front, like that of its close relative *Gastonia*.

Below: The pattern of armour on the back of Polacanthus *is well-known.*

Distribution: England.
Classification: Thyreophora, Ankylosauria, Polacanthidae.
Meaning of name: Many spikes.
Named by: Hulke, 1881.
Time: Barremian stage of the early Cretaceous.
Size: 4m (13ft).
Lifestyle: Low browser.
Species: P. foxii, P. rudgwickensis.

Minmi

This dinosaur is known from an almost complete specimen and several other isolated pieces. The stiffened back skeleton suggests that it may have been a fairly fast runner. A cololite, a fossilized lump of stomach contents, shows that it fed on fruit, seed and soft vegetation. It had been so thoroughly chewed that this ankylosaur must have had cheek pouches to hold and process the food.

Features: This primitive ankylosaur has features that are similar to both the ankylosaurids and the nodosaurids. The legs are quite long and the back stiffened by extensions of the vertebrae – the paravertebrae that give it its species name. Unusually for ankylosaurs, the belly is protected by armour. Conical spines stick out in a rim around the hips, presumably to defend against attack from the rear. A supposed club found on the tail was later found to be merely an effect of fossilization.

Distribution: Australia.
Classification: Thyreophora, Ankylosauria.
Meaning of name: From Minmi Crossing, the place where it was found.
Named by: Molnar, 1980.
Time: Aptian stage of the early Cretaceous.
Size: 2m (6½ft).
Lifestyle: Low browser.
Species: M. paravertebra.

Left: Minmi *was the first armoured dinosaur to be found in the Southern Hemisphere. Its name has the distinction of being the shortest dinosaur name on record.*

Gastonia

Found in the same quarry as *Utahraptor*, *Gastonia* is known from an almost complete skeleton and several individual skulls. It may have fought by head-butting, but its main defence against *Utahraptor* would have been the scything tail with the sideways-pointing blade-like plates, and the long spikes on the shoulders. The plates may have worked like scissors on anything caught between them.

Features: This heavily armoured dinosaur, closely related to the European *Polacanthus*, carries several types of armour on its neck, back and sides. Pairs of spines run along the neck and shoulders, and down the sides. Broad plates stick out sideways from the tail. A solid mass of armour of fused ossicles lie across the hips. Smaller ossicles lie in the spaces between the larger armour pieces. The skull itself is broad and solid.

Distribution: Utah, USA.
Classification: Thyreophora, Ankylosauria, Polacanthidae.
Meaning of name: Gaston's (Robert Gaston, the discoverer) thing.
Named by: Kirkland, 1998.
Time: Barremian stage of the early Cretaceous.
Size: 5m (16½ft).
Lifestyle: Low browser.
Species: *G. burgei*.

ANKYLOSAUR CLASSIFICATION

The classification of the ankylosaurs is constantly undergoing revision. For our purposes we can identify several main evolutionary trends in the group.

Below: Scelidosaurus.

The most basal thyreophorans may be ancestral to both the ankylosaurs and the stegosaurs. These primitive forms include animals such as *Scelidosaurus* and *Minmi*. From these, three main ankylosaur branches develop.

Below: Polacanthus.

The first is the polacanthids, including *Polacanthus* and *Gastonia*. They had broad mouths, and armour consisting of spikes on the back and neck, a buckler over the hip and plates on the tail.

Below: Pawpawsaurus. Next come the nodosaurids, including *Nodosaurus* and *Pawpawsaurus*. Like the polacanthids, they had armour on the neck and shoulders, including sideways-pointing spikes. Unlike the polacanthids, the nodosaurids had a narrow mouth, suggesting more selective feeding.

The final group is the ankylosaurids. This later group is characterized by the club on the end of the tail, used as a weapon. Typical members are the giants *Euoplocephalus* and *Ankylosaurus* from the late Cretaceous. These had the broad mouths of the general-feeding polacanthids.

Above: Euoplocephalus.

Shamosaurus

The earliest-known ankylosaurid is *Shamosaurus*, known from several individuals including a good skull and jaw from Mongolia. It appears that the group evolved in this area and subsequently spread to North America (the two continents were joined at the time). The closely related *Cedarpelta* appeared in North America slightly later.

Features: The beak of the skull is quite narrow – more like that of a nodosaurid than that of an ankylosaurid. There are small horns at each side of the head. As with all ankylosaurids there is a maze of nasal passages inside the skull, probably for increasing the sense of smell, warming the breathed air, or making sounds. The back is well covered by an armour of plates and spines – hence its species name *S. scutatus*, meaning shielded. Otherwise the bones are different enough from the typical ankylosaurids to lead some scientists to put it in its own family.

Distribution: Mongolia.
Classification: Thyreophora, Ankylosauria, Ankylosauridae.
Meaning of name: Lizard from Shamo (the Gobi Desert).
Named by: Tumanova, 1983.
Time: Barremian to early Aptian stages of the Cretaceous.
Size: 7m (23ft).
Lifestyle: Low browser.
Species: *S. scutatus*.

Left: Shamosaurus *had a similar neighbour.* Gobisaurus, *known since the 1950s but not described until 2001, was another ankylosaurid of a similar appearance to* Shamosaurus *and from the same area.*

MOSASAURS

The ichthyosaurs waned and died out at the beginning of the Cretaceous period. Their place in the Cretaceous era, as the fast-swimming marine predators, was taken by a group of animals called the mosasaurs. These were closely related to the modern monitor lizards, but they had adaptations to marine life that made them the veritable sea serpents of their day.

Clidastes

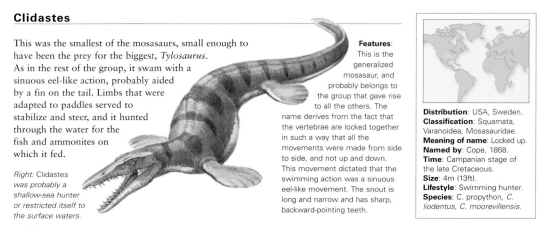

This was the smallest of the mosasaurs, small enough to have been the prey for the biggest, *Tylosaurus*. As in the rest of the group, it swam with a sinuous eel-like action, probably aided by a fin on the tail. Limbs that were adapted to paddles served to stabilize and steer, and it hunted through the water for the fish and ammonites on which it fed.

Right: Clidastes was probably a shallow-sea hunter or restricted itself to the surface waters.

Features: This is the generalized mosasaur, and probably belongs to the group that gave rise to all the others. The name derives from the fact that the vertebrae are locked together in such a way that all the movements were made from side to side, and not up and down. This movement dictated that the swimming action was a sinuous eel-like movement. The snout is long and narrow and has sharp, backward-pointing teeth.

Distribution: USA, Sweden.
Classification: Squamata, Varanoidea, Mosasauridae.
Meaning of name: Locked up.
Named by: Cope, 1868.
Time: Campanian stage of the late Cretaceous.
Size: 4m (13ft).
Lifestyle: Swimming hunter.
Species: C. propython, C. liodentus, C. moorevillensis.

Tylosaurus

One of the biggest and the last of the mosasaurs was *Tylosaurus*. It was so big that it preyed on other sea reptiles including smaller mosasaurs like *Clidastes*. *Plotosaurus*, another mosasaur of a similar size, from California, USA, has been found with impressions of skin, indicating that it was covered in scales, like a snake.

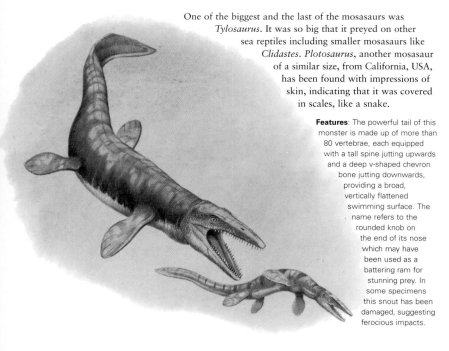

Features: The powerful tail of this monster is made up of more than 80 vertebrae, each equipped with a tall spine jutting upwards and a deep v-shaped chevron bone jutting downwards, providing a broad, vertically flattened swimming surface. The name refers to the rounded knob on the end of its nose which may have been used as a battering ram for stunning prey. In some specimens this snout has been damaged, suggesting ferocious impacts.

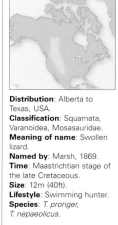

Distribution: Alberta to Texas, USA.
Classification: Squamata, Varanoidea, Mosasauridae.
Meaning of name: Swollen lizard.
Named by: Marsh, 1869.
Time: Maastrichtian stage of the late Cretaceous.
Size: 12m (40ft).
Lifestyle: Swimming hunter.
Species: T. proriger, T. nepaeolicus.

Left: Tylosaurus was found in Kansas in 1868, and described by Louis Agassiz. Cope described it scientifically in 1869, but it was his great rival Othniel Charles Marsh who named it Tylosaurus.

THE FIRST KNOWN MOSASAUR

In 1780 workers in a sandstone quarry near Maastricht (from which the Maastrichtian stage of the Cretaceous is named) in the Netherlands found a set of fossil jawbones and teeth. They were given to a French army surgeon called Hofmann. Dr Goddin, canon of the local cathedral, successfully sued Hofmann for possession, as the quarry was on his land.

In 1794, in the French Revolutionary wars, the French army bombarded Maastricht but spared the suburb in which the fossil was housed as word had spread to the French authorities about the scientific value of the find. Goddin hid the fossil in a cave, but the occupying French bribed the locals, with 600 bottles of wine, into surrendering it. It was recovered and sent by the army to the Jardin des Plantes in Paris, to be studied by the foremost French anatomist Baron Cuvier, and there it has remained.

The fossil was the jaws of the first mosasaur to have been discovered, and it was soon seen to have been a giant aquatic lizard related to the modern monitors. It was named *Mosasaurus*, meaning "lizard from the Meuse", by British geologist William Conybeare.

Below: The fossil jaw of Mosasaurus.

Platecarpus

We know that mosasaurs ate ammonites because several ammonite shells have been found with tooth marks arranged in the distinctive V-shape of a mosasaur's dentition. The mosasaur seems to have bitten into the shell, turned it and bitten into it again, repeating this several times until the shell collapsed. The soft part would then have been swallowed. No shell fragments have been found in a mosasaur's stomach.

Features: This, the most common mosasaur, has a short body and long tail. Well-preserved jaws show how it swallowed its prey. The teeth are generally straight, but at the back of the roof of the mouth there is a series angled back towards the gullet. A joint in the jaws allows the lower jaws to spread and pull the lower jaw backwards along with the food. These upper back teeth held the food and guided it down the throat.

Distribution: Kansas, USA, also Europe, Africa and possibly Australia.
Classification: Squamata, Varanoidea, Mosasauridae.
Meaning of name: Flat wrist.
Named by: Cope, 1869.
Time: Turonian to Maastrichtian stage, of the late Cretaceous.
Size: 7.5m (25ft).
Lifestyle: Fish and ammonite hunter.
Species: *P. ictericus,*
P. tympaniticus, P. bocagei,
P. coryphaeus,
P. planifrons.

Above: Mosasaurs are often portrayed with a fin or a crest along the neck and back. This was due to a misidentification of impressions of the throat structure on a specimen of Platecarpus *found in 1899. Williston, who made the discovery, acknowledged his mistake two years later. There is no proof of such a crest.*

Globidens

The strong blunt teeth and powerful jaws show that the mouth was built for crushing. Diet would have consisted of heavy-shelled animals such as turtles or molluscs like ammonites. Although the mosasaurs as a group lived in shallow seas all over the world, they are best known from the chalk deposits laid down in the shallow sea that covered most of North America in late Cretaceous times.

Features: This specialized mosasaur has a short massive head. The short jaws carry bulbous teeth with narrow bases and rounded tops; their upper surfaces are covered in wrinkles. The teeth on the roof of the mouth, typical of other mosasaurs, are absent. Otherwise the skull and backbone show that it is closely related to *Clidastes*. The paddles, like those of other mosasaurs, consist of five toes, showing polyphalangy, (an increase in the number of joints), as in the primitive ichthyosaurs and the plesiosaurs.

Distribution: Alabama, South Dakota, USA.
Classification: Squamata, Varanoidea, Mosasauridae.
Meaning of name: Bulbous teeth.
Named by: Gilmore, 1912.
Time: Campanian to Maastrichtian stage of the late Cretaceous.
Size: 6m (20ft).
Lifestyle: Shellfish-eater.
Species: *G. alabamaensis.*
G. dakotensis.

Right: Isolated teeth very similar to those of Globidens have been found in Africa, Europe, the Middle East and South America, suggesting that it may have been quite a wide-ranging genus.

PLESIOSAURS

The late Cretaceous seas had as much diversity of plesiosaurs as those of the Jurassic. It seems likely that the different types of plesiosaurs had different hunting areas with, for example, the big whale-like pliosauroids patrolling the open ocean after big prey, and the small-headed, long-necked plesiosauroids preferring the shelf seas and hunting fish and smaller animals.

Hydrotherosaurus

Feeding on fish *Hydrotherosaurus* lived in the shallow waters near the coastlines of continents. We know its diet from the stomach contents of a complete articulated skeleton found in California. The stomach also contained stomach stones, which the animal would have swallowed to help adjust its buoyancy – a technique often used by animals that live and hunt close to the sea floor.

Features: The difference between *Hydrotherosaurus* and the other late-Cretaceous, long-necked plesiosaurs, such as *Elasmosaurus*, lies in the 60 vertebrae in the neck. They are much lower and narrower at the front end, but high towards the shoulders. Early restorations show it with a narrow streamlined body, but this is due to the angle at which the ribs of a complete skeleton were found. It is now thought to have had a squat turtle-like body, with broad shoulder bones and hips like other plesiosauroids.

Distribution: California, USA.
Classification: Plesiosauridae, Plesiosauroidea.
Meaning of name: Fisherman lizard.
Named by: Welles, 1943.
Time: Maastrichtian stage of the late Cretaceous.

Size: 13m (42ft).
Lifestyle: Fish-hunter.
Species: *H. alexandrae*.

Above: The species was named after the famous fossil collector Annie Montague Alexander (1867–1950), who contributed over 20,000 specimens to the Museum of Palaeontology at the University of California, USA, and was even in the field on her 80th birthday.

Elasmosaurus

The neck of *Elasmosaurus* is so long that it was mistaken for a tail by Edward Drinker Cope when he first investigated it. The species name *E. platyurus* means "big flattened tail" and actually referred to the neck. When his rival Othniel Charles Marsh pointed out the mistake the bad feeling was so intense that it sparked off the so-called "bone wars" of the late nineteenth century in which the two tried to outdo each other in the number of fossil vertebrates discovered.

Features: The enormous neck of *Elasmosaurus* has the greatest number of neck vertebrae in any known animal, 72 in all. The undersides of the shoulder girdle and the hip bones are expanded into broad plate-like structures that anchor the powerful muscles driving the paddles. The head is tiny, but has a very wide gape and carries sharp spike-like teeth. Judging by the stomach contents it was capable of catching the fastest of the fish of the time. As with the other plesiosauroids, the long neck of *Elasmosaurus* would have been used to snatch at fast-moving fish unawares, without having to move the great body too quickly.

Distribution: Kansas, USA.
Classification: Plesiosauridae, Plesiosauroidea.
Meaning of name: Metal plate lizard (after the flat bones in its pelvis and shoulder girdle).
Named by: Cope, 1868.
Time: Maastrichtian stage of the late Cretaceous.
Size: 12m (45ft).
Lifestyle: Fish-hunter.
Species: *E. platyurus*, *E. morgani*, *E. serpentines*, *E. snowii*.

Above: There is a large number of species of Elasmosaurus. *However, many scientists, notably Ken Carpenter of Denver, believe many of them to be separate genera rather than just species.*

Dolichorhynchops

The arrangement of teeth and jaw muscles indicated that *Dolichorhynchops* could make very quick bites at its prey, but that these bites would not have been particularly strong. It seems likely that its prey consisted of the soft-bodied squid that spread throughout the seas at this time. It is known from several complete adult skeletons and partial skeletons of youngsters.

Below: The neck of Dolichorhynchops *was short but a little longer than the head.*

Distribution: Kansas, USA.
Classification: Plesiosauridae, Pliosauroidea.
Meaning of name: Long-snouted face.
Named by: Williston, 1902.
Time: Campanian stage of the late Cretaceous.
Size: 5m (17ft).
Lifestyle: Fish-hunter.
Species: *D. osborni.*

Features: The body is short and streamlined, like that of Jurassic *Peloneustes*, but the skull is long and narrow with big eye sockets, like that of an ichthyosaur. The head is much lighter than in the larger pliosauroid genera, and the teeth are small, tightly-packed and all of the same size.

Kronosaurus

Although *Kronosaurus* is often illustrated, and its dimensions are often quoted with confidence, its true nature is more problematic. The name was originally applied to skull bones from Queensland, Australia. A mounted skeleton in Harvard University Museum, USA, on which most restorations are based, actually comes from a different horizon and has been inaccurately assembled.

Features: *Kronosaurus* has the biggest skull of any marine reptile known. It is flat-topped and ends in pointed jaws, and is about 2.7m (9ft) long, accounting for about a third of the entire length of the animal. There are probably about 20 vertebrae in the neck, unlike the 30 or so originally attributed to it, hence the revision of the animal's length. It may have lived rather like a sperm whale.

Below: The first specimen of Kronosaurus *found was a piece of jaw with six teeth, found near Hughenden in Queensland, Australia, in 1901. It was thought to have been an ichthyosaur until more complete skull bones were found in 1924. The famous skull was not found until 1931.*

Distribution: Queensland, Australia and Boyaca, Colombia.
Classification: Plesiosauria, Pliosauridae.
Meaning of name: Kronos' lizard (after the Greek titan).
Named by: Longman, 1924.
Time: Albian stage of the early Cretaceous (but included here because of its importance).
Size: 9m (30ft), as opposed to the 13m (45ft) often quoted.
Lifestyle: Ocean hunter.
Species: *K. queenslandicus, K. boyacensis.*

GIANT PTEROSAURS

The last of the pterosaurs were veritable monsters, some with wingspans greater than those of hang-gliders or small aircraft. The biggest modern flying birds, such as the albatross and the Andean condor, would have been puny beside the biggest of the pterosaurs that ruled the skies at the end of the Age of Dinosaurs. Soon the birds would take over the skies.

Pteranodon

The first specimens to be found were in the chalk of Kansas, USA, by O. C. Marsh, in 1870. The specimens were wing-bone fragments and identified as a species of *Pterodactylus*. In 1876 the first skulls were found and the name was changed to *Pteranodon* because of the lack of teeth. Until the 1970s *Pteranodon* was regarded as the biggest animal that was capable of flight.

Features: *Pteranodon* has a long head with deep toothless jaws and a long crest that extends backwards, giving the skull a kind of a hammerhead appearance. In the largest species, *P. sternbergi*, the crest projects upwards giving a totally different profile. The head is much bigger than the body. The bones are hollow to keep down the weight, and the dorsal vertebrae are fused together with the ribs to form a solid support for the flight muscles. It probably spent more time soaring than in flapping flight.

Distribution: South Dakota, Kansas, Oregon, USA, and Japan.
Classification: Pterosauria, Pterodactyloidea.
Meaning of name: Wing without teeth.
Named by: Marsh 1876.
Time: Santonian to Campanian stages of the late Cretaceous.
Size: 9m (30ft) wingspan.
Lifestyle: Fish-hunter.
Species: *P. longiceps, P. ingens, P. eatoni, P. marshi, P. walkeri, P. oregonensis, P. sternbergi*.

Nyctosaurus

The size of the crest of *Nyctosaurus* is comparable with the area of the wing. Some palaeontologists argue that in life this may have carried a sail of skin, like that of *Tapejara*, used for display or for aerodynamics. There is no direct evidence that such a sail existed, and many palaeontologists think that a sail is impossible for mechanical reasons.

Right: The restoration given here is rendered without the suggested membrane on the head crest.

Features: Since its discovery in the 1870s, *Nyctosaurus* has always been regarded as a small version of *Pteranodon*, but without the crest. It also differs from *Pteranodon* by the shape of the upper arm bone and shoulder joint. Then in 2003 two new skulls were found in the Smoky Hill Chalk at western Kansas, USA, and these bear the most remarkable crests, made up of fine struts of bone like a single deer antler, about three times the length of the skulls themselves.

Right: Tapejara for comparison.

Distribution: Kansas, USA and Brazil.
Classification: Pterosauria, Pterodactyloidea.
Meaning of name: Night lizard.
Named by: Marsh, 1876.
Time: Santonian to Maastrichtian stages of the late Cretaceous.
Size: 2.9m (9½ft) wingspan.
Lifestyle: Fish-hunter.
Species: *N. lamegoi, N. gracilis*.

FOOTPRINTS

Pterosaurs were flying reptiles: we have known that for more than 150 years. However it is only in the last few decades that we have determined that they flew by active flapping as birds do, rather than by gliding like flying squirrels. What they did while on the ground is a mystery.

There were two schools of thought. One was that they sprawled on all fours, rather like frogs. The other was that they were fully bipedal like birds. In the 1980s certain fossil trackways became recognized for what they were – marks made by pterosaurs while they were on the ground. They showed four-toed footprints made by the hind feet as quite a wide trackway. The hands were represented by three-fingered marks, with one of the fingers sweeping away to the side, positioned quite close to the hind prints. The analysis is that the animal had been walking bow-legged on its hind feet, with its body more or less upright, supported by the wings in an action that resembled someone walking on crutches.

A more detailed trackway found in France, in 2004, showed that a pterosaur coming in to land slowed its flight to stalling speed close to the ground, and let down its feet gently. It then dragged its toes for a moment, made a short hop and put down its front legs to walk away on all fours, a very precise landing manoeuvre.

Below: Pterosaur tracks.

Zhejiangopterus

Four articulated specimens of this genus are known. It was regarded as a relative of *Pteranodon* and *Nyctosaurus* when it was discovered, but is now thought to be closer to *Quetzalcoatlus*, largely because of the very long neck vertebrae. Complete pterosaur skulls like these are rare, due to the fact that they are made of very porous lightweight bones with the consistency of expanded polystyrene.

Features: As in other members of the group, *Zhejiangopterus* has short wings, long legs (about half as long again as the arm) and a very big head that is extremely narrow. The only crest is a long one beneath the lower jaw, which may have had a structural significance in supporting the very thin bones of the skull. The jaws are toothless and the eye socket is tiny. It has a very long neck made up of elongated vertebrae.

Distribution: Zhejiang Province, China.
Classification: Pterosauria, Pterodactyloidea.
Meaning of name: Wing from Zhejiang Province.
Named by: Cai and Feng, 1994.
Time: Santonian stage of the late Cretaceous.
Size: 5m (16ft) wingspan.
Lifestyle: Fish hunter.
Species: *Z. linhaiensis.*

Left: The reason that the skulls of these big-headed pterosaurs rarely fossilize is that they were made of very delicate material, which disintegrated rapidly after death.

Quetzalcoatlus

The discovery of *Quetzalcoatlus* in the 1970s gave the world what was thought to have been the biggest flying animal that could possibly have existed. It may have lived by fishing in inland waters or by scavenging from the corpses of dead dinosaurs spotted while soaring on rising air over the arid landscape. Now, however, there is evidence for even bigger genera of pterosaurs emerging.

Below: Despite its great size the skeleton was lightly built and the whole animal may have weighed no more than 100kg (220lbs).

Features: *Quetzalcoatlus* has a very big head with a bony crest along the back portion. Like *Pteranodon* and the other big pterosaurs it is toothless. The weight of the skull is kept to a minimum by the fusion of the nostril and the gap that usually exists between the nostril and the eye socket. Since its discovery the wingspan has been revised from 15m (52ft) to 11m (37ft), still very respectable.

Distribution: Texas, USA.
Classification: Pterosauria, Pterodactyloidea.
Meaning of name: From Quetzalcoatl, the plumed serpent of Aztec mythology.
Named by: Lawson, 1975.

Time: Maastrichtian stage of the late Cretaceous.
Size: 11m (37ft) wingspan.
Lifestyle: Fish- or carrion-eater.
Species: *Q. northropi.* One other unnamed.

BASIC ABELISAURIDS

The abelisaurids were a distinctive group of theropods from the late Cretaceous period. They evolved from the same line as the ceratosaurids, quite distinct from the tetanurans. The abelisaurids were originally thought to have been restricted to South America, but they now seem to have been much more widely distributed across the southern continents.

Masiakasaurus

Distribution: Madagascar.
Classification: Theropoda, Neoceratosauria, Abelisauria.
Meaning of name: Vicious lizard.
Named by: Sampson, Carrano and Forster, 2001.
Time: Maastrichtian stage of the late Cretaceous.
Size: 1.8m (6ft).
Lifestyle: Fisher.
Species: *M. knopfleri.*

This unusual abelisaurid is known from the single, incomplete and disarticulated skeleton, including parts of the jaws, the hind limbs and some of the vertebrae. The lower jaw shows a strange arrangement of teeth, with the front teeth pointing forward and hooked upwards. Those at the rear of the jaw are standard.

Features: The strange forward-pointing teeth on the lower jaw indicate that this was an unusual dinosaur. They are similar to the teeth of some pterosaurs that we know were fishing animals, and by analogy *Masiakasaurus* was probably a fishing animal too. Its long neck also suggests this animal hunted fish. Unfortunately, we do not have the front of the upper jaw, and so it is unclear how the mouth actually worked.

Right: This particular species, M. knopfleri, is named after Mark Knopfler, whose guitar music the team were listening to when they made the discovery.

Noasaurus

The killing claw on the hind foot of *Noasaurus* is a textbook example of convergent evolution. Superficially it is similar to the killing claw of the maniraptorans, but the two animals are only distantly related. Such a claw evolved independently in response to similar needs for similar hunting styles. Possibly they hunted large animals and inflicted deep bleeding wounds, rather than going for a quick kill.

Distribution: Argentina.
Classification: Theropoda, Neoceratosauria, Abelisauridae.
Meaning of name: North-west Argentine lizard.
Named by: Bonaparte and J. E. Powell, 1980.
Time: Maastrichtian stage of the late Cretaceous.
Size: 3m (10ft).
Lifestyle: Hunter.
Species: *N. leali.*

Features: This is a small, active, hunting abelisaurid dinosaur with a deep head, and with a huge killing claw on the second toe of the foot. The claw differs from that of the unrelated maniraptorans by the way the muscles are attached, and by being more sharply curved and more manoeuvrable. Recent research suggests that the big claw could not be used for slashing.

Left: More recent research suggests that this restoration is inaccurate, and that the killing claw belongs not on the foot but on the hand. This would result in a far more conventional-looking animal.

The evolution of the abelisaurids has, for two decades, been held up as one of the proofs of the shifting palaeogeography of the Cretaceous world.

From the observable evidence, the group evolved in the area that is current-day South America, and then spread across the southern continents. The remains have been found in late Cretaceous rocks of South America, Madagascar and India. As the supercontinent of Pangaea split up in Mesozoic times, the southern section, known as Gondwana, remained as a whole for much longer than the northern part, Laurasia. Animals were able to migrate freely across this area, but by that time there was very little connection between Gondwana and Laurasia.

However, the lack of abelisaurids in Africa has always suggested that the African continent broke away from the rest of Gondwana quite early, before the abelisaurids became established. These big animals could not cross the widening ocean areas between Africa and the other southern continents.

Discoveries in the early part of the twenty-first century have shown that the situation is not as simple as this. A large abelisaurid, *Rugops*, has been found in North Africa, and the discovery of *Tarascosaurus*, in France, showed that somehow the group was able to cross the Tethys Ocean that separated Gondwana from Laurasia.

Abelisaurus

This is the dinosaur that gave its name to the whole abelisaurid group – based originally on *Abelisaurus* and *Carnotaurus* – and typifies the late Cretaceous meat-eaters of the Southern Hemisphere. It is only known from a partial skull, but since its discovery there have been more complete specimens of closely related animals, and so we can quite confidently tell what the living animal looked like.

Features: The deep skull and sharp teeth of the basic abelisaurid shows that it was a powerful hunting dinosaur, analogous to the carnosaurs or the tyrannosaurids of the Northern Hemisphere. The skull is distinctly different from either because it has a particularly big gap in front of the eyes.

Distribution: Argentina.
Classification: Theropoda, Neoceratosauria, Abelisauria.
Meaning of name: Abel's (Roberto Abel, the director of the Argentinian Museum of Natural Science) lizard.
Named by: Bonaparte and Novas, 1985.

Time: Maastrichtian stage of the late Cretaceous.
Size: 6.5m (21ft).
Lifestyle: Hunter.
Species: *A. comahuensis*.

Left: A. comahuensis *is named after the Comahue formation, from where it was excavated.*

Tarascosaurus

It was thought that the abelisaurids were confined to the southern continents, evolving there in isolation, until this animal was excavated in France in the 1990s. Although it is only known from a femur and two vertebrae, these are distinctive enough to put it in the group. Its exact provenance is unclear, although it is certainly from a limited sequence of grey limestone of Campanian age.

Features: Since this animal is only known from a few scraps of bone, it is difficult to give a full account of it. It is thought to have had a big head with a blunt snout, and long dagger-like teeth. The body is long and heavy. The arms are small and three-fingered, and the feet have big claws. Much of this description is, of course, based on our knowledge of other large abelisaurids.

Right: The tarasque, after which the dinosaur was named, was a legendary dragon in Provençal legend. We do not know who discovered the dinosaur or who dug it up.

Distribution: France.
Classification: Theropoda, Neoceratosauria, Abelisauria.
Meaning of name: Dragon lizard.
Named by: LeLoeff and Buffetaut, 1991.
Time: Campanian stage.
Size: 10m (33ft).
Lifestyle: Hunter.
Species: *T. salluvicus*.

ADVANCED ABELISAURIDS

Until the 1980s the abelisaurids were not well understood. Since then the burgeoning study of dinosaurs in South America and Madagascar has shown just how wide-ranging and diverse this group actually was. Well-preserved skeletons give us a clear idea of what these animals looked like – mostly big heavy-bodied meat-eaters.

Aucasaurus

Known from an almost complete skeleton, lacking only the end of the tail, *Aucasaurus* was found in lake sediments in Patagonia in 1999. This makes it the best-known abelisaurid skeleton, and it is used as the basis for several other reconstructions. There was damage to the skull of the skeleton found, suggesting that this individual had been involved in a fight shortly before its death.

Features: *Aucasaurus* is similar to its relative, *Carnotaurus*, but only about two-thirds the size. Where *Carnotaurus* has horns on the sides of its head, *Aucasaurus* only has bumps, probably used as sexual display structures. The arms, although tiny, are not as small as those of *Carnotaurus*, and seem to be made up of all humerus, the bones of the lower arm being hardly larger than those of the four fingers.

Distribution: Neuquen Province, Argentina.
Classification: Theropoda, Neoceratosauria, Abelisauria.
Meaning of name: Lizard from Auca Mahuevo.
Named by: Chiappe and Coria, 2001.
Time: Campanian stage of the late Cretaceous.
Size: 5m (16½ft).
Lifestyle: Hunter.
Species: *A. garridoi*.

Carnotaurus

An almost complete skeleton of *Carnotaurus* was extracted with difficulty from the hard mineral nodule in which it was preserved in Argentina. The deep skull suggests that it may have had an acute sense of smell, but the strength of the jaws and neck implied by the muscle attachments seem at odds with the weakness of the lower jaw and the teeth.

Features: The head is very short and squashed-looking with a shallow, hooked lower jaw. Two horns stick out sideways from above the eyes, probably being used for sparring with rivals. The arms are

Distribution: Argentina.
Classification: Theropoda, Neoceratosauria, Abelisauria.
Meaning of name: Flesh-eating bull.
Named by: Bonaparte, 1985.
Time: Campanian to Maastrichtian stage of the late Cretaceous.
Size: 7.5m (25ft).
Lifestyle: Hunter.
Species: *C. sastrei*.

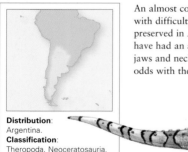

Right: The skull of Carnotaurus *has an enormous hole in front of the eye sockets; this is known as the antorbital fossa. All theropods possess this, but only in the abelisaurids is it so large.*

extremely short with no apparent forearms, even shorter than the tiny arms of *Tyrannosaurus*. They form mere stumps with four miniscule fingers. The skin texture, the best-known of any theropod, has a groundmass of small, pebbly scales but with large, conical scutes forming rows along the sides.

Rugops

The fossil skull of *Rugops* was found in 2000 by a team from *National Geographic* magazine, led by the Chicago Field Museum's Paul Sereno. Several big sauropods were found in the same area, so there was no shortage of food. It was the first abelisaurid to be found in Africa, all others having been found in South America, Madagascar and India. Evidently there was some land connection between Africa and the rest of the continents at the time.

Features: The wrinkled face of *Rugops* was the result of the bones being riddled with arteries and veins, leaving grooves etched across the skull. This implies that the head was covered in skin or armour. Holes along the snout suggest the presence of some kind of fleshy display structure. The skull is short and has a rounded snout. The teeth are those of a meat-eater, but are small and weak, suggesting that *Rugops* was not a hunter but a scavenger of dead animals. The skull bones were found lying on the surface of the rock where they had been weathered by the wind and flying sand.

Above: The presence of Rugops *in north Africa suggests that there was a connection between South America and the northern part of Africa as late as 100 million years ago – a good 20 million years later than originally thought.*

Distribution: Niger.
Classification: Theropoda, Neoceratosauria, Abelisauria.
Meaning of name: Wrinkled face.
Named by: Sereno, Wilson and Conrad, 2004.
Time: Cenomanian stage of the late Cretaceous.
Size: 9m (30ft).
Lifestyle: Scavenger.
Species: *R. primus*.

HUNTER OR SCAVENGER?

Cannibalism, as observed in *Majungatholus*, is not unusual in the animal world. Nowadays there are at least 14 species of mammal, and many species of reptiles and birds, that are known to kill and eat members of their own kind when conditions become harsh. In Cretaceous times, the environment of the area that is now northern Madagascar had a seasonal climate with no steady supply of food or water, and times of extreme dryness. Conditions such as this are conducive to cannibalistic behaviour among animals.

We do not know whether *Majungatholus* killed its own kind or scavenged from the corpses of those that had already succumbed to starvation and drought. The *Majungatholus* bones in question show sets of tooth marks that match the size and spacing of the teeth in a *Majungatholus* jaw, and also the pattern of serrations on the individual teeth.

The only other evidence of dinosaur cannibalism is in the story of the *Coelophysis* discovery at Ghost Ranch, in Arizona, where a pack of *Coelophysis* appear to have perished in a drought, but not before devouring at least one of their young.

Left: Cannibalism.

Majungatholus

The first specimen of *Majungatholus* to be discovered was a small skull fragment. The thickening on the top of this led early researchers to classify this animal as a pachycephalosaurid, hence the name. The term *tholus* means "dome", and its presence in a dinosaur name suggests that the animal was a pachycephalosaurid; and *Majunga* is the province in which it was found. Since the first specimen, several better fossils, including an almost complete skull, have been found, proving it to be an abelosaurid. It seems to have been a cannibal, judging by the toothmarks on some of the bones.

Distribution: Madagascar and possibly India.
Classification: Theropoda, Neoceratosauria, Abelisauria.
Meaning of name: Dome from Majunga (the district in which it was found).
Named by: Sues and Taquet, 1979.

Features: The head is short and broad, broader than in most other theropods, and the snout deep and blunt with thickened bones around the nostrils. The dome-like bulges on top of the skull, the cause of confusion in the original specimen, may have been the bases of horns. The head ornamentation is a single spike rising from above the eyes, and was used for display.

Time: Campanian stage of the late Cretaceous.
Size: 7–9m (23–30ft).
Lifestyle: Hunter and scavenger.
Species: *M. atopus*.

Left: Majungatholus is known to have been a cannibal, attacking its own species.

SUNDRY THEROPODS

Almost weekly the list of meat-eating dinosaurs becomes longer, as new specimens representing completely new species come to light. The 1990s was a particularly fruitful decade in the discovery of new theropods, especially in South America and Africa. In some instances the new discoveries represented animals that had been found but whose original specimens had been lost or forgotten.

Deltadromeus

This theropod resembled the late Jurassic *Ornitholestes* in its anatomy, but it was very much bigger. It was a late-surviving member of this primitive group of meat-eating dinosaurs that are more usually associated with North America. Of the dinosaurs excavated from the Cretaceous beds of North Africa in the 1990s, this would have been the fiercest hunter.

Features: A surprising feature of this animal is the extraordinarily long and delicate limbs. The leg bones are half the thickness of those of a similar-size *Allosaurus*, and the lower limbs approach the proportions of one of the fleet-footed ostrich mimics. The arms are very long for a theropod, and the slim legs make it a very fast runner. The teeth are thin and adapted for stripping flesh rather than for crushing bone. They have very fine serrations but this feature does not seem to be significant in the classification.

Distribution: Kem Kem region, Morocco.
Classification: Theropoda, Tetanurae, Coelurosauria.
Meaning of name: Delta runner.
Named by: Sereno, Duthiel, Iarochene, Larsson, Lyon, Magwene, Sidor, Variccio and J. A. Wilson, 1996.
Time: Late Cretaceous.
Size: 8m (26ft).
Lifestyle: Hunter.
Species: *D. agilis*.

Above: As in many other dinosaurs, the best skeleton found is incomplete. However the cast of a complete skeleton appears in several museums, based largely on guesswork. The head, significantly, is totally speculative.

Carcharodontosaurus

Bones of *Carcharodontosaurus* found in Egypt were described by Stromer in the 1920s, but were lost when an air raid destroyed the Bavarian State Museum, Munich, Germany, in which they and other valuable palaeontological specimens were housed. At the time it had not been clear just how big an animal *Carcharodontosaurus* had been. Half a century later, new specimens were found in Morocco by a team from the Field Museum, Chicago, USA. The skull found was one of the biggest known, and the animal must have been one of the biggest meat-eaters that ever lived.

Distribution: Morocco, Tunisia, Algeria, Libya, Niger.

Classification: Theropoda, Tetanurae, Carnosauria, Allosauridae.
Meaning of name: Great white shark lizard.
Named by: Stromer, 1931.
Time: Albian to Cenomanian stages of the Cretaceous.
Size: 14m (46ft).
Lifestyle: Hunter.
Species: *C. saharicus*.

Features: The skull is missing the lower jaw and the snout, but the full length has been estimated at 1.53m (5ft), which is longer than any *Tyrannosaurus* skull yet found. The brain cavity, however, is tiny, much smaller even than that of *Tyrannosaurus*. The upper jaw has cavities for 14 blade-like teeth at each side, and they are curved and finely serrated at the edges, with grooves across the sides to allow blood to flow away.

Right: Carcharodontosaurus was orignally named in 1925 by Deperet and Savornin, who described it as a megalosaur.

Giganotosaurus

For a century we have regarded *Tyrannosaurus* as the biggest meat-eater that ever lived. That was before the discovery of *Giganotosaurus* in Patagonia in 1993, by amateur collector Ruben Carolini. The subsequent excavation unearthed a skull that was bigger than the skull of any *Tyrannosaurus* known. Then they found the jawbone of another individual that was even bigger. Unlike its close relative, *Carcharodontosaurus*, from Africa, whose remains consisted only of the skull and a few scraps of bone, the skeleton of *Giganotosaurus* is 70 per cent complete. It lived at the time that the big titanosaurs were flourishing on the South American plains, and it probably found its prey among them.

A mounted skeleton of *Giganotosaurus* stands in the entrance hall of the Academy of Natural Sciences in Philadelphia, USA.

Features: In the legs, the tibia and the femur are about the same length. This indicates that *Giganotosaurus* was not a running animal. With a principal prey of big sauropods, such as titanosaurs, it would not need speed to help it hunt. It weighed between four and eight tonnes.

Below: Giganotosaurus,
Carcharodontosaurus and
Tyrannosaurus were the biggest
known meat-eating dinosaurs.
Studies carried out in 1999 at the
North Carolina State Univeristy,
USA, showed that Giganotosaurus
and Tyrannosaurus *had some form*
of warm-blooded metabolism.

Distribution: Neuquén Province, Argentina.
Classification: Theropoda, Tetanurae, Carnosauria, Allosauridae.
Meaning of name: Giant southern lizard.
Named by: Coria and Salgado, 1995.
Time: Albian stage of the middle Cretaceous.
Size: 15m (49½ ft).
Lifestyle: Hunter.
Species: *G. carolini.*

THE DINOSAURS OF BAHARIYA OASIS

Early in the twentieth century, pieces of bone and tooth were discovered in Egypt. They were collected and described by E. S. von Richenbach, of the Bavarian State Collection of Palaeontology and Historical Geology, in Munich. He named them as *Carcharodontosaurus*, seeing a resemblance to the lifestyle of a killer shark with such teeth. The remains of some of the biggest meat-eating dinosaurs ever were uncovered by German palaeontologists in the early years of the century and housed in the museum, including the famous sail-backed *Spinosaurus*.

Unfortunately, this bone collection, together with some of the biggest meat-eating dinosaur fossils ever uncovered, was destroyed when the museum was caught in a World War II bombing raid. Then, in the 1980s, an expedition from the Field Museum, in Chicago, USA, found a skull in the deserts of Morocco that matched the published descriptions of the *Carcharodontosaurus* remains.

Stromer's original site in Egypt has been pinpointed, and has become the focus of a great deal of palaeontological activity, including the discovery of the enormous sauropod *Paralititan*.

TROODONTS

The troodonts were a group of active little theropods that seem to fall somewhere between the long-legged ostrich mimics, the ornithomimids, and the killer-clawed dromeosaurids. The skeletons were bird-like and all had big eyes and large brains. They were probably warm-blooded and covered in feathers but, as yet, there is no direct evidence for this.

Troodon

The teeth of *Troodon* were the first part of this animal to be found, and they were thought to have come from a lizard or pachycephalosaur, or even a carnivorous ornithopod, something of an absurdity. They were then seen to have come from a subsequently discovered dinosaur that had been called *Stenonychosaurus*. As the name *Troodon* was invented first, *Stenonychosaurus* had to be dropped.

Distribution: Alberta, Canada, Montana, Wyoming and perhaps Alaska, USA.
Classification: Theropoda, Tetanurae, Coelurosauria, Troodontidae.
Meaning of name: Tearing tooth.
Named by: Leidy, 1856.
Time: Campanian stage of the late Cretaceous.
Size: 2m (6½ft).
Lifestyle: Stealthy crepuscular hunter.
Species: *T. formosus*.

Features: The long head contains the biggest brain for its body size of any dinosaur, being comparable to that of a modern emu. The hands are long and slim, with three-clawed fingers, and could grasp objects palm-to-palm. The legs are particularly long and each foot has a big killing claw, like that of a *Velociraptor*, on its second toe. Its big eyes, arranged stereoscopically, suggest that it was a hunter of small prey during darkness or at dusk.

Left: The big brain of Troodon *led Canadian palaeontologist Dale Russell to suggest that, had the dinosaurs not become extinct,* Troodon *would have evolved into an intelligent humanoid form by today.*

Saurornithoides

When the scattered remains of *Saurornithoides* were discovered on the Central Asiatic Expedition by the American Museum of Natural History, they were thought to have been the remains of a toothed bird (the skeleton of the whole troodontid group is particularly bird-like). *Saurornithoides* hunted by stealth and ambush rather than by pursuit. The teeth, like those of other troodonts, were small and sharp, adapted for gripping small prey-like lizards or mammals, rather than shearing flesh.

Distribution: Mongolia.
Classification: Theropoda, Tetanurae, Coelurosauria, Troodontidae.
Meaning of name: Bird-shaped lizard.
Named by: Osborn, 1924.
Time: Campanian to Maastrichtian stages of the late Cretaceous.
Size: 2m (6½ft).
Lifestyle: Stealthy hunter.
Species: *S. mongoliensis, S. junior, S. asiamericanus, S. isfarensis.*

Features: In general, the skull is shorter than that of *Troodon* and similar to that of *Velociraptor*. However, it has many more teeth, with 38 in each upper jaw compared with 30 in *Velociraptor*. The big brain seems to be enlarged in the auditory region, suggesting that it had a very good sense of hearing, and the eyes face forwards giving good stereoscopic vision, a valuable aid to hunting small prey. The middle foot bone is a mere splint, a feature that the troodonts share with the ostrich mimics and even the tyrannosaurs. This may have lessened the stress on the foot during running.

Above: The brain of Saurornithoides was the largest known among dinosaurs. It was six times larger than that of a crocodile of a similar weight.

Borogovia

This troodontid is known only from the hind limbs, which were at first attributed to *Saurornithoides*. There are several other small theropods known from the area, including *Saurornithoides*, and there must have been enough food types to enable them all to survive. The name derives from a fictitious animal, the borogove, in Lewis Carroll's poem *Jabberwocky*.

Features: *Borogovia* was a much slimmer animal than *Saurornithoides*, as suggested by the very delicate toe bones. The killing claw on the second toe is much straighter than that found on any of the other troodontids, and is quite small. These killing claws seem to have become smaller as the group evolved. Otherwise there is no reason to doubt that the rest of the skeleton is typically troodontid. The leg bones are so much like those of *Saurornithoides* that some palaeontologists regarded it as a specimen of *S. junior*.

Above: Borogovia, Saurornithoides *and* Tochisaurus *have been found at the same site, leading to the suggestion that they are all the same animal.*

Distribution: Bayankhongor, Mongolia.
Classification: Theropoda, Tetanurae, Coelurosauria, Troodontidae.
Meaning of name: From borogove.
Named by: Osmólska, 1987.
Time: Campanian to Maastrichtian stages of the late Cretaceous.
Size: 2m (6½ft).
Lifestyle: Hunter.
Species: *B. gracilicrus*.

EGG MOUNTAIN

In 1979 the famous hadrosaur nesting sites in Montana were being excavated and studied by a team from Princeton University, USA, when a seismic company began a survey of the area on behalf of an oil company. As this involved drilling and explosions, the team did a rapid hunt for fossils across the whole area before any damage was done. Right beside one of the boreholes they discovered a whole nesting site and a total of 52 eggs from a small dinosaur. These were totally different from the hadrosaur sites they had been studying. Along with the nests there were the remains of lizards, small mammals and dinosaur bones.

The bones belonged to the small hypsilophodont *Orodromeus*, and it was assumed that the site represented an *Orodromeus* nesting colony. However, when the remains were studied properly, it was found that the eggs contained *Troodon* babies and it was actually a *Troodon* nesting site. The original site had been on an island in an alkaline lake, and a parent *Troodon* had been bringing *Orodromeus* remains back to feed its babies. The seismologists appreciated the importance of the site and diverted the line of the survey in order to leave it alone. As the site was located on the top of a knoll, it eventually came to be known as Egg Mountain. *Left*: Troodon *at Egg Mountain.*

Byronosaurus

The smallest of the troodontids was *Byronosaurus*. It is known from one of the best-preserved troodontid skulls ever found. It was discovered in 1994, and from bits of bone found by expeditions during the two subsequent years in the very rich fossil locality of Ukhaa Tolgod, in the Gobi Desert. All troodontids, except for *Troodon* itself, have been found in Asia.

Features: What distinguishes *Byronosaurus* from all other troodontids is the unserrated teeth. As a rule the theropods have teeth that were serrated like steak knives, with other rare exceptions being among the spinosaurids and primitive ostrich mimics. Primitive toothed birds also have teeth that are not serrated. The mouth has a palate separating it from the nasal passages, and the structure of the snout suggests that *Byronosaurus* had a very sensitive nose.

Distribution: Ukhaa Tolgod, Mongolia.
Classification: Theropoda, Tetanurae, Coelurosauria, Troodontidae.
Meaning of name: Byron lizard (after a sponsor of the American Museum of Natural History's Palaeontological Expeditions).
Named by: Norell, Mackovicky and Clark, 2000.
Time: Campanian stage of the late Cretaceous.
Size: 1.5m (5ft).
Lifestyle: Hunter.
Species: *B. jaffei*.

Left: By the time Byronosaurus *was found there were eight troodonts known, seven of which were found in Asia. The group probably evolved in Asia.*

ORNITHOMIMIDS

*The ornithimimids were famous as the ostrich mimics. In general structure they were similar to modern,
ground-living birds with a lightly built skeleton, compact body, long neck and small skull.
Although all the ostrich mimics have a very similar body plan, they are distinguished by details of the
beak, the hands and the body proportions.*

Archaeornithomimus

Originally described by Gilmore in 1933 as a species of *Ornithomimus*,
Archaeornithomimus is known from limb bones and vertebrae.
Many scientists regard it as a *nomen dubium* since there is so
little to study, and it is possible that the remains represent not
an ornithomimid but quite a different
kind of theropod dinosaur.

Features:
Archaeornithomimus
appears to be very similar
to both *Struthiomimus* and
Gallimimus, but it lived
about 30 million years
before either of
them and is
slightly smaller.
Its fingers are much
smaller than those of the other
ornithomimids, and the third finger is
particularly short. It has straight claws
on its fingers, a rather primitive
feature. Like its bigger relatives, it
used speed as a means of escaping
from predators.

*Left: Whatever the true
classification of
Archaeornithomimus,
it was evidently a fast
runner, suggesting that it
was an active predator,
hunting small reptiles
and mammals.*

Distribution: Erenhot City,
Inner Mongolia.

Classification: Theropoda,
Tetanurae, Coelurosauria,
Ornithomimosauria.
Meaning of name: Early
Ornithomimus.
Named by: Russell, 1972.
Time: Early to late
Cretaceous.
Size: 3.5m (11½ft).
Lifestyle: Omnivore or hunter.
Species: *A. asiaticus*,
A. bissektensis.

Garudimimus

Although all the ornithomimids were built for speed,
Garudimimus, a fairly early form, was not in the same
league. The relatively short leg and heavy foot show that it
was not as speedy as the more advanced forms. The foot
had the vestige of the first toe, whereas
all other ornithomimids were
purely three-toed with
the first and fifth
toes lost.

Distribution: Bayshin Tsav,
Mongolia.
Classification: Theropoda,
Tetanurae, Coelurosauria,

Ornithomimosauria.
Meaning of name: Garuda
(an Indian deity) mimic.
Named by: Barsbold, 1981.
Time: Coniacian to Santonian
stages of the late Cretaceous.
Size: 4m (13ft).
Lifestyle: Omnivore.
Species: *G. brevipes*.

*Right: The crest on the skull
of Garudimimus was tiny.
However, in life it may have
been covered in horn and would
have appeared much larger. If so,
it would have been used for
display and communication.*

Features: The skull has a
more rounded snout than
others of the group, and larger
eyes, and the side
of the skull is
more like that of a
primitive theropod. It
has a small crest in
front of its eyes, something no other
known ornithomimid possesses. The
leg is different too, having a much
shorter lower section and foot bones
than the other ornithomimids, and
four toes rather than three. The short
ilium bone in the hip suggests that
the musculature of the legs was quite
weak, an argument against its being a
powerful runner.

Gallimimus

The ornithomimid featured in the film *Jurassic Park* was *Gallimimus*. It was a fairly good representation, except that it is now thought that these animals were covered in feathers, which would make sense if they were to be as active as they were portrayed in the film. There are skeletons of juveniles that have allowed scientists to study the growth pattern of ornithomimids in general.

Distribution: Bayshin Tsav, Mongolia.

Classification: Theropoda, Tetanurae, Coelurosauria, Ornithomimosauria.
Meaning of name: Chicken mimic.
Named by: Osmólska, Roniewicz, Barsbold, 1972.
Time: Maastrichtian stage of the late Cretaceous.
Size: 6m (19½ft).
Lifestyle: Omnivore.
Species: *G. bullatus*, *G. mongoliensis*.

Right: Like the other ostrich mimics and modern birds Gallimimus *had hollow bones. This device allowed for a reduction of weight in the body, without reducing the strength, and enabled the animal to move quickly.*

The main difference between the two known Gallimimus *species is the shape of the fingers.* G. mongoliensis *had shorter hands and would not have grasped as well.*

Features: *Gallimimus* is the largest known type of ornithomimid, but it has shorter arms in proportion to the other species. The hands, too, are quite small and the fingers are not very flexible. The head is quite long and graceful and, as in nearly all ornithomimids, the jaws have no teeth. The beak of the lower jaw is shovel-shaped, and the big eyes are situated on the sides of the head, so it did not have binocular vision.

HERD OR SOLITARY ANIMAL?

The ornithomimids were made famous by their appearance in the film *Jurassic Park* when a whole herd of *Gallimimus* weaved across the countryside, acting as a unified mass. At the time and for a long while after, this behaviour was regarded as good cinema but poor science because there was no evidence that *Gallimimus*, or any other ornithomimid, lived in a herd.

Then, in 2003, a paper was published by Yoshitsugu Kobayashi and Jun-Chang Lu that described the discovery of a bed of bones in Inner Mongolia consisting of 14 ornithomimid individuals, 11 of them youngsters. A new genus, *Sinornithomimus*, was established based on these specimens, and lay somewhere between *Archaeornithomimus* and *Anserimimus* on the evolutionary scale. The important aspect of the find was that it seemed to show evidence that ornithomimids lived in large groups, with adults protecting the young. The bone bed may have been the result of either an accident befalling a big herd in which a high proportion of the youngsters died, or a catastrophic event that killed the whole herd, which consisted of many youngsters and a few adults. Studies of the juvenile bones suggested that the adults could run faster than the young.

Anserimimus

Anserimimus is known from a partial skeleton that includes an incomplete forelimb. The strong arms of this ornithomimid suggest that it may have dug in the ground for food such as roots, insects or dinosaur eggs. It would have been a fast runner, giving the lie to its name of "goose mimic".

Features: What set this species apart from other ornithomimids is the size of the muscle attachments of the upper arm. This must have meant that the forelimbs were particularly strong. The bones of the hand are bound closely together to give a rigid structure, and claws on the hand are flat and hoof-like. Otherwise the skeleton is very much like that of the other ornithomimids.

Distribution: Mongolia.
Classification: Theropoda, Tetanurae, Coelurosauria, Ornithomimosauria.
Meaning of name: Goose mimic.
Named by: Barsbold, 1988.
Time: Campanian to Maastrichtian stages of the late Cretaceous.
Size: 3m (10ft).
Lifestyle: Omnivore.

Species: *A. planinychus.*

Left: "Goose mimic," "chicken mimic," "ostrich mimic" and "emu mimic" – all evocative names that emphasize these animals' resemblance to modern ground-dwelling birds. However, their skulls were more like those of the extinct ground-dwelling birds from New Zealand, the moas, in being sturdily constructed and strong.

ADVANCED ORNITHOMIMIDS

The late ornithomimids were the fastest dinosaurs known. In fact, the known fossils of this group of advanced ostrich-mimics seem to reflect the known fossils of horse ancestors, a sequence leading from small generalized beasts to large, elegant, long-legged creatures adapted for speed. The presence of feathers is suggested by pits in the arm bones of a specimen in Tyrrell Museum, Alberta, Canada.

Struthiomimus

This is the animal that gave rise to the term ostrich mimic, which is often used instead of the slightly more formal ornithomimid. It was the first complete ornithomimid skeleton to be found. It was a fast runner on the late Cretaceous plains of North America.

Its main predators would have been sickle-clawed dromeosaurids and the tyrannosaurid *Albertosaurus*, from which it would have escaped by a sudden turn of speed.

Features: The small head, lack of teeth, long neck, compact body and long legs are the ostrich-like features of this dinosaur. Its non-ostrich-like features are its long arms with three-fingered hands, and the long tail. It is very similar to its close relative *Ornithomimus*, the main differences being its slightly smaller size and longer tail. Nevertheless many scientists regard it, along with many other genera of ornithomimids, as merely a species of *Ornithomimus*. Stomach stones have been associated with the skeleton. Usually only plant-eating animals have them, so this find indicates that *Struthiomimus* was partly vegetarian.

Left: Although the specimen of Struthiomimus *was fairly complete, it was quite badly damaged, resulting in the ongoing confusion as to whether or not it is really a specimen of* Ornithomimus.

Distribution: Alberta, Canada.
Classification: Theropoda, Tetanurae, Coelurosauria, Ornithomimosauria.
Meaning of name: Ostrich-mimic.
Named by: Osborn, 1917.
Time: Campanian stage of the late Cretaceous.
Size: 3–4.3m (10–14ft).
Lifestyle: Omnivore.
Species: *S. sedens*.

Ornithomimus

The image of *Ornithomimus* took many decades to compile after the first very fragmentary specimen was found in the 1880s. It was not until 1917 that a good skeleton of *O. edmontonicus*, complete, but lacking the skull, was discovered in Canada, and the true bird-like nature was appreciated. Three species of *Ornithomimus* existed together, their slight differences in beak shapes suggesting that they ate different foods, some preferring insects, with others opting for small reptiles or plants.

Features: The head is small and carries a fluted toothless beak. The neck and tail are long, and the body is more compact than that of other ornithomimids. The legs are very long for the size of the body, although not quite as long as those of other ornithomimids, and show that it was a running animal. The arms are quite long and slender, and carry three-clawed fingers, the first longer than the others.

Above: Ornithomimus *is the best-known of the ornithomimids, and the remains are quite widely dispersed. Several other ornithomimid genera are regarded by some scientists as species of* Ornithomimus.

Distribution: Alberta, Canada, to Texas, USA.
Classification: Theropoda, Tetanurae, Coelurosauria, Ornithomimosauria.
Meaning of name: Bird mimic.
Named by: Marsh, 1890.
Time: Campanian to Maastrichtian stages of the late Cretaceous.
Size: 4.5m (15ft).
Lifestyle: Omnivore.
Species: *O. antiquus, O. edmontonicus, O. velox, O. lonzeensis* and *O. sedens*.

Dromiceiomimus

The big eyes and the shape of the beak and hands suggest that *Dromiceiomimus* specialized in hunting small prey in the twilight hours. The discovery of an adult skeleton with two young indicates that there was some form of family structure. There is a suggestion, that because of the wide hips, *Dromiceiomimus* did not lay eggs but gave birth to live young.

Above: Dromiceiomimus was originally described by William A. Parks, a Canadian palaeontologist, in 1926, as a species of Struthiomimus.

Features: The shins are long, longer in proportion than those of any other ornithomimid, and this indicates a very fast runner. It was probably the fastest dinosaur known, with the ability to reach speeds of 73kph (45mph). The eyes are bigger than those of other ornithomimids. The shape of the muzzle and the weak jaw muscles suggest a diet of insects, and the hands seem more adapted to scraping in the ground than clutching big prey.

Distribution: Alberta, Canada.
Classification: Theropoda, Tetanurae, Coelurosauria, Ornithomimosauria.
Meaning of name: Emu mimic.
Named by: Russell, 1972.
Time: Campanian to Maastrichtian stages of the late Cretaceous.
Size: 3.5m (12ft).
Lifestyle: Omnivore or hunter of small animals.
Species: *D. brevitertius*, *D. samuelli*.

VEGETARIAN OR CARNIVORE?

The way of life of the ornithomimids seems to have been different from that of all other theropods. In general, the typical theropod was the hunter of the time, whether it was a small animal preying on mammals and small dinosaurs, or a gigantic, dragon-like monster preying on the biggest plant-eaters of the day.

In contrast, the lack of teeth in the jaw precluded the average ornithomimid from such a lifestyle. Instead of teeth it had a beak, like that of a bird. This has led palaeontologists to assume that they were omnivorous animals, feeding on both plant and animal food. There is, however, a strong argument against the vegetable part of this diet; there seems to have been no means of processing such food. Modern plant-eating birds, lacking teeth for chewing, swallow grit and stones to help them grind up the food. This is true of some of the sauropod plant-eaters that lacked chewing teeth. Gastroliths have been found in the stomach areas of skeletons of some of them. However, no gastroliths have ever been found associated with an ornithomimid skeleton, with the exception of *Struthiomimus*.

The fluting on the beak has led to a suggestion that the beak was used for straining shrimps and other tiny animals from lake water, but this is not a generally accepted view. The consensus is that a typical ornithomimid fed on small animals, such as insects or lizards.

Deinocheirus

The only thing that is known of this supposed ornithomimid is a pair of enormous arms, with huge, clawed hands. Since the description in 1970, scientists have speculated on whether the arms were unusually large for the size of animal, or whether they were in the same proportions as in the other ornithomimids, which would have meant an animal as big as *Tyrannosaurus*.

Features: Each arm is 2.6m (8½ft) long. The lower arm is about two-thirds the length of the upper arm, and the hand with the three equal-length fingers is about the same length as the lower arm.

Distribution: Mongolia.
Classification: Theropoda, Tetanurae, Coelurosauria, Ornithomimosauria.
Meaning of name: Terrible hand.
Named by: Osmólska and Roniewicz, 1970.
Time: Maastrichtian stage of the late Cretaceous.
Size: 7–12m? (23–39ft?).
Lifestyle: Unclear.
Species: *D. mirificus*. One other, unnamed.

Left: The fingers have strongly curved claws that are 25cm (10in) long. In life these were covered in horn. We know nothing more about this mysterious animal.

OVIRAPTORIDS

Oviraptorids were extremely bird-like dinosaurs, not only in their general build but also in the presence of the beak and the fact that the shoulders were strengthened by a collarbone. It has even been suggested that the accepted classification of oviraptorids is wrong, and that they should be classed as birds instead of dinosaurs.

Oviraptor

Distribution: Mongolia.
Classification: Theropoda, Tetanurae, Coelurosauria, Oviraptorosauria.
Meaning of name: Egg stealer.
Named by: Ostrom, 1924.
Time: Campanian stage of the late Cretaceous.
Size: 1.8m (6ft).
Lifestyle: Specialist feeder.
Species: *O. philoceratops*, *O. mongoliensis*.

The genus name *Oviraptor* derives from the belief that the first *Oviraptor* found had been eating the eggs of a ceratopsian dinosaur. The mouth seems to have evolved for a crushing action, and the current diet suggestions point to shellfish or nuts. There seems to be a variation in size and shape of the skull crest, maybe a sign of different stages of growth and maturity, or different species. The skull, on which most restorations, including this one, are based, is thought to have belonged to the related oviraptorid *Citipati*.

Features: As with all other oviraptorids the head is short and carries a heavy toothless beak at the end of its well-muscled jaws. A hollow crest, like that of a cassowary, which sticks up on the head was probably used for display and intimidation. There is a pair of teeth on the palate. The skull is extremely lightweight and has very large eye sockets.

Right: Oviraptor hands are very long. Its eggs are about the size of a hot-dog bun.

Khaan

The three almost complete articulated skeletons that have been excavated give a good idea of what this oviraptorid looked like. It was first thought to have been a specimen of the related oviraptorid, *Ingenia*, showing just how little variation there was between animals of this group. Several oviraptorids lived in the same area at the same time. The specific name of the type species *K. mckennai* honours American palaeontologist Malcolm McKenna.

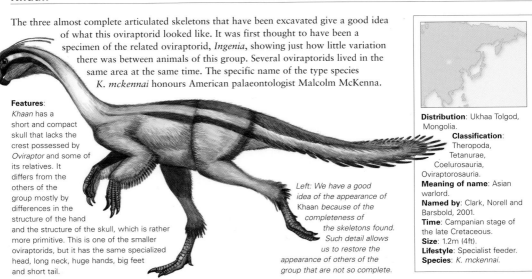

Features: *Khaan* has a short and compact skull that lacks the crest possessed by *Oviraptor* and some of its relatives. It differs from the others of the group mostly by differences in the structure of the hand and the structure of the skull, which is rather more primitive. This is one of the smaller oviraptorids, but it has the same specialized head, long neck, huge hands, big feet and short tail.

Left: We have a good idea of the appearance of Khaan because of the completeness of the skeletons found. Such detail allows us to restore the appearance of others of the group that are not so complete.

Distribution: Ukhaa Tolgod, Mongolia.
Classification: Theropoda, Tetanurae, Coelurosauria, Oviraptorosauria.
Meaning of name: Asian warlord.
Named by: Clark, Norell and Barsbold, 2001.
Time: Campanian stage of the late Cretaceous.
Size: 1.2m (4ft).
Lifestyle: Specialist feeder.
Species: *K. mckennai*.

OVIRAPTOR DIET

With the oviraptorids' absurdly short heads, huge, toothless beaks and pair of little teeth on the palate, their diet was clearly understood when specimens were first discovered but has since become a mystery. The American Museum of Natural History expeditions to the Gobi Desert in the 1920s produced a whole host of new dinosaurs. The most abundant were the little ceratopsian, *Protoceratops*, found with nesting sites full of eggs. The first *Oviraptor* was found close to one of these nests, and it was believed to have been raiding the *Protoceratops* nest when it was overcome and killed by a sandstorm. This made perfect sense, since the mouth of *Oviraptor* was ideal for breaking into hard-shelled eggs, and its hands were perfectly shaped to clutch eggs of that size.

Then, in the 1990s, another American Museum of Natural History expedition found the remains of an identical nest, but this time with an oviraptorid sitting on it brooding the eggs. So the original *Oviraptor* had not been raiding a *Protoceratops* nest at all, it had been on its own nest. But we still do not know what *Oviraptor* could have eaten with its very specialized mouth parts.

Nomingia

The remarkable pygostyle on the end of the tail of *Nomingia* must have supported a fan of feathers. Since this was manifestly not a flying animal, the feathers must have been used for display purposes. The fan's effect was probably reinforced by display feathers on the arms, but the forelimbs and the head are missing from the only known specimen.

Features: This is the only dinosaur known with a pygostyle, a structure formed of the fusion of tail vertebrae. In birds, the entire tail is formed of a pygostyle and it is used as a mounting platform for the fan of tail feathers. In *Nomingia* it is on the end of a short length of tail formed of stubby vertebrae, but it probably represents the base of a tail fan as well. The legs are proportionally longer than in other oviraptorids. It is possible that the second toe bears a killing claw.

Distribution: Bugin Tsav, Mongolia.
Classification: Theropoda, Tetanurae, Coelurosauria, Oviraptorosauria.
Meaning of name: From the Nomingiin Gobi, part of the Gobi Desert.
Named by: Barsbold, Osmólska, Watabe, Currie and Tsogtbataar, 2000.
Time: Maastrichtian stage of the late Cretaceous.
Size: 1.8m (6ft), but this takes into account a shorter tail than is usual in the oviraptorids.
Lifestyle: Specialist feeder.

Left: Despite the specialization of the tail, Nomingia *seems to be quite a primitive oviraptorid in other respects.*

Chirostenotes

This animal is assembled from bits of skeleton excavated over a period of time. The hand was found in 1924, the feet (named *Macrophalangia*) in 1932, and the jaws (named *Caenagnathus*) in 1936. It was not realized that they all came from the same genus of animal until, in 1988, an unprepared skeleton was found in a museum collection where it had lain unstudied for 60 years.

Right: Chirostenotes was closely related to another caenagnathid, called Elmisaurus, from Mongolia, showing that this was quite a wide-ranging group.

Features: Each hand has three narrow, clawed fingers with the middle finger bigger than the others. *Chirostenotes* belongs to a side branch of the oviraptorid family known as the caenagnathids. They have toothless jaws, but the head is not as extremely specialized as we see in the oviraptorids proper, although it does have a spectacular crest.

Distribution: Alberta, Canada.
Classification: Theropoda, Tetanurae, Coelurosauria, Oviraptorosauria.
Meaning of name: Slim hand.
Named by: Gilmore, 1924.
Time: Campanian stage of the late Cretaceous.
Size: 2.9m (9½ft).
Lifestyle: Specialist feeder.
Species: *C. sternbergi*, *C. pergracilis*.

THERIZINOSAURIDS

The therizinosaurids represent a rare, exclusively Cretaceous group of dinosaurs, which is still undergoing revision. They may not represent a coherent group. The biggest is Therizinosaurus, which could be so different from the rest that the others have to be put into a different family altogether, and called the segnosaurids. They seem to show a mixture of plant-eating and meat-eating features.

Therizinosaurus

Distribution: Mongolia, Kazakhstan and Transbaykalia.
Classification: Theropoda, Tetanurae, Coelurosauria, Therizinosauria.
Meaning of name: Scythe lizard.
Named by: Maleev, 1954.
Time: Campanian stage of the late Cretaceous.
Size: 8–11m (26–36ft).
Lifestyle: Unclear.
Species: *T. cheloniformis*.

Therizinosaurus was first thought to have been a turtle-like animal (hence the species name). That was the best that could be made of the original remains, which consisted of some flattened ribs and a set of enormous arms and claws. Over the last half century more pieces have come to light and these give a better picture of the whole animal.

Features: The claws of this animal are its most remarkable feature. The hand has three of them, roughly the same length, with the longest measuring 71cm (28in). In life they would have been covered with a horny sheath perhaps half as long again. These massive hands are supported by arms that are 2.1m (7ft) long. Most of the rest of the animal has been restored on the basis of other better-known relatives, but this is the giant of the group.

Segnosaurus

Distribution: Mongolia.
Classification: Theropoda, Tetanurae, Coelurosauria, Therizinosauria.
Meaning of name: Slow lizard.
Named by: Perle, 1979.
Time: Cenomanian to Turonian stages of the late Cretaceous.
Size: 4–9m (13–30ft).
Lifestyle: Unclear.
Species: *S. galbinensis*.

We know of only a few remains of *Segnosaurus*, but the whole animal can be restored by comparison with its close relatives. The head is based on the skull of the closely related *Erlikosaurus*, and the feathery covering comes from an early Cretaceous form, *Beipiaosaurus*, found perfectly preserved in the Liaoning sediments, in China.

Features: The fragmentary remains of this genus show the typical down-turned jaw with the leaf-shaped teeth, and the hip bones with the swept-back pubis that give the impression of an ornithischian dinosaur. These are important details in establishing the make-up of this whole line of dinosaur. It is such an important animal that the name *Segnosauria* has been proposed as an alternative name for the group. Fossil eggs about the size of duck eggs have been attributed to these animals.

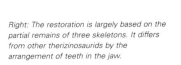

Right: The restoration is largely based on the partial remains of three skeletons. It differs from other therizinosaurids by the arrangement of teeth in the jaw.

Nothronychus

The first therizinosaurid to be found outside Asia is the most complete. It lived at the edge of the shallow sea that covered most of central North America at the time, in swampy deltas in the area of the Arizona/New Mexico borderlands. It is one of the few dinosaurs to be found from the early part of the late Cretaceous. Its name comes from the similarity between it and the giant ground sloths that existed until a few million years ago.

Right: The resemblance to a ground sloth lies in its upright stance and the enormous claws on its hands. The claws were probably used for pulling down vegetation in the swampy forests where it lived.

Features: The almost complete skeleton of this animal forms the basis for most modern restorations of therizinosaurids. It has a small head on a long neck, leaf-shaped teeth (suggesting a herbivorous habit), a heavy body with broad hips (also suggesting a partial diet of plant material), heavy hands and a short, stumpy tail. It carried itself with a more upright stance than other theropods, and its hind legs are relatively short. The hip girdle has the swept-back bird-like ischium bone that is usually only seen in plant-eating dinosaurs.

Distribution: New Mexico, USA.
Classification: Theropoda, Tetanurae, Coelurosauria, Therizinosauria.
Meaning of name: Sloth claw.
Named by: Kirkland and Wolfe vide Stanley, 2001.
Time: Turonian stage of the late Cretaceous.
Size: 4.5–6m (15–18½ft).
Lifestyle: Unclear.
Species: *N. mckinleyi*.

THERIZINOSAUR DIET

With heads containing leaf-shaped, plant-shredding teeth, cheek pouches (indicated by the depressions at the side of the head), sharp cutting beaks, huge ripping claws and heavy, pot-bellied bodies, the therizinosaurids' lifestyle has always been a mystery. The hip bones, although in most respects saurischian, have the swept-back pubis typical of the ornithischians. This could be an adaptation to a plant-eating diet, to accommodate the big plant-processing digestive system. However, we see this feature also in the maniraptorans – as unambiguous a family of meat-eaters as we can find.

One intriguing interpretation of the huge claws is a comparison with the big claws of modern ant-eating animals, such as ant-eaters, aardvarks, armadillos and the like. In these animals the claws are an adaptation to ripping into nests and logs to reach tiny insects. However, it seems unlikely that such a diet would support an animal as big as *Therizinosaurus*.

The most likely interpretation is that the therizinosaurs were plant-eaters, the huge claws being used for ripping branches down from trees. What environmental pressures induced a family of meat-eating theropods to evolve into such a lifestyle is still a mystery.

Neimongosaurus

Neimongosaurus is known from two partial skeletons, one of which has most of the backbone and nearly all the limb bones. We know of only part of the skull, including the brain box. Mysterious arm and claw bones found in the same area in 1920, and named *Alectrosaurus*, may belong to *Neimongosaurus* and shed light on the missing hand bones.

Features: The long neck and the short tail, as well as the air spaces in the vertebrae and the arrangement of the shoulder muscles, suggest that these animals are closely related to the oviraptorids. Unfortunately, the hand bones are absent and we cannot compare them with the rest of the group. The jaw is deep and markedly down-turned, and bears a broad beak. The teeth are very similar to those of some ornithischians, suggesting a plant diet.

Distribution: Nei Mongol, China.
Classification: Theropoda, Tetanurae, Coelurosauria, Therizinosauria.
Meaning of name: Lizard from Nei Mongol.
Named by: Xu, Sereno, Kuang and Tan, 2001.
Time: Cenomanian to Campanian stages of the late Cretaceous.
Size: 2.3m (7½ft).
Lifestyle: Unclear.
Species: *N. yangi*.

Left: The shoulder girdle is very much like that of an oviraptorid, as are the vertebrae, which are full of air spaces.

ALVAREZSAURIDS

Bird or dinosaur? The fleet-footed, compact-bodied, pointy-snouted alvarezsaurids represent one of those enigmatic groups that could be classed as either. Their bird-like features include specialized forelimbs with the breast bone, fused ankle bones and narrow skull. The non-bird-like features include the huge claw and long tail.

Alvarezsaurus

The skull and forelimbs of *Alvarezsaurus* (the most important features of the group) are missing from the only skeleton found, and it was not until other members of the group were discovered that scientists could appreciate just how unusual this animal was. It was originally restored with the three-fingered hands of a typical coelurosaurid: this restoration still appears today.

Features: The lack of spines on the back vertebrae, resulting in a compact body with no ridge down its back, show it to have been very bird-like. The tail is flattened from side to side and is very long, about twice the length of the body and neck. The neck is long and flexible. It had the long, lightweight feet of a running animal. *Alvarezsaurus* appears to be the most primitive of the Alvarezsaurid group.

Distribution: Neuquen, Argentina.
Classification: Theropoda, Tetanurae, Coelurosauria, Alvarezsauria.
Meaning of name: Alvarez's (historian Don Gregorio Alvarez) lizard.
Named by: Bonaparte, 1991.
Time: Coniacian to Santonian stages of the late Cretaceous.
Size: 2m (6½ft).
Lifestyle: Insectivore.
Species: *A. calvoi*.

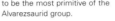

Right: The marked contrast between this alvarezsaurid and the bird Archaeopteryx *has been noted.* Archaeopteryx *was obviously evolved to fly, but it had very few bird-like features.* Alvarezsaurus, *on the other hand, had many bird-like features but was certainly not evolved to fly. Evidently the various bird-like features and the various features adapted for flight evolved over and over again.*

Patagonykus

Distribution:
Neuquen Province, Argentina.
Classification: Theropoda, Tetanurae, Coelurosauria, Alvarezsauria.
Meaning of name:
Patagonian claw.
Named by: Novas, 1996.
Time: Turonian stage of the late Cretaceous.
Size: 2m (6½ft).
Lifestyle: Insectivore.
Species: *P. puertai*.

The discovery of *Patagonykus* in western Argentina in 1996 was important in establishing the alvarezsaurids as a group. The fragmentary skeleton was similar to *Alvarezsaurus*, but it had the distinctive short, powerful forelimb with the single claw that was such an important feature of the more complete Asian relative, *Mononykus*. *Patagonykus* and *Alvarezsaurus* both lived in South America, about as far away as it was possible to get in the late Cretaceous world from Mongolia where the rest of the group have been found. The alvarezsaurids represented a very widespread group.

Features: The arm is very distinctive, with quite a short, thin humerus, but a massive ulna that projects well back from the elbow joint, suggesting a very powerful leverage. The claw bone is almost as big as this ulna. The joint between the first tail vertebra and the hips is very flexible, indicating a great deal of movement here. This may show an ability to sit down on the heavy pubis bone, with the tail out of the way.

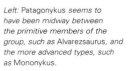

Left: Patagonykus *seems to have been midway between the primitive members of the group, such as* Alvarezsaurus, *and the more advanced types, such as* Mononykus.

Mononykus

The alvarezsaurid *Mononykus* was found in 1923 by an expedition from the American Museum of Natural History, but its significance was not realized at the time. It was referred to as an "unidentified bird-like dinosaur". Another expedition from the Museum found better examples in the 1990s. It was originally named *Mononychus*, but unfortunately that name had already been given to a kind of beetle.

Features: The upper and lower arms and the single claw are of equal length, and support a very strong musculature, attached to a keeled breast bone. This is interpreted as a digging adaptation. The long legs are very bird-like, with the fibula reduced to a vestige, and this has led some palaeontologists to think that *Mononykus*, and the rest of the alvarezsaurids, should be regarded as birds.

Distribution: Bugin Tsav, Mongolia.
Classification: Theropoda, Tetanurae, Coelurosauria, Alvarezsauria.
Meaning of name: Single claw.
Named by: Perle, Norrell, Chiappe and Clark, 1993.
Time: Campanian stage of the late Cretaceous.
Size: 0.9m (3ft).
Lifestyle: Insectivore.
Species: *M. olecranus*.

ALVAREZSAURID LIFESTYLE

The alvarezsaurids have always been a mystery. In their adaptations they were quite unlike any other dinosaur. The strange, stubby arms and the single massive claw were obviously an adaptation to some very specialized lifestyle. The current thinking is that these forelimbs were adapted for digging. The bird-like breast bone suggests powerful arm muscles, but they were certainly not used for flying. They were probably used for digging into the mounds of termites and other colonial insects, breaking up the concrete-like walls so that the living chambers could be reached by the long, narrow prehensile jaws and, perhaps, a long tongue. A similar lifestyle has been suggested for the therizinosaurids, but the large size of these animals seems to rule this out. It is a much more realistic prospect for the alvarezsaurids, which were considerably smaller animals and could possibly have subsisted on such a diet. Termite mounds are known from as far back as the Triassic period.

The particularly long legs would have been used for escape from predators, since they lacked any other form of defence. Indeed an alvarezsaurid ankle bone found by Othniel Charles Marsh in the 1880s was thought to have come from one of the fleet-footed ostrich mimics.

Right: *Mononykus* arm bones.

Shuvuuia

This dinosaur is known from two very well-preserved skulls. In fact it is the only member of the alvarezsaurids for which the skull is known. The appearance of the head of all the other alvarezsaurids shown in these restorations is based on these specimens. Some fossils originally attributed to *Mononykus* are thought to have come from *Shuvuuia*, the two are so similar.

Features: The notable feature of the skull is the joint between the long, pointed snout and the bone in front of the eye. This means that the mouth could open very widely, hinged upwards in front of the brain case. The teeth, set in a continuous groove, are numerous and small, very similar to those of primitive birds. Analysis of the fossils shows traces of a chemical beta keratin only found in feathers, indicating that it had a feathered covering of some sort.

Right: *Shuvuuia* and *Mononykus* were so similar that many fossil bones previously attributed to *Mononykus* are now regarded as having belonged to *Shuvuuia*. However, *Shuvuuia* had two additional tiny claws on the hand that *Mononykus* lacked.

Distribution: Ukhaa Tolgod, Mongolia.
Classification: Theropoda, Tetanurae, Coelurosauria, Alvarezsauria.
Meaning of name: Bird.
Named by: Chiappe, Norell and Clark, 1998.
Time: Campanian stage of the late Cretaceous.
Size: 1m (3ft).
Lifestyle: Insectivore.
Species: *S. deserti*.

DROMAEOSAURIDS

Without a doubt the dromaeosaurids were the most important active predators of late Cretaceous times. With their clawed hands that could grasp prey between their palms, their killing claw on the hind foot, and their mental agility that would have enabled them to balance and slash at the same time, they represented a fearsome group of animals.

Dromaeosaurus

Distribution: Alberta, Canada, and Montana, USA.
Classification:
Theropoda, Tetanurae, Coelurosauria, Deinonychosauria.
Meaning of name:
Running lizard.
Named by: Matthew and Brown, 1922.
Time: Campanian stage.
Size: 1.8m (6ft).
Lifestyle: Hunter.
Species: *D. albertensis, D. cristatus, D. gracilis, D. explanatus.*

Dromaeosaurus was the first of the group to have been discovered, and led to the establishment of the family. It is surprisingly poorly known, although a cast of a complete mounted skeleton, prepared by the Tyrrell Museum in Alberta, Canada, appears in several museums throughout the world. Its construction was made possible by knowledge of others of the group that have been discovered more recently.

Left: The famous dinosaur hunter Barnum Brown found the first and best of the Dromaeosaurus remains on the banks of the Red Deer River in Alberta, Canada, in 1914, naming it eight years later.

Features: The jaws are long and heavily built, and the neck is curved and flexible. The snout is deep and rounded. The tail is stiff and straight, articulated only at the base, stiffened by bony rods growing backwards from above and below the individual vertebrae. This would have helped the animal to balance while hunting prey. Its large eyes gave it excellent vision, and the size of the nasal cavities suggest that it could hunt by smell as well. The killing claw on the second toe is smaller than that of others of the group, but still efficient.

Saurornitholestes

This hunting dinosaur is known from the remains of three individuals. One remarkable occurrence is of a *Saurornitholestes'* tooth embedded in a pterosaur bone. It is not impossible to imagine an active predator like this snatching a pterosaur from the sky, but it is more likely that it scavenged the carcass of a pterosaur that had died already.

Right: Saurornitholestes seems to be an amalgam of different animals. The head is very much like that of Velociraptor, while the rest of the skeleton (what has been found of it) is more like that of Deinonychus.

Features: The shape of the skull suggests a bigger brain than many of its relatives, but a poorer sense of smell. The teeth are also different from those of *Dromaeosaurus*, but otherwise the two animals are very similar, having grasping hands with sharp claws and a killing claw on the second toe. It was originally classed among the troodontids, but now some palaeontologists regard it as a species of *Velociraptor*.

Distribution: Alberta, Canada.
Classification: Theropoda, Tetanurae, Coelurosauria, Deinonychosauria.
Meaning of name: Lizard bird thief.
Named by: Sues, 1978.
Time: Campanian stage of the late Cretaceous.
Size: 2m (6½ft).
Lifestyle: Hunter.
Species: *S. langstoni.*

A theory put forward by American palaeontologist Greg Paul, in 2000, suggests that the dromaeosaurids actually descended from a flying ancestor. Their ancestor would have been *Archaeopteryx*, or some relative of *Archaeopteryx*, back in late Jurassic times. It would have possessed the *Archaeopteryx* jaws and teeth, the clawed wings and the long, bony tail. At some subsequent time the descendants would have lost their powers of flight, and taken up a ground-dwelling existence. The wings would have atrophied, but the jaws and teeth, the claws, and the long tail would have been retained. They would have kept their plumage and their warm-blooded metabolism, but not their flight feathers. Like modern flightless birds, such as

ostriches and emus, they would have become bigger and heavier, and eventually they would have looked as though they had never flown, so well would they have adapted to ground-living. This would account for the bird-like skeletons of these animals, and the suggestion of bird-like musculature in animals such as *Unenlagia*.

Left: Archaeopteryx (bottom) and Velociraptor (top).

Unenlagia

This dinosaur caused some confusion when it was discovered in the 1990s. It was so bird-like that it was initially regarded as a kind of bird. At one time it was thought to have been a juvenile specimen of the contemporary *Megaraptor*, but that seems unlikely as *Megaraptor* now appears to have been a totally different type of dinosaur.

Below: The only specimen found consists of 20 bones found in river-laid sedimentary rocks in Argentina by Fernando Novas of the Museum of Natural History in Buenos Aires. Its name is a mixture of Latin and the local Mapache language.

Features: The shoulder joint of this dinosaur would allow for flapping, as though it had wings rather than the more normal dinosaurid arms. It was too big for a flying animal, and it is possible that wing-like arms could have been used for stabilization and steering while the dinosaur was running at speed. This

Distribution: Argentina.
Classification: Theropoda, Tetanurae, Coelurosauria, Deinonychosauria.
Meaning of name: Half bird.
Named by: Novas, 1997.
Time: Turonian to Coniacian stages of the late Cretaceous.
Size: 2–3m (6½–10ft).
Lifestyle: Hunter.
Species: *U. comahuensis*.

appears to add weight to the theory that dromaeosaurids evolved from flying ancestors. Certainly it shows how closely these animals are related to birds.

Velociraptor

Perhaps the best-known of the dromaeosaurids, *Velociraptor* is known from several specimens, the first found by the American Museum of Natural History expedition to Mongolia in the 1920s. Found in 1971, a famous fossil consisted of a complete *Velociraptor* skeleton wrapped around that of a *Protoceratops*. The two had been preserved in the middle of a fight, possibly engulfed in a sandstorm.

Features: The 80 very sharp curved teeth in a long snout, flattened from side to side, the three-fingered hands, each finger equipped with eagle-like talons, and the curved killing claw, 9cm (3½in) long, on the second toe of each foot show this to have been a ferocious hunter. Its long, stiff tail functioned as a balance while running and making sharp turns. A covering of feathers would help to keep the animal insulated, a necessity for its active, warm-blooded lifestyle.

Distribution: Mongolia, China and Russia.
Classification: Theropoda, Tetanurae, Coelurosauria, Deinonychosauria.
Meaning of name: Fast hunter.
Named by: Osborn, 1924.
Time: Campanian stage of the late Cretaceous.
Size: 2m (6½ft).
Lifestyle: Hunter.
Species: *V. mongoliensis*.
Several other genera, such as *Deinonychus*, *Saurornitholestes* and *Bambiraptor*, have been regarded as species of *Velociraptor* in the past.

TYRANNOSAURIDS

If the dromaeosaurids represented the ultimate small and agile killers, then the tryannosaurids were undoubtedly the typical killer giants, especially of Asia and North America. They were among the biggest meat-eaters that ever walked the Earth. The tyrannosaurids are classed with the dromaeosaurids in the Coelurosauria, a division that formerly encompassed only the smallest of the meat-eating dinosaurs.

Albertosaurus

The earliest dinosaur remains to be found in Alberta, Canada, were *Albertosaurus* bones. They were found in 1884 by J. B. Tyrrell, after whom the world-famous dinosaur museum in Drumheller was named. It was one of the most abundant predators of the North American Cretaceous plains, and despite its great weight it was probably a fast runner, running down its prey, which would have consisted of duckbills.

Features: *Albertosaurus* is very similar to its later cousin *Tyrannosaurus*, but is only about half the size. It is much better known as its remains are more numerous. Its skull is heavier, with smaller gaps in it surrounded by thicker struts of bone, the muzzle is longer and lower and also much wider, and the jaw is considerably shallower. The arms, although small, are a little larger than those of *Tyrannosaurus*.

Distribution: Alberta, Canada; Alaska, Montana, Wyoming, USA.
Classification: Theropoda, Tetanurae, Coelurosauria, Tyrannosauroidea.
Meaning of name: Lizard from Alberta.
Named by: Osborn, 1905.
Time: Campanian to Maastrichtian stages of the late Cretaceous.
Size: 8.5m (28ft).
Lifestyle: Hunter.
Species: *A. sarcophagus*, *A. grandis*.

Nanotyrannus

Distribution: Montana, USA.
Classification: Theropoda, Tetanurae, Coelurosauria, Tyrannosauroidea.
Meaning of name: Little tyrant.
Named by: Bakker, Currie and Williams, 1988.
Time: Maastrichtian stage of the late Cretaceous.
Size: 5m (16½ft).
Lifestyle: Hunter.
Species: *N. lancensis*.

This is significant in being the smallest tyrannosaurid yet recovered from the late Cretaceous. It is known from a skull discovered in 1942 and studied in the 1980s, when its significance was realized. Modern analytical techniques such as CAT scanning, a medical technique that produces 3D images of body interiors, were brought to it, revealing features inside that had not been noted in any other tyrannosaurid skull.

Features: The skull is 57.2cm (22½in) long. It is long and low, with a narrow snout, broadening towards the rear. The eye sockets point forward, giving stereoscopic vision. Turbinals (scrolls of bone inside the nose) are present, showing an enhanced sense of smell or a cooling device. Some palaeontologists regard *Nanotyrannus* as a juvenile specimen of something better known, or even a dwarf species of *Albertosaurus* or *Gorgosaurus*.

Left: A new skeleton found in Montana in 2000 by the Burpee Museum in Illinois, USA, was thought to have been a second specimen of Nanotyrannus. It was later established that it was a juvenile Tyrannosaurus.

Alioramus

The Asian tyrannosaurid *Alioramus* is poorly known, and only from the jaw bones, part of the skull and some foot bones. Its name derives from the fact that this dinosaur represents a distinct branch of tyrannosaurid evolution that left the main evolutionary line early in the late Cretaceous. The animal was lightly built, and well adapted to chasing prey over open landscapes.

Right: The only specimen of Alioramus *found consisted of a partial skull and some foot bones found in Ingenii Höövör Valley, in Mongolia. The name, meaning "different branch," refers to the fact that it diverged early from the main tyrannosaurid line.*

Features: The long, rugged snout of *Alioramus* is distinctive. A ridge down the nose shows six prominent bumps that may be horn cores, two side by side, with four in a single row in front of them. The long jaws have many more teeth – 18 in the front of each lower jaw alone – than any other tyrannosaurid, suggesting that this is a very primitive member of the group. The eyes are large and well adapted for hunting.

Distribution: Mongolia.
Classification: Theropoda, Tetanurae, Coelurosauria, Tyrannosauroidea.
Meaning of name: Different branch.
Named by: Kurzanov, 1976.
Time: Maastrichtian stage of the late Cretaceous.
Size: 6m (19½ft).
Lifestyle: Hunter.
Species: *A. remotus*.

TYRANNOSAURID ARMS

The small arms of a tyrannosaurid have always been a puzzle. Of what use are a pair of arms that are too small to reach the mouth, or anything else? The arm bones of a 12m (39ft) tyrannosaurid are about the same length as those of a human, but about three times as thick. The muscle scars show that the upper arm was very heavily muscled. There are only two fingers on each hand. The second finger is much larger than the first, and the third has disappeared altogether. The elbow joint was not very flexible. The hands were positioned so that the palms faced towards each other.

There are three main theories about the purpose of these arms. First, that they were used for clutching prey to the chest so that the jaws and teeth could reach it. Second, that they were used for grasping the female while mating. And third, that they were used to stabilize the body as the great animal stood from a lying position. We don't know for sure.

Left: Bones and tendons of the tyrannosaurid arm.

Appalachiosaurus

The finding of a tyrannosaurid in eastern North America in the early 1980s, and its scientific publication in 2005, was a surprise. Until then all members of the group had been found in western North America and Asia – separated from eastern North America by the Cretaceous inland sea. The ancestors must have spread across the North American continent before the inland sea cut the landmass in two.

Features: The single skeleton of *Appalachiosaurus* consists of most of the skull, the hind legs and parts of the tail and hips. This is enough to show that it is a typical primitive medium-size tyrannosaurid. The skull is similar to that of *Albertosaurus* and the ornamentation on it is small – unlike the bumps and knobs of *Alioramus* and other primitive tyrannosaurids. The most remarkable feature of this animal is the fact that it was found outside the expected geographic range of the group.

Distribution: Alabama, USA.
Classification: Theropoda, Tetanurae, Coelurosauria, Tyrannosauroidea.
Meaning of name: Lizard of the Appalachians.
Named by: Carr, Williamson and Schwimmer, 2005
Time: Campanian stage of the late Cretaceous.

Size: 7m (23ft).
Lifestyle: Hunter.
Species: *A. montgomeriensis.*

Left: The skeleton of Appalachiosaurus *was found in marine mudstones. The body had been washed out to sea by river currents and deposited on the sea bed.*

THE LAST TYRANNOSAURIDS

At the end of the Age of Dinosaurs, the biggest of the meat-eaters of the Northern Hemisphere were the tyrannosaurids. By then, their body shape had settled into a consistent design, with each of the late examples becoming almost indistinguishable from the others. Most of the descriptions here concentrate on the small departures from the basic Tyrannosaurus *anatomy.*

Tyrannosaurus

Perhaps the best-known of all dinosaurs, *Tyrannosaurus* held the record for the biggest and most powerful land-living predator of all time for a century, until the discovery of the big allosaurids, such as *Carcharodontosaurus* and *Giganotosaurus*, in the 1990s. About 20 skeletons of *Tyrannosaurus* are known, some articulated and some scattered, and so the appearance of this dinosaur is known with confidence.

Features: The skull is short and deep, and solid compared with that of other big meat-eaters. The teeth are 8–16cm (3–6in) long and about 2.5cm (1in) wide. Those at the front are D-shaped, built for gripping, while the back teeth are thin blades, evolved for shearing meat. The eyes are positioned so that they give a stereoscopic view forward. The ear structure is like that of crocodiles, which have good hearing.

Distribution: Alberta to Texas, USA.
Classification: Theropoda, Tetanurae, Coelurosauria, Tyrannosauroidea.
Meaning of name: Tyrant lizard.
Named by: Osborn, 1905.
Time: Maastrichtian stage of the late Cretaceous.
Size: 12m (39ft).
Lifestyle: Hunter or scavenger.
Species: *T. rex*, although *Daspletosaurus*, *Gorgosaurus* and *Tarbosaurus* are sometimes regarded as species of *Tyrannosaurus*.

Tarbosaurus

Tyrannosaurid *Tarbosaurus* is the largest Asian predator known, a close cousin of *Tyrannosaurus*. Indeed some regard it as a species of *Tyrannosaurus*, named *T. bataar*. Three skeletons were found by a Russian expedition to the Nemegt Formation, in the Gobi Desert, in the 1940s. Since then there have been almost as many *Tarbosaurus* skeletons as *Tyrannosaurus* skeletons found.

Features: *Tarbosaurus* is very similar to *Tyrannosaurus*, but it is less heavily built. It has a larger head with a shallower snout and lower jaw, and slightly smaller teeth. The other differences are in minor points of the shape of the

individual skull bones. These features are slightly more primitive in *Tarbosaurus*, and so the early evolution may have taken place in Asia. Had it been found in North America, *Tarbosaurus* would have been regarded as a species of *Tyrannosaurus*.

Right: The two Tarbosaurus *skeletons found in Mongolia in the 1940s are currently mounted in the Palaeontological Institute of the Russian Academy of Sciences in Moscow, Russia.*

Distribution: China and Mongolia.
Classification: Theropoda, Tetanurae, Coelurosauria, Tyrannosauroidea.
Meaning of name: Alarming lizard.
Named by: Maleev, 1955.
Time: Maastrichtian stage of the late Cretaceous.
Size: 12m (39ft).
Lifestyle: Hunter or scavenger.
Species: *T. efremovi*, *T. bataar*.

Daspletosaurus

When C. M. Sternberg found the first skeleton of *Daspletosaurus* in 1921 he regarded it as a species of *Gorgosaurus*. However, it was revealed to be a much heavier animal. It is so similar to the slightly later *Tyrannosaurus* that it is sometimes regarded as its ancestor. Bonebeds of *Daspletosaurus* found in Montana, USA, suggest that it may have hunted in packs.

Features: The difference between *Daspletosaurus* and *Tyrannosaurus* is in the teeth. *Daspletosaurus* has even larger teeth than *Tyrannosaurus*. The neck and back are stockier, and the foot slightly shorter and heavier. Although slightly smaller, *Daspletosaurus* has a stronger build than *Tyrannosaurus*, and was much

Distribution: Alberta, Canada and Montana, USA.
Classification: Theropoda, Tetanurae, Coelurosauria, Tyrannosauroidea.
Meaning of name: Frightful lizard.
Named by: D. A. Russell, 1970.
Time: Campanian stage of the late Cretaceous.
Size: 9m (30ft).
Lifestyle: Hunter or scavenger.
Species: *D. torosus* and one other, unnamed.

Right: There have been approximately six good specimens of Daspletosaurus *found, as well as quite a few scattered remains. The best mounted skeleton is in the Royal Tyrrell Museum, in Alberta, Canada.*

stronger than its contemporary *Albertosaurus*. It possibly fought the big, slow-moving, horned dinosaurs, while its more agile relatives preyed on the swift-footed duckbills.

HUNTER OR SCAVENGER?

A long-running debate among scientists concerns the lifestyle of the tyrannosaurids, and *Tyrannosaurus* in particular. Was it the fearsome hunter, the terror of the Cretaceous plains and forests, as it has always been portrayed? Or was it a slow-moving animal, eking out a living as a scavenger, subsisting only on the corpses of animals that had already died or been killed by more active hunters?

Evidence for the former includes the position of the eyes. These pointed forward giving stereoscopic vision, essential for a hunter of swift prey. The nostrils contained turbinal bones, thin sheets of bone that would have carried moist, sensitive tissue. This would have enhanced the sense of smell, which would be useful for either a hunter or a scavenger.

On the other hand, a *Tyrannosaurus* was a very big animal, perhaps too big to be capable of much sustained speed or activity. Once up to speed, the slightest stumble would have resulted in a crash that might have been fatal. The *Tyrannosaurus* tooth marks found on ceratopsian bones show signs of flesh being scraped off a dead animal, rather than chunks bitten off a live one, but this is not to say that the ceratopsian had not been killed by the *Tyrannosaurus* in the first place. A bite mark in the backbone of a duckbill was made by a *Tyrannosaurus* on a living animal. It is, in fact, very likely that both lines of evidence are valid, and that *Tyrannosaurus* and its relatives were active hunters, but did not pass up the chance of devouring any corpse that they came across.

Gorgosaurus

Gorgosaurus is known from more than 20 skeletons, since the first was found and named by Lambe in 1914. In the 1970s, a study of the tyrannosaurids indicated that *Gorgosaurus* and *Albertosaurus* were the same, and the name *Gorgosaurus* was dropped. Another study in 1981 showed that they were different, however, and so the name was reinstated.

Features: The several skulls found have different numbers of teeth, but are thought to belong to animals of different growth stages. There is a pair of short horns above the eyes which appears in two types, one in which the horns point up and forwards, and another in which they are longer and much more horizontal. It may be that this indicates two sexes or more than one species.

Distribution: Alberta to New Mexico, USA.
Classification: Theropoda, Tetanurae, Coelurosauria, Tyrannosauroidea.
Meaning of name: Gorgon lizard.
Named by: Lambe, 1914.
Time: Campanian stage of the late Cretaceous.
Size: 9m (30ft).
Lifestyle: Hunter or scavenger.
Species: *G. libratus*.

Left: The original specimen of Gorgosaurus *consisted of a compete skeleton with a crushed skull. Subsequent finds have shown different numbers of teeth in the jaws, but this is probably due to the ages of the different individuals.*

BIG TITANOSAURIDS

As the late Cretaceous period dawned, the sauropods appeared to have had their day, their role as the principal herbivores having been taken over by the ornithopods. However, the surviving titanosaurid group, a branch of the sauropod family, flourished, spreading out mostly over the southern continents, with some examples on the northern lands. They developed into the biggest land animals known.

Andesaurus

This very big animal is known from only a handful of vertebrae and some leg bones. It lived in an area of river plains and enclosed lagoons, a landscape clothed by a thick covering of conifers and ferns, in what is now the Comahue region of Patagonia, in Argentina. The remains were found by Alejandro Delgado while swimming in a lake.

Features: *Andesaurus* must be one of the biggest sauropods that ever lived. The few tail bones that have been found have ball-and-socket joints, rather than the usual flat faces like other sauropods. This suggests strength and flexibility in the tail. It comes from the same period of history as another titanosaurid giant, *Argentinosaurus*, but there are enough differences to show that they are separate genera. The back

Distribution: Argentina.
Classification: Sauropoda, Macronaria, Titanosauria.
Meaning of name: Lizard of the Andes.
Named by: Calvo & Bonaparte, 1991.
Time: Albian stage of the Cretaceous.
Size: 40m (131ft).
Lifestyle: Browser.
Species: *A. delgadoi*.

Right: The beds in which the only specimen was found contain many iguanodont footprints. It seems that the ornithopods were the main herbivores of the area at that time, and the titanosaurids did not become really important until later.

vertebrae have high spines, giving the animal a tall ridge down its back. The specimen is incomplete, and consists of some vertebrae, the hind legs and a few ribs.

Paralititan

When *Paralititan* was discovered in 2000, it was seen to be the biggest dinosaur ever to have been found in Africa. Its discovery came as a surprise, as its presence in North Africa suggests that even in the late Cretaceous period there was some land link to the continent of South America, the place where the gigantic titanosaurids then flourished. It was found in rocks that were deposited in vegetated tidal flats and channels.

Distribution: Egypt.
Classification: Sauropoda, Macronaria, Titanosauria.
Meaning of name: Giant of the beaches.
Named by: Lamanna, Lacovara, Dodson, Smith, Poole, Giegengack and Attia, 2001.
Time: Albian or Cenomanian

Features:
Paralititan is very similar in its build to the huge titanosaurids of South America, Although the only specimen known consists of about 100 fragments of 16 different bones, we can judge the size of the beast by the fact that the upper arm bone, the humerus, is 1.69m (5½ft) long, the longest complete humerus known. The animal's weight has been estimated at a massive 70–80 tonnes. The bones are titanosaurid, and so its size can be calculated by comparing it with other titanosaurids.

Left: Paralititan's specific name honours Ernst Stromer, a German palaeontologist who worked in this area of Africa a century before Paralititan was found.

stage of the Cretaceous.
Size: 24–30m (79–98ft).
Lifestyle: Browser.
Species: *P. stromeri*.

THE TITANOSAURID SKELETON

Titanosaurids are known from very incomplete remains. Few titanosaurid skulls have been found but the skull fragments that we know show a wide, steeply sloping head and thin, peg-like teeth with tapering crowns. As a rule the neck is relatively short, the front legs are about three-quarters of the length of the back legs, and there is a shortish tail.

For all that the titanosaurids are the last group of sauropods to appear, the backbones are very primitive in aspect, lacking the deep hollows and the weight-bearing flanges and plates of the earlier groups. Likewise, the pelvis seems to have been different. It is fused to the backbone by six vertebrae rather than five as in the diplodocids. It is less robust than that of the earlier relatives and is much wider. This suggests that the living animals walked with a different gait from that of the diplodocids or the brachiosaurids. Titanosaurid footprints are easily recognized by their wide gait, as though they are spreading their legs further to the side than other sauropods. This contrasts with the narrow trackways of diplodocids and brachiosaurids that seem to show the animals putting the left foot just in front of the right foot in a rather mincing manner. Since titanosaurid trackways are known from middle Jurassic times, the group is older than it appears from the body fossils that have been preserved.

Left: Prints of a diplodocid (above), titanosaurid (below).

"Bruhathkayosaurus"

When the remains of this dinosaur were first discovered in the 1980s it was thought they belonged to some kind of monstrous theropod. However, Chatterjee reclassified it as a titanosaurid in 1995. All in all, it is something of a *nomen dubium*, but the remains are important in belonging to the biggest animal found in India so far.

Features: Despite the difficulties in actually identifying this huge animal, it seems that the tibia is about 25 per cent longer than that of South American *Argentinosaurus*. This would make it far bigger than *Argentinosaurus*, currently accepted as the biggest dinosaur. However, the uncertainty about the identity of *'Bruhathkayosaurus'* keeps it out of the record books. Its classification as a sauropod is based on the fact that nothing else is as big.

Distribution: Tamilnadu, India.
Classification: Sauropoda, Macronaria, Titanosauria.
Meaning of name: Huge body lizard.
Named by: Yadagiri and Ayyasami, 1989.
Time: Maastrichtian stage of the late Cretaceous.
Size: 40m (131ft).
Lifestyle: Browser.
Species: "*B.*" *matleyi*.

Antarctosaurus

The original *Antarctosaurus* skeleton found is very complete for such a big animal, and consists of the skull and jaws, the shoulders and parts of the legs and hips. There is some doubt as to whether all this belonged to the same animal. Although it was found in South America, questionable material referred to *Antarctosaurus* has also been unearthed in India and Africa.

Features: *Antarctosaurus* has slim legs for the size of animal, and a small head with big eyes and a broad snout with only a few peg-shaped teeth at the front. The jaws are rather squared off at the front, like *Bonitasaura*, suggesting a low browsing or grazing feeding habit. Although it is one of the better-known southern sauropods, there is still a great deal of confusion about whether all the pieces belong to the same beast. One specimen has a thigh bone that measures 2.3m (7ft 8in), one of the biggest dinosaur bones ever found.

Distribution: South America.
Classification: Sauropoda, Macronaria, Titanosauria.
Meaning of name: Southern lizard.
Named by: von Huene, 1929.
Time: Campanian to Maastrichtian stages of the late Cretaceous.
Size: 40m (131ft).
Lifestyle: Browser.
Species: *A. wichmannianus*, *A. jaxarticus*, *A. brasiliensis*.

Right: The squared-off jaw appears in several other titanosaurids and also in unrelated sauropods. It, and the peg-shaped teeth, would have been used for feeding indiscriminately on groundcover.

MISCELLANEOUS TITANOSAURIDS

As with other widespread animal groups, there were many different types of titanosaurid in different parts of the world. They ranged from the true titanosaurids of South America to the dwarf species of the European islands. Although a complete skull was not identified until recently, the variation of shapes in isolated skull bones suggests that different titanosaurids had different head shapes.

Hypselosaurus

Distribution: France and Spain.
Classification: Sauropoda, Macronaria, Titanosauria.
Meaning of name: High ridge lizard.
Named by: Matheron, 1869.
Time: Maastrichtian stage of the late Cretaceous.
Size: 12m (39ft).
Lifestyle: Browser.
Species: *H. priscus*.

The dinosaur *Hypselosaurus* is known from the scattered remains of at least ten individuals. Fossilized eggs, approximately 30cm (12in) in diameter and lying in groups of five, found near Aix in southern France (eggs-en-Provence, as some wit put it), have been attributed to *Hypselosaurus*, although this has not been scientifically confirmed. Another theory is that the eggs actually belong to a contemporary flightless bird, *Gargantuavis*.

Right: Hypselosaurus' eggs are spherical, more than twice the size of ostrich eggs and have a volume of 2 litres (½ gallon).

Features: It is difficult to restore *Hypselosaurus*. It is a large, four-footed, long-necked plant-eater, of a typical sauropod shape, and is known only from disarticulated remains. Comparing it with other titanosaurids, it seems to have had more robust limbs than its relatives. The teeth are weak and peg-shaped. We do not know whether it was covered in armour like some titanosaurids.

Magyarosaurus

The smallest known adult sauropod, *Magyarosaurus*, was found in Romania and Hungary. In late Cretaceous times this area of Europe was an island chain, and it seems likely that dwarf dinosaurs evolved on these islands to make the best use of limited food supplies. Other contemporary dwarf dinosaurs from this area include the duckbill, *Telmatosaurus*, and the ankylosaur, *Struthiosaurus*.

Below: Ampelosaurus, a relative of Magyarosaurus.

Features: *Magyarosaurus* is probably related to *Ampelosaurus*, and is probably one of the armoured forms. Its unusual stature makes it difficult to classify. Some scientists believe it to be a small species of '*Titanosaurus*', or even of *Hypselosaurus*, rather than a genus in its own right. In fact, the several specimens known are not consistent, and there are those with slender humeri and those with robust humeri. This may be a sexual difference, or there may be more than one genus of dwarf sauropod represented.

Distribution: Romania, Hungary.
Classification: Sauropoda, Macronaria, Titanosauria.
Meaning of name: Lizard of the Magyars, an ancient European tribe.
Named by: von Huene, 1932.
Time: Maastrichtian stage of the late Cretaceous.
Size: 6m (19½ft).
Lifestyle: Low browser.
Species: *M. dacus*, *M. transylvanicus*.

Gondwanatitan

The single, partial skeleton known of *Gondwanatitan* consists of bones from all over the body. It was discovered in the mid-1980s by Fausto L. de Souza Cunha, palaeontologist of the National Museum of Rio De Janeiro, and researcher Jose Suarez. It lived by the sides of lakes, marshes and rivers that were common in central Brazil in late Cretaceous times.

Features: *Gondwanatitan* is a relatively small and lightly built titanosaurid. The skeletal features, particularly the articulation of the tail vertebrae, show it to have been quite different from, and more highly evolved than, any of the better known of the titanosaurids. The tibia is straight whereas in other titanosaurids it is curved. The spines on the tail bones point forward, suggesting that it is most closely related to *Aeolosaurus*.

Distribution: São Paulo state, Brazil.
Classification: Sauropoda, Macronaria, Titanosauria.
Meaning of name: Giant from Gondwana.
Named by: Kellner and de Azevedo, 1999.
Time: Maastrichtian stage of the late Cretaceous.
Size: 8m (26ft).
Lifestyle: Browser.
Species: *G. faustoi*.

Left: Gondwanatitan is named after the supercontinent Gondwana, which encompassed the whole landmass of the Southern Hemisphere at the time that this dinosaur lived.

DWARF SPECIES

The flora and fauna of islands have always evolved differently from those of mainland areas in response to local conditions. One of the results of this is dwarfism, the development of small versions of animals found elsewhere. During the last Ice Age, elephants evolved into dwarf forms that roamed the rugged outcrops of Malta. Miniature mammoths existed on islands off the coast of California, USA. On Caribbean islands the 4m (13ft)-high South American giant ground sloth evolved into a form that was no bigger than a domestic cat. Modern equivalents include the Shetland pony of the Scottish islands. Until a few thousand years ago a dwarf human, *Homo floresiensis*, inhabited the island of Flores in Indonesia.

In these circumstances there may be no predators. As there is only a limited amount of food in such areas, large populations of small individuals make the best use of the resources. Miniature dinosaurs roamed the island chains that lay across the shallow seas of late Cretaceous Europe.

Below: Size comparison of a dwarf species.

Aeolosaurus

The dinosaur *Aeolosaurus* was discovered in 1987 when much of the skeleton, including the backbone and leg bones, was found. However, an additional find in 1993 by Salgado and Coria produced evidence of back armour. It seems likely that *Aeolosaurus* inhabited the swampy lowlands and coastal plains of late Cretaceous Argentina, while other sauropods fed on the vegetation of the surrounding uplands. Eggs were also found associated with some of the bones.

Distribution: Patagonia.
Classification: Sauropoda, Macronaria, Titanosauria.
Meaning of name: Windy lizard (from the windswept plains of Patagonia).
Named by: Powell, 1987.
Time: Campanian or Maastrichtian stage of the late Cretaceous.
Size: 15m (49ft).
Lifestyle: High browser.
Species: *A. rionegrinus*.

Features: *Aeolosaurus* is known from many skeletal parts, including pieces of armour about 15cm (6in) in diameter. The forward-pointing spines on the tail vertebrae close to the hips are regarded as proof of its ability to rise on its hind legs, propped by its tail, to feed from the high branches of conifers. Other sauropods would have been able to do this.

MORE MISCELLANEOUS TITANOSAURIDS

Titanosaurids belonged to the sauropod group. In general, although sauropods were more typical of Jurassic and early Cretaceous landscapes, the dinosaur Alamosaurus, a titanosaurid, was among the last of the dinosaurs to have existed at the end of the Age of Dinosaurs. In all their era spanned 150 million years from the end of the Triassic to the end of the Cretaceous.

Bonitasaura

When found, the remains of *Bonitasaura* were thought to have been a mixture of a titanosaurid skeleton and that of a diplodocid. The shape of the skull, with its broad-fronted jaws, is extremely similar to that of the diplodocid, *Nigersaurus*. This indicates a rapid evolutionary expansion of the titanosaurids once the diplodocids, their evolutionary cousins, had died out, with the former taking over all the niches of the latter.

Features: The amazing feature of this titanosaurid is the skull. The front of the jaw is broad and square, evidently adapted to cropping plants near the ground. The front teeth are short and pencil-like, rather like those of a diplodocid. Directly behind them, where the skull narrows, the jaws are equipped with horny blades evolved for slicing vegetation. Apart from this, the rest of the animal is similar to its relative, *Antarctosaurus*. The species is named to honour Leonardo Salgado, the famous Argentinian sauropod specialist.

Distribution: Patagonia.
Classification: Sauropoda, Macronaria, Titanosauria.
Meaning of name: Female lizard from La Bonita Hill, where it was discovered.
Named by: Apesteguia, 2004.
Time: Maastrichtian stage of the late Cretaceous.
Size: 9m (30ft).
Lifestyle: Browser.
Species: *B. salgadoi*.

OTHER TITANOSAURIDS

The titanosaurid *Pellegrinisaurus*, and other titanosaurids that were once assigned to the now defunct genus "*Titanosaurus*", probably inhabited the hilly areas of the South American island continent. They fed on the upland flora of conifers and ferns, and were menaced by the theropod *Abelisaurus*, while other titanosaurids such as *Aeolosaurus* and the relatively few ornithopods present lived on the nearby coastal lowland plains.

Since *Pellegrinisaurus* is known from a single specimen consisting of part of a backbone and a femur, little can be said about its appearance. It was first thought to have been a species of *Epachthosaurus* when it was discovered in 1975, but work 20 years later showed that it was sufficiently different to have been a separate genus.

NESTING IN PATAGONIA

One of the most spectacular dinosaur nesting sites in the world was discovered, and studied, in the late 1990s in Campanian rocks in Patagonia. Called Auca Mahuevo, the site stretches for several square kilometres. It was used regularly by hundreds or thousands of dinosaurs that returned to the site each year. Each kicked out a shallow hole, laid the eggs in clusters and covered them with vegetation.

The site represents the floodplain of a river that periodically overflowed its banks. Mud deposited each time it flooded buried and suffocated the eggs that were exposed there, and preserved them. Eggs discovered in subsequent layers showed that dinosaurs were not discouraged by random disasters.

Each egg is almost spherical, measuring 13–11.5cm (5–4½in), and the best examples preserve the structure of the eggshell, the imprint of the internal membrane and even the baby complete with lizard-like skin. It is clear that the eggs were laid by titanosaurids, but the exact genus has not been identified. The embryonic skin lacks armour, which suggests that it developed after the animal hatched and was growing, as in modern armoured lizards and crocodiles.

Left: Bonitasaurus.

Alamosaurus

After a gap of 35–40 million years when there were no sauropods in North America, the genus returned with *Alamosaurus*. Until the discovery of an as yet unnamed titanosaurid in the late Cretaceous period of Chihuahua Province in Mexico in 2002, this was the sole representative of the group in North America. It migrated from South America across the land-bridge of Central America established in late Cretaceous times.

Features: *Alamosaurus* is larger, but more lightly built than its close relative *Saltasaurus*, and does not seem to have possessed armour. Several partial skeletons are known, from various parts of North America. A site in Texas has produced the remains of an adult and two well-grown juveniles, suggesting that it lived in a family structure. A study has suggested that the total population of *Alamosaurus* in Texas at any one time would have been about 350,000. That would represent a density of one per 2km².

Distribution: New Mexico, Utah and Texas, USA.
Classification: Sauropoda, Macronaria, Titanosauria.
Meaning of name: Lizard from the Ojo Alamo Sandstone (alamo being the local Spanish name for the cottonwood tree).
Named by: Gilmore, 1922.
Time: Maastrichtian stage of the late Cretaceous.
Size: 21m (69ft).
Lifestyle: Browser.
Species: *A. sanjuanensis.*

Left: Remains of Alamosaurus have been found in the latest Cretaceous rocks from New Mexico to Utah, USA. However none has been found in Alaska, where similar aged deposits yield other dinosaurs. Perhaps climate was a factor in this distribution.

Epachthosaurus

When it was found, *Epachthosaurus* was thought to have been buried in Cenomanian deposits, from the base of the late Cretaceous. This fitted well with the primitive aspect of the bones. However, later studies showed it to have come from the end of the period, making it a primitive throwback to an earlier type. Primitive forms must have existed alongside the more advanced.

Features: This primitive titanosaurid shares its features with many earlier types. It differs from other titanosaurids in the structure of the back, the shape of the articulations between the vertebrae, and the presence of a very strong linkage between the hip and the backbone giving it a very strong back structure. The single skeleton known is almost complete except for the neck and head, and the tip of the tail, but shows no sign of armour plates. The join between the hips and the backbone consists of six vertebrae that are completely fused.

Right: An articulated skeleton of Epachthosaurus has been found, with its hind legs crumpled beneath it and its front legs splayed out. It lay on its stomach in death.

Distribution: Argentina.
Classification: Sauropoda, Macronaria, Titanosauria.
Meaning of name: Heavy lizard.
Named by: J. E. Powell, 1990.
Time: Maastrichtian stage of the late Cretaceous.
Size: 15–20m (49–65½ft).
Lifestyle: Browser.
Species: *E. sciuttoi.*

SMALL ORNITHOPODS

The late Cretaceous period was the time of the big ornithopods. However, there were still many smaller types that scampered around the feet of the giants. Some retained their primitive features from earlier times, but some were quite advanced. Their narrow beaks were able to select food from the newly evolved flowering plants on the ground.

Thescelosaurus

The classification of *Thescelosaurus*, known from complete and articulated remains, has been moved between the hypsilophodontids and the iguanodontids for most of the last century, and the authorities are still not sure where it sits. One specimen, nicknamed Willo, has the remains of the internal organs, including the heart, preserved. Its rather primitive structure is surprising considering it is among the last of the dinosaurs to have lived.

Above: The specific name, Thescelosaurus neglectus, refers to the fact that the skeleton was not studied for 22 years after it was discovered and collected.

Features: *Thescelosaurus* is more heavily built than other medium-sized ornithopods, and the legs are relatively shorter. This is not an animal built for speed. As with its relatives, it has a beak at the front of the mouth but an unusual arrangement of three different kinds of teeth – sharp in the front, canine-like at the side and molar-like grinding teeth at the back. The four-toed foot is one of the primitive features of *Thescelosaurus*.

Distribution: Alberta, Canada, Saskatchewan, Colorado, Montana, South Dakota and Wyoming, USA.
Classification: Ornithopoda.
Meaning of name: Wonderful lizard.
Named by: Gilmore, 1913.
Time: Campanian to Maastrichtian stages of the late Cretaceous.
Size: 4m (13ft).
Lifestyle: Browser.
Species: *T. neglectus.* Several others, all unnamed.

Orodromeus

Skeletons of this small ornithopod were found in Montana with the remains of dinosaur eggs and nests at a site known as Egg Mountain. For some time it was believed that *Orodromeus* was the builder of the nests, but later it was found that they actually belonged to the meat-eater *Troodon* and that *Orodromeus* represented the prey.

Features: This is a fairly primitive small ornithopod. Its legs show it to have been a speedy animal, hence its name. The head is small, with a beak and cheek pouches. The neck is long and flexible, and the tail is stiffened by bony rods that helped to keep it straight for balance while running. There are three toes on the hind foot, and four fingers on the hand.

Distribution: Montana, USA.
Classification: Ornithopoda.
Meaning of name: Mountain runner.
Named by: Horner and Weishampel, 1988.
Time: Campanian stage of the late Cretaceous.
Size: 2.5m (8ft).
Lifestyle: Low browser.
Species: *O. makelai.*

Left: Fossil finds of Orodromeus suggest that it lived in small groups.

Parksosaurus

This is known from an incomplete skeleton and a skull. It was originally thought by Parks to have been a species of *Thescelosaurus* when he studied it in 1927. However, when Sternberg studied the remains ten years later, he found that it was quite different. For a long time it was regarded as a hypsilophodont, but is now thought, once more, to be closer to *Thescelosaurus*. An extra bone lies above the shoulder blade, an unusual feature.

Features: In appearance *Parksosaurus* is very similar to other small ornithopods, such as *Hypsilophodon*. There are, however, a number of differences in the skull, and the tail carries a system of tendons that have turned to bone, presumably to make it stiff and straight. There is also an extra bone in the shoulder girdle, which is still puzzling the palaeontologists. Like *Hypsilophodon*, it is built as a fast runner, using speed to escape enemies.

Left: The only specimen of this dinosaur found is of the left-hand side of the skeleton. On death it sank into river mud, and the exposed bones of the right-hand side were broken up and washed away.

Distribution: Alberta, Canada.
Classification: Ornithopoda.
Meaning of name: William Parks' lizard.
Named by: Sternberg, 1937.
Time: Maastrichtian stage of the late Cretaceous.
Size: 2.4m (8ft).
Lifestyle: Browser.
Species: *P. warrenae*.

DISCOVERY OF THE HEART

The internal organs of the *Thescelosaurus* specimen nicknamed Willo, named after the wife of the rancher on whose property it was found, consist of cartilaginous ribs and plates, tendons attached to the vertebrae, and most importantly, the heart.

The grapefruit-size lump in the chest was suspected of being the heart but there was a great deal of doubt. Finally it was scanned using a medical CAT scanning device, which is used in hospitals to build up a three-dimensional image of a patient's internal organs, and this proved without a doubt that it was a heart. The heart has four chambers, a double pump and a single aorta, making it more like that of a bird or mammal than that of a reptile, and suggests a very high metabolic rate. The work was done at North Carolina State University and the North Carolina Museum of Natural Sciences.

Below: The heart on the left is that of a crocodile, while that on the right is that of a dinosaur.

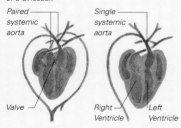

Paired systemic aorta

Single systemic aorta

Valve

Right Ventricle

Left Ventricle

Gasparinisaura

Ornithopods are rare in South America when compared with the sauropods, and so the discovery of *Gasparinisaura* was quite important. This small plant-eater is known from a complete juvenile specimen and about 15 other incomplete skeletons. It is classed as a small iguanodontid, but recent studies show that it may have been closer to the hypsilophodontids.

Features: As in the hypsilophodontids, *Gasparinisaura* has very narrow front bones in the hips, which suggest that it may be related to this group, but the feature may have developed independently. It is a rather dainty-looking dinosaur, but with quite sturdy legs for the size of animal and very small arms. The head is short, ending in a beak, and the mouth contains diamond-shape chopping teeth. The eyes are large, but this may be because the specimen is a juvenile.

Distribution: Neuquen Province, Argentina.
Classification: Ornithopoda, Iguanodontia.
Meaning of name: Dr Zulma Gasparini's lizard.
Named by: Coria and Saldago, 1996.
Time: Coniacian to Santonian stages of the late Cretaceous.
Size: 0.8m (2½ft), but this may be a juvenile.

Lifestyle: Low browser.
Species: *G. cincosaltensis*.

Left: The discovery of parts of several individuals in a small area suggests that Gasparinisaura *may have been gregarious.*

THE IGUANODONTID-HADROSAURID CONNECTION

The most important plant-eaters in the Northern Hemisphere at the end of the dinosaur era were the hadrosaurids, or duck-billed dinosaurs. They evolved from iguanodontids halfway through the Cretaceous. At the time many ornithopods had features from both groups, suggesting a transition stage.

Telmatosaurus

We know *Telmatosaurus* from several fragmentary skulls from individuals of different ages, as well as pieces of the rest of the skeleton. There are also clutches of two to four eggs that have been attributed to *Telmatosaurus*. It is one of the few hadrosaurids to have been found in Europe. Its relative smallness may be due to the fact that it lived on islands.

Features: What we know of *Telmatosaurus* suggests that it was very similar to *Iguanodon*, despite the fact that it occured at the end of the Cretaceous period. Like *Equijubus*, the appearance of its skeleton supports the idea that the hadrosaurids evolved from the iguandontid line.
It has a deep skull like *Gryposaurus*, but lacks the duck-like bill possessed by other members of the group; instead it has a long, drawn-out snout.

Right: Telmatosaurus *is very primitive, despite its late appearance, which suggests that its island habitat acted as a refugium – an isolated area where animals continued to exist despite dying out elsewhere.*

Distribution: Romania, France and Spain.
Classification: Ornithopoda, Iguanodontia, Hadrosauroidea.
Meaning of name: Marsh lizard.
Named by: Nopcsa, 1899.
Time: Maastrichtian stage of the late Cretaceous.
Size: 5m (16½ft).
Lifestyle: Browser.
Species: *T. transylvanicus*, *T. cantabrigiensis*.

Gilmoreosaurus

One of the earliest Asian hadrosaurids, *Gilmoreosaurus* was originally thought to have been a species of *Mandschurosaurus*. The confusion is typical of its history. Fossils that were originally thought to be from *Bactrosaurus* now seem to be *Gilmoreosaurus*. Another genus, *Cionodon*, from Asia, is now regarded as *Gilmoreosaurus* as well.

Features: *Gilmoreosaurus* is quite a lightweight animal for its size, but has particularly strong legs. Its feet are rather like those of *Iguanodon*, emphasizing the close relationship between the two groups. As with other medium-size ornithopods it had the ability to walk either on its hind legs, balanced by the heavy tail, or on all fours, taking the weight of its forequarters on its strengthened fingertips.

Distribution: Mongolia.
Classification: Ornithopoda, Iguanodontia, Hadrosauroidea.
Meaning of name: Charles Whitney Gilmore's lizard.
Named by: Brett-Surman, 1979.
Time: Cenomanian to Maastrichtian stages of the late Cretaceous.
Size: 8m (26ft).
Lifestyle: Browser.
Species: *G. mongoliensis*, *G. atavus*, '*G. arkhangelskyi*'.

Left: Although it was found in 1923, and partially studied in 1933, Gilmoreosaurus *was not properly studied until 1979. One species,* G. arkhangelskyi, *is a* nomen dubium, *as the partial skeleton seems to be made of different animals.*

Equijubus

Although *Equijubus* was found in late early Cretaceous beds, it is included here because of its significance in hadrosaurid evolution. It consists of a complete skull with an articulated jaw, combining features found in the iguanodontids and the hadrosaurids. Its presence suggests that the hadrosaurids evolved from the iguanodontids somewhere in Asia at the end of the early Cretaceous, or the beginning of the late Cretaceous.

Features: The significance of *Equijubus* is that it had the long jaws of the iguanodontids, but the elaborate tooth arrangement consisting of a large number of small, tightly packed grinding teeth in very mobile jaws seen in the hadrosaurids. This shows a transition between the two groups, but *Equijubus* is

Distribution: Gansu Province, China.
Classification: Ornithopoda, Iguanodontia, Hadrosauroidea.
Meaning of name: Horse's mane (local name of the site of discovery).
Named by: You, Luo, Shubin, Witmer, Zhi-lu Tang and Feng Tang, 2003.
Time: Albian stage.
Size: 5m (16½ft).
Lifestyle: Browser.
Species: *E. normanii*.

Right: The genus name Equijubus *is derived from the Chinese translation of Ma Zong, the name of the mountain range where this dinosaur was found. The specific name comes from David Norman (the British expert on iguanodontids).*

regarded as the earliest and most primitive known of the hadrosaurids. The rest of the skeleton could have come from the group of ornithopod dinosaurs.

EVOLUTIONARY PROGRESS

The hadrosaurids were the most abundant of the plant-eaters at the end of the age of dinosaurs. They evolved with the development of flowering plants, and may have diversified in response to the spread of this new food source.

There is a theory, proposed by Bob Bakker in the 1980s, that flowering plants evolved in response to heavy grazing by hordes of herbivorous dinosaurs. If low-growing plants were being eaten at a great rate, then evolution would favour the ability of a plant to regenerate itself quickly, and hence the seed of a flowering plant, with its self-contained food supply feeding its fertilized embryo, became more efficient than the spore of the fern which is cast off and fertilized by a hit-and-miss lottery.

The shape of a hadrosaurid neck is consistent with a low feeder. The distinctive S-shape is very similar to that of a modern horse or a buffalo, bringing the snout and the mouth down to where the low plants are growing. There is a suggestion that the traditional restoration, with the snaky neck, is wrong. Perhaps the curve of the neck was filled with muscle, like that of a horse. The two types of hadrosaurid, the lambeosaurines and the hadrosaurines, have differently-shaped duckbills that must reflect a different feeding strategy. The narrow-beaked lambeosaurines would be selective about what plants they ate, while the broad-beaked hadrosaurines would have taken great mouthfuls indiscriminately. They would both also have been able to rear up on their hind legs to scrape leaves from trees. Neither type would have eaten grass, which did not evolve significantly until after the dinosaurs were extinct.

Protohadros

This dinosaur is based on a skull and some bones from the rest of the body. When it was found in Texas, USA, by Gary Bird, it was hailed as the most primitive hadrosaurid yet discovered, even though it was not the earliest, hence its name. Scientific opinion has changed and it is now regarded as a very specialized member of the iguanodontid group.

Features: The lower jaw of *Protohadros* is very large. The snout is turned down at the front, which suggests a habit of grazing low-growing vegetation, rather than browsing from bushes or overhanging branches. It does not have the flexible mouth that produced the food-grinding action seen in either iguanodontids or hadrosaurids. Its food would have been the aquatic plants of the delta streams in its habitat, scooped up by the broad, down-turned mouth.

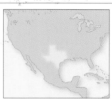

Distribution: Texas, USA.
Classification: Ornithopoda, Iguanodontia.
Meaning of name: First hadrosaurid.
Named by: Head, 1998.
Time: Cenomanian stage of the late Cretaceous.
Size: 6m (19½ft).
Lifestyle: Low browser.
Species: *P. byrdi*.

Right: The location of the Protohadros discovery in Texas, USA, confounded the accepted idea that the hadrosaurids evolved in Asia. However, its reclassification as an iguanodontid has removed this ambiguity.

LAMBEOSAURINE HADROSAURIDS

The lambeosaurine hadrosaurids were those with ornate hollow crests on their heads. Each genus could be distinguished from the next by the shape of this crest, and the crests would have been brightly coloured to enhance this function. Their duck-like bills were narrower than those of the other group of hadrosaurids – the hadrosaurines.

Tsintaosaurus

The characteristic crest of *Tsintaosaurus* has always been a bit of a mystery, being so unlike the crests of other hadrosaurids. At times the crest was thought to have been just the way the skull was preserved, with a sliver of bone out of place, and that it was just a damaged specimen of *Tanius*. Later discoveries of more specimens show that the crest did actually exist.

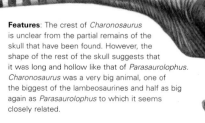

Features: The obvious feature of *Tsintaosaurus* is the thin, hollow crest jutting directly upward from the middle of the head, like the horn of a unicorn. It may have been a display feature, but it is also possible that the crest supported a flap of skin that could be inflated with air from the nostrils, to make some kind of a sound signal. This would make sense in view of the resonating chambers that have been found inside the skull.

Left: The name Tsintaosaurus *is an anglicized version of Ch'ing-tao, or Qingdao, which means "green island". The species name denotes the fact that it had a spine on its nose.*

Distribution: Shandong Province, China.
Classification: Ornithopoda, Iguanodontia, Hadrosauroidea, Lambeosaurine.
Meaning of name: Lizard from Tsintao, the city close to the place where it was found.
Named by: Young, 1958.
Time: Campanian to Maastrichtian stages of the late Cretaceous.
Size: 10m (36ft).
Lifestyle: Browser.
Species: *T. spinorhinus.*

Charonosaurus

Charonosaurus comes from late Maastrichtian deposits, the Yuliangze Formation at Jiayin, China, at the end of the Age of Dinosaurs and after most of the other lambeosaurines had died out. The best specimen is a partial skull, but many other pieces of bone were found in a bonebed suggesting that this animal lived in big herds close to a river that flooded frequently.

Distribution: China.
Classification: Ornithopoda, Iguanodontia, Hadrosauroidea, Lambeosaurine.
Meaning of name: Charon's lizard (after the ferryman in the underworld of ancient Greek legend).
Named by: Godefroit, Zan and Jin, 2000.
Time: Maastrichtian stage of the late Cretaceous.
Size: 13m (42½ft).
Lifestyle: Browser.
Species: *T. jiayinensis.*

Features: The crest of *Charonosaurus* is unclear from the partial remains of the skull that have been found. However, the shape of the rest of the skull suggests that it was long and hollow like that of *Parasaurolophus*. *Charonosaurus* was a very big animal, one of the biggest of the lambeosaurines and half as big again as *Parasaurolophus* to which it seems closely related.

Right: The long forelimbs suggest that Charonosaurus *was habitually a quadruped, but it would have moved about on its hind legs from time to time.*

Parasaurolophus

Probably the most familiar lambeosaurine hadrosaurid, and the one with the most flamboyant crest, was *Parasaurolophus*. Three species are recognized, and each has a different curve of the crest. Different crest sizes may indicate different sexes. Computer studies in the New Mexico Museum of Natural History and Science have produced a deep trombone-like sound that would have been produced by *Parasaurolphus* snorting through its crest.

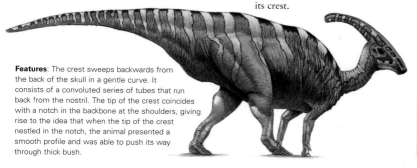

Features: The crest sweeps backwards from the back of the skull in a gentle curve. It consists of a convoluted series of tubes that run back from the nostril. The tip of the crest coincides with a notch in the backbone at the shoulders, giving rise to the idea that when the tip of the crest nestled in the notch, the animal presented a smooth profile and was able to push its way through thick bush.

Distribution: New Mexico, USA, to Alberta, Canada.
Classification: Ornithopoda, Iguanodontia, Hadrosauroidea, Lambeosaurinae.
Meaning of name: Almost Saurolophus.
Named by: Parks, 1922.
Time: Campanian to Maastrichtian stages of the late Cretaceous.
Size: 10m (33ft).
Lifestyle: Browser.
Species: *P. walkeri, P. cyrtocristatus, P. tibicen.*

Lambeosaurus

This is a well-known dinosaur, and gives its name to the whole group. Its remains were discovered in 1889, but it was not recognized as a distinct genus until 1923. More than 20 fossils have been found. The wide geographical range of the finds suggests that it lived all along the western shore of the late Cretaceous inland sea of North America.

Features: The hollow crest on the top of the head is in the shape of an axe, with a squarish blade sticking up and a shaft pointing backwards. The square portion is the hollow part with the convoluted nasal passages, while the spike is solid. The crest of the larger species, *L. magnicristatus*, has a larger hollow portion, bigger than the skull itself, and a very small spike. The skin is thin and covered in small polygonal scales.

Distribution: Alberta, Canada, and Montana to New Mexico, USA.
Classification: Ornithopoda, Iguanodontia, Hadrosauroidea, Lambeosaurinae.
Meaning of name: Lawrence M. Lambe's lizard.
Named by: Parks, 1923.
Time: Campanian stage of the late Cretaceous.
Size: 9–15m (30–49ft).
Lifestyle: Browser.
Species: *L. lambei, L. laticaudus, L. magnicristatus.*

MORE LAMBEOSAURINE HADROSAURIDS

We know that lambeosaurines spread across the whole of the northern continent, since their remains have been found in Russia, China, Canada and the USA. This shows that for at least some of the late Cretaceous period, the northern continents were continuous across the area where the Bering Strait now separates them.

Hypacrosaurus

Several good specimens of *Hypacrosaurus* have been found, the first by famous American dinosaur hunter Barnum Brown in 1910. Subsequent discoveries of another species of *Hypacrosaurus* included eggs and youngsters in various stages of growth, allowing palaeontologists to chart the growth pattern and family life of the animal.

Features: Tall spines along the backbone gave *Hypacrosaurus* a deep ridge, or a low fin, along its back, which was probably used for display. Alternatively they may have supported a fatty hump used as a food store for lean seasons, as with modern camels. The crest is shorter than that of other lambeosaurines. It is semicircular like that of *Corythosaurus* but smaller and thicker, and it has a short spike jutting backwards.

Distribution: Alberta, Canada, and Montana, USA.
Classification: Ornithopoda, Iguanodontia, Hadrosauroidea, Lambeosaurinae.
Meaning of name: Below the top lizard.
Named by: Brown, 1913.
Time: Maastrichtian stage of the late Cretaceous.
Size: 9m (30ft).
Lifestyle: Browser.
Species: *H. stebingeri*, *H. altispinus*.

Left: Hypacrosaurus is the most primitive known of the lambeosaurine hadrosaurids.

Corythosaurus

One of the best-known hadrosaurids is *Corythosaurus*. The remains of more than 20 individuals are known, many with complete skulls and some with pebbly skin impressions. The abundance of the remains suggest that *Corythosaurus* travelled in groups. As in other hadrosaurids, it would have spent much of its time on all fours, but could also run on its hind legs.

Below: The scaly texture of the skin is known, and is similar to that of other hadrosaurids. However, there is evidence of three rows of broad scales along the belly. This may have protected its underside while it moved through thick undergrowth.

Distribution: Alberta, Canada, and Montana, USA.
Classification: Ornithopoda, Iguanodontia, Hadrosauroidea, Lambeosaurinae.
Meaning of name: Corinthian lizard (referring to the crest, which is shaped like a Corinthian helmet).
Named by: Brown, 1914.
Time: Campanian stage of the late Cretaceous.
Size: 10m (33ft).
Lifestyle: Browser.
Species: *C. casuarius*.

Features: The crest of *Corythosaurus* is distinctive, being half a disc, about the size of a dinner plate. As in other lambeosaurines, the nasal passages within this disc are convoluted and are connected to the nostrils. The part of the brain dealing with the sense of smell is also close to the crest. The crest is in two different sizes, with the smaller crest probably belonging to the female of the species.

LAMBEOSAURINE CRESTS

The flamboyant crests of the labeosaurines consisted of hollow tubes that were connected to the nasal passages. It is possible that their function was to modify the air that was breathed. If the tubes were lined with moist membrane (something that could never be proved), cold dry air would be warmed and moistened before travelling to the lungs.

In some instances the part of the brain that dealt with the sense of smell was located close to the crest. This suggests that the huge volume of nasal passage would have had something to do with this sense, helping the animal to detect slight variations in scents in the air. A third possibility is that the crest was a communication device. Air snorted through these tubes would have made distinctive honking noises like the sound of a bass trombone. Such distinctive sounds would carry for long distances through forests and across swamps, and would be a valuable means of keeping a herd together. There is no reason why all these functions could not have been combined.

Left: The different Lambeosaurine crests.

Olorotitan

Excavated in 1999–2000 on the banks of the Amur River in far eastern Russia, this is the most complete Russian dinosaur skeleton ever found, and the most complete lambeosaurine outside North America. The skeleton, including the head with the spectacular crest, is now on display in the Amur Natural History Museum in Blagoveschensk, Russia. It is very closely related to the North American *Corythosaurus* and *Hypacrosaurus*.

Features: The crest is huge, sweeping up and back, with a hatchet-like blade on its rear edge. This presumably carried some kind of skin ornamentation. The neck is longer than that of other lambeosaurine hadrosaurids, and so is the pelvis. The tail is even more rigid than that of other hadrosaurids, but that may be due to disease affecting the tail vertebrae of the one individual that has been found.

Below: Despite being very closely related to the North American Corythosaurus *and* Hypacrosaurus, Olorotitan *is different in a number of ways. The neck is longer than in the other genera, and the joints of the tail show that the tail was even more rigid.*

Distribution: Kundur, Russia.
Classification: Ornithopoda, Iguanodontia, Hadrosauroidea, Lambeosaurinae.
Meaning of name: Giant swan.
Named by: Godefroit, Bolotsky and Alifanov, 2003.
Time: Maastrichtian stage of the late Cretaceous.
Size: 12m (39ft).
Lifestyle: Browser.
Species: *O. arharensis*.

CRESTED HADROSAURINE HADROSAURIDS

The hadrosaurine hadrosaurids lacked the flamboyant hollow crests of the lambeosaurines.
Instead they either had flat, crestless heads or heads ornamented with bony bumps or solid crests
formed of spikes of bone. They also tended to have broader jaws, and longer and slimmer limbs
than their hollow-crested relatives.

Aralosaurus

This Asian hadrosaurine hadrosaurid is known from the back half of an articulated skull with the distinctive nasal bone from a subadult specimen, along with some limb bones and vertebrae from several other individuals unearthed from the same deposits. Despite its current classification, it may have been a lambeosaurine.

Features: The skull of *Aralosaurus* has a peculiar arched area on the nasal region, swept back to a kind of point just in front of its eyes, suggesting that it may have actually been one of the crested lambeosaurines. Otherwise the skull is similar to that of *Gryposaurus*. It appears to have had different teeth in the upper and lower jaws, but it is difficult to see the significance of this.

Above: Like other hadrosaurines Aralosaurus would have had a broad beak. The skull would have been tall at the back and broad at the front.

Distribution: Kazakhstan.
Classification: Ornithopoda, Iguanodontia, Hadrosauroidea, Hadrosaurinae.
Meaning of name: Lizard of the Aral Sea, central Asia.
Named by: Rozhdestvensky, 1968.
Time: Turonian to Coniacian stages of the late Cretaceous.
Size: 6–8m (19½–26ft).
Lifestyle: Browser.
Species: *A. tuberiferus*.

Gryposaurus

This dinosaur was once thought to have been the same genus as *Kritosaurus*, the original species being *K. notabilis*. *Gryposaurus* lived much further north than *Kritosaurus*, and this was probably the only thing that distinguished them (the two were certainly related). It is known from several specimens, including ten complete skulls and 12 fragmentary skulls, along with skin impressions.

Features: In appearance *Gryposaurus* is similar to *Kritosaurus*, with its narrow, deep skull and highly arched nostrils. The tooth shape is slightly different though, suggesting a slightly different diet. The skin is covered in smooth scales less than 4mm (⅛in) in diameter, and there are cone-shape plates 1.3cm (½in) in diameter, spaced 5–7cm (2–2⅘in) apart on the tail.

Distribution: Alberta, Canada.
Classification: Ornithopoda, Iguanodontia, Hadrosauroidea, Hadrosaurinae.
Meaning of name: Hook-nosed lizard.
Named by: Lambe, 1914.
Time: Campanian stage of the late Cretaceous.
Size: 8m (26ft).
Lifestyle: Low browser.
Species: *G. notabilis*, *G. incurvimanus*, *G. laltidens*.

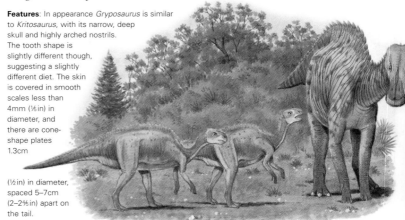

Kritosaurus

The original description of *Kritosaurus* is based on a poorly preserved skull. It is traditionally shown as a flat-headed hadrosaurine hadrosaurid of a subgroup that lacked crests. However, it now seems more likely to have belonged to the subgroup with the small solid crests, such as *Saurolophus*. Other hadrosaurines such as *Anasazisaurus* and *Naashoibitosaurus* may be species of *Kritosaurus*.

Distribution: New Mexico and Texas, USA, and a species in Argentina.
Classification: Ornithopoda, Iguanodontia, Hadrosauroidea, Hadrosaurinae.
Meaning of name: Separated lizard.
Named by: Brown, 1910.
Time: Campanian to Maastrichtian stages of the late Cretaceous.
Size: 10m (33ft).
Lifestyle: Browser.
Species: *K. australis*, *K. navajovius*.

Features: *Kritosaurus* has a flat skull with a ridge of bone just below the eyes, giving the snout a "Roman nose" appearance. The high ridge on the nose probably supported a flap of skin at each side, and this may have cooled the inhaled air, or helped make noises for signalling to one another. Alternatively, it may have been a horn used for battling rival mates.

HADROSAURINE COMMUNICATION

If the lambeosaurines could signal to one another by using their hollow crests as musical instruments, then how did their close relatives the hadrosaurines manage? The hadrosaurines had much broader beaks than the lambeosaurines, and the tops of their skulls presented a broad platform of bone around the nasal openings. It is quite possible that this platform was covered by a flap of skin that surrounded the nostrils. Such a flap of skin would have been inflated as the animal breathed, and could have been controlled to produce a sound, very much like that produced by the inflatable throat pouch of many modern frogs.

This would explain the solid crests of some of the hadrosaurines, especially among the Maiasaurini and the Saurolophini. The flap of skin could have been supported by this crest, and the various crest shapes in the many species would give a different shape of skin flap, and so a different sound would be created to identify all the herds.

Right: An inflated hadrosaurine nose flap.

Kerberosaurus

The skull of *Kerberosaurus* was found in a river-deposited bonebed along with many indeterminate skeletal parts. The presence of this hadrosaurid, similar to *Prosaurolophus* and *Saurolophus*, the type of animal common in North America but unusual for Russia, shows the exchange of fauna that could take place between Asia and North America in late Cretaceous times.

Features: The 1m- (3ft-) long skull, the only part of the animal to have been found, looks very much like that of *Prosaurolophus* or *Saurolophus*, but it differs in the detail of the individual bones and how they are articulated with one another. The significant thing about this animal is the fact that it lived in eastern Asia at this time. Its ancestors migrated there from North America, where the hadrosaurines had evolved.

Distribution: Blagoveshchensk, Russia.
Classification: Ornithopoda, Iguanodontia, Hadrosauroidea, Hadrosaurinae.
Meaning of name: Lizard of Cerberus (a monster dog in Greek mythology).
Named by: Bolotsky and Godefroit, 2004.
Time: Maastrichtian stage of the late Cretaceous.
Size: 10m (33ft).
Lifestyle: Browser.
Species: *K. manakini*.

Left: The single species is named in honour of Colonel Manakin, a pioneer of dinosaur discovery in the Amur region of Russia.

MORE CRESTED HADROSAURINES

The Maiasaurini and the Saurolophini represent advanced subgroups of the Hadrosaurinae, distinguished by the possession of solid crests, formed by the backward extension of the nasal bones. These crests were quite unlike the hollow crests of the Lambeosaurinae, and may have served as a support for some skin structure, whether a ballooning nose flap or the end of a frill down the neck.

Brachylophosaurus

This was quite a rare hadrosaurine hadrosaurid, but it is known from several remains including one, nicknamed "Leonardo", that preserves the skin, the muscles of the neck and the gastric tract that includes the remains of its last meal. The solid nature of the skull suggests that it may have taken part in head-butting behaviour.

Features: The short crest implied in the name consists of a flat plate on the top of the head with a short spine sticking back from the rear of the skull. The skull itself is quite tall for a hadrosaurine hadrosaurid, and has a steep face. The forelimbs are relatively long. The skin is covered in fine scales, the biggest being on the lower limbs. The two species known are so similar that they may represent a male and female of the same species, the only difference being that the skull of *B. goodwini*, the only part known, has a depression before the crest.

Distribution: Alberta, Canada, to Montana, USA.
Classification: Ornithopoda, Iguanodontia, Hadrosauroidea, Hadrosaurinae, Maiasaurini.
Meaning of name: Short-crested lizard.
Named by: C. M. Sternberg, 1953.
Time: Santonian to Campanian stages of the late Cretaceous.
Size: 7m (23ft).
Lifestyle: Browser.
Species: *B. canadensis, B. goodwini.*

Right: The term duckbill becomes something of a misnomer when dealing with Brachylophosaurus. Rather than being broad and duck-like, the bill is flattened from side to side and downturned.

Maiasaura

This hadrosaurine is known from a nesting colony found in Montana in the 1970s. It lived in big herds and nested in groups, with possibly as many as 10,000 returning to the same area every year. They probably did so for protection. More than 200 skeletons, embryos, hatchlings, immature and mature adults were found.

Features: The crest is a short, broad projection above the eyes, and is solid, distinguishing it from the hollow-crested lambeosaurines. The crest, and a pair of triangular projections on the cheekbones, form the basis for the definition of the Maiasaurini. They have batteries of self-sharpening teeth, and a jaw mechanism that allowed the surfaces to grind past each other to chew up tough vegetation.

Distribution: Montana, USA.
Classification: Ornithopoda, Iguanodontia, Hadrosauroidea, Hadrosaurinae, Maiasaurini.
Meaning of name: Good mother lizard.
Named by: Horner and Makela, 1979.
Time: Campanian stage of the late Cretaceous.
Size: 9m (30ft).
Lifestyle: Browser.
Species: *M. peeblesorum.*

Saurolophus

Confusingly, *Saurolophus* is only very distantly related to the much more popular *Parasaurolophus*. The type species *S. osborni* is known from the remains of at least three individuals. Another species, *S. angustirostris*, is known from the Gobi desert. Some palaeontologists think it should be the same species as *S. osborni*, but others think it is a different genus altogether.

Right: The uniting feature between the species of Saurolophus *is the presence of the backward-pointing spike above the eye.*

Features: The distinguishing feature of *Saurolophus* is the prominent spine that rises above the eyes and projects backwards. This is formed from the nasal bones that extend backwards and may have been associated with some sound-producing mechanism. The skull is quite narrow for a hadrosaurid, especially across the snout where we would expect to see the duck-like bill. The original species was the most complete to have been found in Canada at the time (1911). The Asian species is much larger.

Distribution: Alberta, Canada, with a species in Mongolia.
Classification: Ornithopoda, Iguanodontia, Hadrosauroidea, Hadrosaurinae, Saurolophini.
Meaning of name: Lizard crest.
Named by: Brown, 1912.
Time: Maastrichtian stage of the late Cretaceous.
Size: 9–12m (30–39ft).
Lifestyle: Browser.
Species: *S. angustirostris, S. osborni, "S. krischtofovici"*.

DINOSAUR NESTS

The nesting site of *Maiasaura* in Montana gives us a unique insight into the family life of hadrosaurids. Each nest was about 2m (6½ft) across and formed of a low mound of mud. The nests were situated at least an adult's length from the next, just the right distance to prevent the adults from pecking at one another while they rested on the eggs. The eggs were laid in a hollow in the top of the mound and covered with a layer of plant material which provided warmth as it rotted, like a compost heap. The remains of the nests lie in successive layers of rock, suggesting that the herds returned year after year to use the same nesting ground. Bones from all age groups were discovered on the site, suggesting that the hatchlings remained in the nest for some time and were looked after by their parents until they were well grown. For the rest of the year the great mass of *Maiasaura* would have split up into smaller bands and migrated to more productive feeding grounds. The presence of the remains of egg-eating lizards and of nest-infecting beetles reveal some of the hazards that the nesting *Maiasaura* had to face.

Prosaurolophus

As its name suggests, this is possibly the ancestor of *Saurolophus* which appeared a little later in the geological succession. It is known from the remains of more than 30 individuals of varying ages. It inhabited the forest plains of late-Cretaceous Canada, and its diet consisted of the conifers, cycads, ginkgoes and flowering plants that grew there.

Features: The crest is a bony lump just in front of the eyes, rising out of a mass of small knobs and developing into a backward-pointing spike. It is not nearly as big as that of *Saurolophus*. The face is sloping and the muzzle broad and flat, but not as widely flared as on other hadrosaurids. The difference in the shape of the skull between the species may be due to crushing during the fossilization process, and it may be that there is only one species known.

Distribution: Alberta, Canada, to Montana, USA.
Classification: Ornithopoda, Iguanodontia, Hadrosauroidea, Hadrosaurinae, Saurolophini.
Meaning of name: Before *Saurolophus*.
Named by: Brown, 1916.
Time: Coniacian to Campanian stages of the late Cretaceous.
Size: 8–9m (26–30ft).
Lifestyle: Browser.
Species: *P. maximus, P. blackfeetensis, P. breviceps*.

Above: Although the official length of Prosaurolophus *is given as 8–9m (26–30ft), there is a suggestion that it may actually have reached lengths of 15m (50ft).*

FLAT-HEADED HADROSAURINES

The flat-headed hadrosaurines were the Edmontosaurini, and were among the last of the dinosaurs to evolve. It seems likely that they evolved from the solid-crested types. It is possible that they signalled to each other by bellowing, using flaps of skin on their flat heads to amplify the sound, but as yet this has not been proved.

Anatotitan

Often seen in old books as *Anatosaurus* and even as *Trachodon*, *Anatotitan* is known from two good skeletons including the skulls. It rested on all fours, but walked on its powerful back legs with the three hoof-like toes on each foot taking the weight. It was among the last of the dinosaurs to have existed.

Features: This is the most duck-like of the duckbilled dinosaurs, with a particularly broad and flat beak, and jaws that are toothless for over half their length. However, the hardened horny bill that covered the front of the mouth is actually quite different from the sensitive organ found on a duck. There are slight knobs above the eyes, but apart from that there is no sign of a crest. Although bigger than *Edmontosaurus*, it is more lightly built.

Distribution: Montana to South Dakota, USA.
Classification: Ornithopoda, Iguanodontia, Hadrosauroidea, Hadrosaurinae, Edmontosaurini.
Meaning of name: Giant duck.
Named by: Brett-Surman, 1990.
Time: Maastrichtian stage of the late Cretaceous.
Size: 10–13m (33–43ft).
Lifestyle: Browser.
Species: *A. copei*, *A. longiceps*.

Edmontosaurus

The most abundant of the herbivores at the end of the Cretaceous period, *Edmontosaurus* is known from many skeletons. One skull shows signs of theropod tooth marks suggesting an attack to the neck, while another has a chunk bitten out of the top of the tail. The bite subsequently healed. The size suggests it was bitten by a tyrannosaurid mouth.

Features: The skeleton of *Edmontosaurus* is regarded as the benchmark, the shape to which all other hadrosaurids are compared. The tail is deep and heavy, used as a balance when it walked on hind legs, and the hands have weight-bearing pads on the fingers to support it while it stood on all fours. The neck is flexible, allowing the duckbilled head to reach the food growing all around it.

Left: We know that Edmontosaurus had leathery skin with non-overlapping scales, as mummified remains of it have been found associated with two skeletons.

Distribution: Alberta, Canada, to Wyoming and maybe Alaska, USA.
Classification: Ornithopoda, Iguanodontia, Hadrosauroidea, Hadrosaurinae, Edmontosaurini.
Meaning of name: Edmonton lizard.
Named by: Lambe, 1917.
Time: Maastrichtian stage of the late Cretaceous.
Size: 13m (43ft).
Lifestyle: Browser.
Species: *E. annectens*, *E. regalis*, *E. saskatchewanensis*.

Shantungosaurus

Shantungosaurus appears to have been the biggest ornithopod known, an adult weighing several tonnes. It was probably the biggest animal ever to walk on two legs, although it would have spent most time on all fours. It is so similar to *Edmontosaurus* that it could be a species of that Canadian hadrosaurid. *Shantungosaurus* is based on an almost complete skeleton in the Beijing Geological Museum, built from the remains of five individuals.

Features: The features that distinguish *Shantungosaurus* from *Edmontosaurus* are those that relate to size – bigger limbs, stronger bones, a more powerful back to take the weight, and so on.

Distribution: Shandong Province, China.
Classification: Ornithopoda, Iguanodontia, Hadrosauroidea, Hadrosaurinae, Edmontosaurini.
Meaning of name: Lizard from Shandong.
Named by: Hu, 1973.
Time: Maastrichtian stage of the late Cretaceous.
Size: 12–15m (39–49ft).
Lifestyle: Browser.
Species: *S. giganteus.*

Right: All the specimens that went into the mounted skeleton came from the same quarry. The bones were all disarticulated, but the number of jaw bones indicates that there were at least five individuals present.

Its estimated weight was around 3–3½ tonnes. The head is 1.5m (5ft) long. The fact that the single, complete skeleton is made up of several individuals, picked out of 30 tonnes of fossilized material, suggests that it was a herd-dweller moving about the plains in large numbers to avoid the tyrannosaurs.

TRACHODON AND HADROSAURUS

Many old books refer to a duckbilled dinosaur called *Trachodon*. The name was first given to some teeth found in Montana in 1855 by Joseph Leidy. The first good dinosaur remains in North America were found two years later in New Jersey. They were of a hadrosaurid that was named *Hadrosaurus* by Leidy. Unfortunately the later find lacked the skull, and was thought to have been an iguanodontid. The first hadrosaurid skull, with beak and teeth, was found by Cope in 1883. The teeth were similar to Leidy's original, and so *Trachodon* became the genus to which all duckbilled dinosaurs belonged – an early example of a "wastebasket taxon".

Soon all sorts of other hadrosaurids came to light, with all shapes of skulls, and many other genera were established. The original *Trachodon* teeth could not confidently be assigned to any of them, and so the name fell into disuse. For a while a hadrosaurine hadrosaurid called *Anatosaurus* became the most likely owner of the *Trachodon* teeth but later, due to the complexity of allocating scientific names, *Anatosaurus* became *Anatotitan*. The name *Trachodon* is no longer used.

Incidentally, the first good North American dinosaur, *Hadrosaurus*, has fallen into disfavour too. The lack of a skull means that it cannot be classified with certainty, which is why, important as it is, it is not featured here.

Tanius

The Chinese geologist H. C. Tan (or Tan Xi-zhou) collected most of the original specimen of *Tanius* in 1923 at Laiyang, in Shandong Province, China. Other specimens that have been assigned to this genus have since been identified as other hadrosaurids.

Features: As the skull is incomplete, it is not entirely certain that the skull found belonged to the Hadrosaurinae. It is possible that it is a *Tsintaosaurus* that has lost its distinctive crest. In either case the body is large, with a deep tail. The specialized, self-sharpening teeth with mobile jaws allow it to feed on tough vegetable matter. The hind legs are longer than the front, and it could walk either on its hind legs or on all fours. Most of the front of the head is missing, hence the uncertainty about its identity.

Distribution: China.
Classification: Ornithopoda, Iguanodontia, Hadrosauroidea, Hadrosaurinae, Edmontosaurini (probably).
Meaning of name: After Tan.
Named by: Wiman, 1929.
Time: Undetermined stage of the late Cretaceous.
Size: 9m (30ft).
Lifestyle: Browser.
Species: *T. sinensis.*

Right: Whatever it was, Tanius had the appearance of other big ornithopods.

BONEHEADS

Most of what we know about the pachycephalosaurids is based on the range of skulls that have been found; there are very few skeletons. For a time most members of the group were regarded as species of Stegoceras, for want of any evidence to the contrary. Current knowledge suggests the subtle differences in skull pattern indicate that there are more genera of pachycephalosaurid than was first appreciated.

Stegoceras

Distribution: Alberta, Canada, and Montana, USA.
Classification: Marginocephalia, Pachycephalosauria.
Meaning of name: Valid roof horn.
Named by: Lambe, 1902.
Time: Campanian stage of the late Cretaceous.
Size: 3m (10ft).
Lifestyle: Low browser.
Species: *S. validum*, *S. browni* (formerly *Ornatotholus browni*).

This is the best known of the pachycephalosaurids, with dozens of skull fragments known and also a partial skeleton. Most of the reconstructions of other pachycephalosaurid genera are based on this skeleton. The structure of the bone in the head dome was such that the bone fibres aligned to absorb impact from the top. The vertebrae of the neck and back were very strong, lashed together with strong tendons that prevented twisting, and aligned to absorb shocks emanating from the head end. The hips were particularly wide and solid. All this is consistent with the idea that the dome was used as a weapon, like a battering ram.

Below: The muzzle is wide and the teeth set quite far apart, suggesting that Stegoceras had a different diet from other boneheads.

Features: The dome on the head of *Stegoceras* is high, but not as high as that of others in the group, and is surrounded by a frill of little horns and knobs. The teeth at the front of the jaw are very widely set and the muzzle is broad compared with other pachycephalosaurids. This may indicate a less selective feeding strategy. The very broad hips suggest that the pachycephalosaurids gave birth to live young, (this is not widely accepted).

Colepiocephale

This dinosaur was discovered by fossil-hunter L. M. Sternberg in 1945 and regarded as a species of *Stegoceras*, but it was studied again by palaeontologist Robert Sullivan in 2003 who found enough features of the dome to distinguish it, and put it in a genus of its own. It is the oldest definite pachycephalosaurid found in North America (although another has been found in slightly earlier strata but has not yet been studied).

Features: The dome has quite an oblique slope to it, producing a flattened aspect to the front of the head, and is somewhat triangular in top view. The bones of the side of the skull are not as complex as those of other pachycephalosaurids. The range in dome shapes among the different genera of pachycephalosaurids suggests that they functioned as display

Distribution: Alberta, Canada.
Classification: Marginocephalia, Pachycephalosauria.
Meaning of name: Knucklehead.
Named by: Sullivan, 2003.
Time: Campanian stage of the late Cretaceous.
Size: 1m (3ft).
Lifestyle: Low browser.
Species: *C. lambei*.

devices, enabling individuals to distinguish one herd from another. In this way they would be similar to modern birds that have big spectacular beaks or flamboyant feather crests to distinguish one species from another.

Above: The "knucklehead" of the name refers to the shape of the dome head that has the appearance of a finger joint.

The popular image of a pachycephalosaurid is of a pair of rival males, excited by the prospect of leading the herd and mating with the females, head-butting one another furiously. The

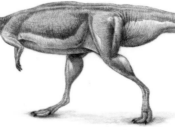

Above: Pachycephalosaurids probably butted one another on the flanks.

vision of mountain goats doing the same thing is difficult to dispel. The idea of this behaviour was put forward by American palaeontologist Ed Colbert in 1955.

However, there are flaws to this evocative scene, as pointed out by Mark Goodwin of the University of California, at Berkeley. For one thing, the shapes are wrong. Two dome-heads on a collision course need to have pinpoint accuracy to be effective, otherwise they would just glance off one another. Also, of all the fossils of bonehead skulls found so far, none of them shows the kind of damage that would have been sustained in such an engagement. It now seems more likely that the dome-head of the pachycephalosaurid was used as a battering ram, but not for head-on attacks.

Hanssuesia

This dinosaur was identified as *Stegoceras sternbergi* by Brown and Schlaikjer in 1943. They gave it a new species name because they found it to be different from *S. validus*. In the study and reclassification of the pachycephalosaurids, conducted by Robert Sullivan of the State Museum of Pennsylvania in 2003, enough differences were found to put it in a brand new genus of its own.

Features: This pachycephalosaurid has a very low and round dome. The joins between the bones that make up the dome are different from those in other pachycephalosaurids. The bones at the side of the skull are smaller too. These features are significant enough to determine that this is a new genus of dinosaur. The most obvious difference is that there is no shelf around the back of the dome as with other members of the group.

Distribution: Alberta, Canada.
Classification: Marginocephalia, Pachycephalosauria.
Meaning of name: From Hans-Dieter Sues, the Canadian pachycephalosaurid expert.
Named by: Sullivan, 2003.
Time: Cenomanian stage of the late Cretaceous.
Size: 2.5m (8ft).
Lifestyle: Low browser.
Species: *H. sternbergi.*

Left: When the bones were found they were thought to have belonged to Troodon, but that was before Troodon *was known to have been a meat-eater. Later they were reassigned to the bonehead group.*

Tylocephale

Knowledge of *Tylocephale* is based on a single damaged skull. It is closely related to *Prenocephale* and may well have been a new species of this genus. The distribution suggests that pachycephalosaurids evolved in Asia and migrated to North America, and then as a group, represented by *Tylocephale*, migrated back again.

Features: *Tylocephale* has the tallest dome of any pachycephalosaurid. The highest portion of it is further back than in others of the group, and it is quite narrow from side to side. There are small spikes around the back of the skull, similar to those of *Stegoceras*. The teeth are quite large for a pachycephalosaurid.

Distribution: Mongolia.
Classification: Marginocephalia, Pachycephalosauria.
Meaning of name: Swollen head.
Named by: Maryańska and Osmólska, 1974.
Time: Campanian stage of the late Cretaceous.
Size: 2.5m (8ft).
Lifestyle: Low browser.
Species: *T. gilmorei, T. bexelli.*

Left: Tylocephale *and the other boneheads may have lived in inland or even mountainous areas.*

LATE BONEHEADS

The pachycephalosaurids were among the last of the dinosaurs to evolve and flourish. Representatives from this group existed until the end of the Age of Dinosaurs. They began as small rabbit-sized animals, some with a slight thickening to the skull, but towards the end of the Cretaceous they became quite large and the skull ornamentation developed to spectacular proportions.

Pachycephalosaurus

Distribution: Montana, South Dakota and Wyoming, USA.
Classification:
Marginocephalia,
Pachycephalosauria,
Pachycephalosaurini.
Meaning of name: Thick-headed lizard.
Named by: Brown and Schlaikjer, 1943, based on Gilmore, 1931.
Time: Maastrichtian stage of the late Cretaceous.
Size: 8m (26ft).
Lifestyle: Low browser.
Species: *P. wyomingensis*, plus one other unnamed species.

Although this animal is often shown in restorations, little is known about it except for the skull. It was the biggest pachycephalosaurid known, hence its fame. Its three kinds of teeth suggest that it fed on a mixed diet of leaves, seed, fruit and insects. *Pachycephalosaurus* and *Stygimoloch* belonged to a family called the Pachycephalosaurini, which has a notable defining feature, which is the presence of horns on the muzzle and cheeks.

Features: The muzzle of *Pachycephalosaurus* is quite long and narrow, and carries a number of tall spikes. Around the rear of the dome is a complex array of nodules and lumps. The length given is based on the assumption that the rest of the body is in the same proportion to the skull as in the others of the group, and the restoration is based on the bodies of other pachycephalosaurids.

Below: The dome on the head of Pachycephalosaurus was 20cm (8in) thick. Estimates of its total length vary from 4.5–8m (15–26ft).

AN ALTERNATIVE EXPLANATION
The head-butting activities of the pachycephalosaurids came under scrutiny with research published in 2004 by Mark Goodwin of the University of California, at Berkeley, and Jack Horner of Montana State University. They found that the radiating bone structure that was thought to have given strength to the head dome and which made it a powerful battering ram was only present in juvenile specimens, and not in adults, when it was assumed the head-butting would most likely have taken place. The head dome continued to grow as the animal grew, with the bone structure altering all the time. In addition, the bone carried blood vessels that suggest that the dome was covered with a horny cap while the dinosaur was alive. This would have been used for species recognition. As a result we cannot put together an accurate restoration of a pachycephalosaurid without knowing what this horn ornamentation was like.

Right: The dome may have been the base of a tall horn.

Stygimoloch

The name of this pachycephalosaurid derives from its frightful appearance. Moloch was a horned devil in Hebrew mythology, and in Greek legends the river Styx was the river that the dead had to cross to reach the Underworld. The fossils were found in the Hell Creek formation in Montana, and this was a further inspiration for the name. The first *Stygimoloch* horn core was found in 1896 and regarded as part of a *Triceratops* skull. In the 1940s, when pachycephalosaurids were recognized, it was classed as a species of *Pachycephalosaurus*.

Features: The most obvious feature of *Stygimoloch* is the array of horns projecting from the rim of the dome. The head is quite long and the dome is high, narrow and thin. From the front this presents a startling apparition of ornamentation, with long horns surrounded by clusters of more stubby spikes, that would evidently have been very effective as a threat or defence display, very much like those of some of the horned ceratopsians.

Distribution: Montana to Wyoming, USA.
Classification: Marginocephalia, Pachycephalosauria, Pachycephalosaurini.

Meaning of name: Horned devil from the river of death.
Named by: Galton and Sues, 1983.
Time: Maastrichtian stage of the late Cretaceous.
Size: 3m (10ft).
Lifestyle: Low browser.
Species: *S. spirifer*.

Above:
Stygimoloch *is known mostly from the skull. There have been five partial skulls found, but there have been other parts of the skeleton found in remains from North and South Dakota, USA.*

Sphaerotholus

Sphaerotholus is known from two skulls, one of which is one of the most complete pachycephalosaurid skulls to have been found. Despite this, the animal's actual identity is rather unclear. The remains are very similar to those of the Mongolian pachycephalosaurid *Prenocephale* and it may be that they are all the same genus. If that is the case, *Prenocephale* will be the official name, since it was invented ("erected" is the term used by palaeontologists) in 1974, long before *Sphaerotholus* was named.

Distribution: Montana and New Mexico, USA.
Classification: Marginocephalia,

Pachycephalosauria.
Meaning of name: Spherical dome.
Named by: Williamson and Carr, 2003.
Time: Maastrichtian stage of the late Cretaceous.
Size: 2m (6ft).
Lifestyle: Low browser.
Species: *S. goodwini, S. edmontonense*.

Features: The domes of the specimens of *Sphaerotholus* are particularly spherical, and the arrangement of knobs around the back of the skull in a single row is quite distinctive, but apart from that there is not enough of the preserved skull to distinguish it from other pachycephalosaurids, such as *Prenocephale*. There is, however, a partial jawbone very similar to that of *Stegoceras* which has been attributed to *Sphaerotholus*.

Right: The dome of Sphaerotholus *is the most ball-like of all the pachycephalosaurids. There is a row of small knobs around the back of the skull.*

PRIMITIVE ASIAN CERATOPSIANS

The main division of the Marginocephalian group is represented by the ceratopsians, or ceratopians as some palaeontologists prefer. These are the horned dinosaurs. Their origin can be traced back into early Cretaceous times, but it was in the late Cretaceous period that they really came into their own. The early forms were quite graceful little animals but they soon evolved into pig-sized beasts.

Graciliceratops

When *Graciliceratops* was discovered in 1975, the skeleton was referred to as a specimen of *Microceratops gobiensis*, but Paul Sereno of Chicago subsequently identified it as the juvenile of something quite different. The name derives from its small size and light build, and its bipedal stance shows that the group had its origins in the two-footed plant-eaters.

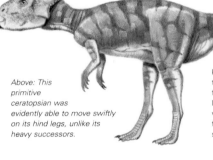

Above: This primitive ceratopsian was evidently able to move swiftly on its hind legs, unlike its heavy successors.

Features: Although this dinosaur is only known from a juvenile skeleton, there is enough to show that it is basically a bipedal animal with a front limb that is smaller than the hind limb. The hind limbs show that it was capable of running swiftly. As with all primitive ceratopsians, it has a beak at the front of its mouth and a ridge of bone, not quite a shield, around the back of the skull.

Distribution: Omnogov, Mongolia.
Classification: Marginocephalia, Ceratopsia, Neoceratopsia.
Meaning of name: Graceful horned face.
Named by: Sereno, 2000.
Time: Santonian to Campanian stages of the late Cretaceous.
Size: 0.9m (3ft), but this is immature. The adult was probably 2m (6½ft).
Lifestyle: Low browser.
Species: *G. mongoliensis*.

Protoceratops

There have been dozens of skeletons of *Protoceratops* found, both adult and juvenile, and so the whole growth pattern is known. It was found by the expeditions to the Gobi Desert undertaken by the American Museum of Natural History in the 1920s. It seems to have lived in herds, and its remains are so abundant that it has been termed the "sheep of the Cretaceous".

Features: *Protoceratops* is a heavy animal with short legs, a deep tail and a heavy head. Although a member of the horned dinosaurs, it does not have true horns. Two forms of adult are known, a lightweight form with a low frill, and a heavier form with a big frill and a bump on the snout where a horn would have been. These probably represent the two sexes, with the males having the heavier head.

Distribution: China and Mongolia.
Classification: Marginocephalia, Ceratopsia, Neoceratopsia.
Meaning of name: Before the horned heads.
Named by: Granger and Gregory, 1923.
Time: Santonian and Campanian stages of the late Cretaceous.
Size: 2.5m (8ft).
Lifestyle: Low browser.
Species: *P. andrewsi*.

CERATOPSIAN DEVELOPMENT AND DISTRIBUTION

The basal ceratopsians used to consist merely of *Protoceratops*, found in the 1920s. Then in the 1960s and 1970s a Polish-Mongolian expedition turned up all kinds of other small primitive ceratopsians at various places in the Gobi Desert.

It now seems likely that the ceratopsians evolved in Asia, and developed into fairly small, compact animals. Some time later they migrated eastwards, crossing the land bridge that is now the Bering Strait, entering North America where they evolved into the huge shield-necked, multi-horned, rhinoceros-like animals that have been known for more than a century. Strangely enough, the primitive forms also existed in North America, and continued largely unchanged until the end of the Age of Dinosaurs.

Below: Asia was once joined to America.

Asia

North America

Bagaceratops

Bagaceratops is known from about two dozen skulls, five of them complete, and some bits of the rest of the skeleton from several juveniles and adults. It is closely related to *Protoceratops*, but is smaller, hence the name. Like *Protoceratops* this is a desert-dweller, and is often found preserved in desert sandstone, having been overwhelmed by sandstorms or collapsing dunes.

Features: Unlike its hornless relatives, *Bagaceratops* sports a small horn on the snout. It also has a distinctive triangular frill around the neck forming a shield. Another distinction is that it lacks the pointed teeth at the front of the upper jaw, relying instead on its beak to gather food. Despite these features, and the fact that it came later, *Bagaceratops* is considered to be a more primitive animal than *Protoceratops*.

Distribution: Mongolia.
Classification: Marginocephalia, Ceratopsia, Neoceratopsia.
Meaning of name: Small horned head.
Named by: Maryańska and Osmólska, 1975.
Time: Campanian stage of the late Cretaceous.
Size: 1m (3ft).
Lifestyle: Low browser.
Species: *B. rozhdestvenskyi.*

Left: So many specimens of Bagaceratops, *young and old, are known that scientists have a good idea of how the individuals grew and developed.*

Breviceratops

Some scientists think that *Breviceratops* is a synonym for *Bagaceratops*. Indeed the two must have looked very much like one another. They lived at about the same time and had the same lifestyle, grazing the low-growing scrubby plants on a bleak desert landscape. In size, *Breviceratops* was intermediate between *Bagaceratops* and *Protoceratops*.

Features: Although some scientists think *Breviceratops* was the same as *Bagaceratops*, or even *Protoceratops* (it was at first thought to have been a species of *Protoceratops*), there are a number of important differences. Unlike *Bagaceratops*, *Breviceratops* has no sign of a horn and has two upper front teeth. It has a straight, lower jaw, unlike the curved jaw of *Protoceratops*, and the bones that form the frill do not flare outwards as widely as in *Protoceratops*. Notable though these differences are, there is still a possibility that *Breviceratops* may be a juvenile *Protoceratops*, and that the differences just represent different growth stages.

Distribution: Khulsan, Mongolia.
Classification: Marginocephalia, Ceratopsia, Neoceratopsia.
Meaning of name: Short horned face.
Named by: Maryańska and Osmólska, 1990.
Time: Santonian to Campanian stages of the late Cretaceous.
Size: 2m (6½ft).
Lifestyle: Low browser.
Species: *B. kozlowskii.*

Left: Classification of Breviceratops *is not helped by the fact that the most important specimen was stolen from the Palaeontological Institute of the Russian Academy of Sciences, in Moscow, in 1996.*

NEW WORLD PRIMITIVE CERATOPSIANS

The basal ceratopsians were mostly found in Asia, but several were found in North America too. Those in North America were mostly from a slightly later date, in fact one was from the end of the Age of Dinosaurs, suggesting that they evolved in Asia and later migrated eastwards. There some evolved into the big horned dinosaurs, while others remained in a small and primitive state.

Montanoceratops

Montanoceratops is the state fossil of Montana. It is known from two partial skeletons, the first studied by Brown and Schlaikjer in 1942, and named *Leptoceratops cerorhynchus*. C. M. Sternberg later realized that it was different enough to be a separate genus. There are two known specimens, the first found in 1916 and the other 80 years later in 1996, just a few metres away from the first.

Features: As noted by Sternberg in the 1950s, in almost every respect *Montanoceratops* is more advanced than *Protoceratops*, and much more so than *Leptoceratops*, both of which it superficially resembles. The first three vertebrae of the neck are fused into a solid lump, presumably as an adaptation to carrying the heavy skull. The long spines on the vertebrae of the tail give it a deep profile until mid-length, after which it tapers rapidly to a point.

Distribution: Montana, USA.
Classification: Marginocephalia, Ceratopsia, Neoceratopsia.
Meaning of name: Horned head from Montana.
Named by: Brown and Schlaikjer, 1942.
Time: Maastrichtian stage of the late Cretaceous.
Size: 3m (10ft).
Lifestyle: Low browser.
Species: *M. cerorhynchus*.

Zuniceratops

The earliest known ceratopsian to have carried a pair of brow horns is *Zuniceratops*. It is the oldest-known ceratopsian from North America. Its presence in New Mexico at such an early stage of the late Cretaceous suggests that the ceratopsians with the brow horns evolved in North America rather than Asia. A bonebed shows that it may have lived in herds.

Features: This is quite a lightweight horned dinosaur, rather like *Protoceratops* or *Montanoceratops* but less heavily built, with a well-developed neck shield containing large openings, and a pair of horns over the eyes. There is no horn on the snout.

Distribution: New Mexico, USA.
Classification: Marginocephalia, Ceratopsia, Neoceratopsia.
Meaning of name: Horned face of the Zuni tribe.
Named by: Wolfe and Kirkland, 1998.
Time: Turonian stage of the late Cretaceous.
Size: 3.5m (11½ft).
Lifestyle: Low browser.
Species: *Z. christopheri*.

An interesting point is that in the young the teeth have single roots, the double roots typical of ceratopsians only developing with age. The brow horns also grow considerably as the individual become older.

Above:
Zuniceratops *is the earliest of the brow-horned ceratopsians.*

Turanoceratops

Despite the fact that all its relatives lived in North America *Turanoceratops* is included here. So far this is the only brow-horned ceratopsian to have been found in Asia. However, the material is very fragmentary and it is difficult to make any definitive statement about it. It does, however, appear to have a pair of horns above the eyes, and the double-rooted teeth of the more advanced ceratopsians.

Features: The double-rooted teeth so distinctive of the later North American ceratopsians are present here, long before they appeared in North American forms. This interesting fact is about the only thing we can say about *Turanoceratops*. A fragmentary skull with evidence of a pair of horns above the eyes, and bits of vertebrae and shoulder bone are all that are known, unfortunately not enough to tell us anything more.

Distribution: Kazakhstan.
Classification: Marginocephalia, Ceratopsia, Neoceratopsia.
Meaning of name: Turanian horned head.
Named by: Nessov and Kaznyshkina, 1989.
Time: Cenomanian to Turonian stages of the late Cretaceous.
Size: 2m (6½ft).
Lifestyle: Low browser.
Species: *T. tardabilis*.

Right: Apart from Turanoceratops, brow-horned ceratopsians are unknown from Asia. It is possible that it returned to Asia after evolving in North America, and that the migration was two-way.

CERATOPSIAN DIET

The food of ceratopsians has always been a puzzle. They were herbivores, we know that, but what exactly were the plants that they ate?

The basal ceratopsians had relatively simple teeth which worked by sliding past one another in a kind of a scissor movement. Their muzzles were quite delicate, suggesting that they were quite selective about what they ate. The heavy ridge at the back of the skull probably evolved first as a base for strong jaw muscles. (The later development into a neck shield was for quite another purpose, such as defence or display.) Their food of choice was probably the shoots and leaves of cycads. These would have been pecked off with the beak and then chopped up in the mouth, the cheeks holding it while the powerful jaws chopped it into bits. Later ceratopsians used the same action but with whole batteries of slicing teeth, with four or five growing upwards to replace each one that was wearing away.

Below: Ceratopsians ate shoots and leaves.

Leptoceratops

Leptoceratops may have been bipedal like the early forms, or it may have walked on all fours. The surprising feature of this animal is its extreme primitiveness, despite which it is among the last of the dinosaurs to have existed, at the end of the Cretaceous period, sharing the North American landscape with the biggest and most advanced ceratopsian, *Triceratops*.

Features: *Leptoceratops* has a slender body with short forelimbs. The skull is deep and the jaws carry primitive, single-rooted teeth. The teeth are adapted for crushing rather than for slicing as in other ceratopsians. There are no horns on the head. The neck shield is flattened from side to side, and carries a tall central ridge and smooth rear border. The front foot has five toes, with claws rather than hooves. All in all, it is a very primitive animal.

Below: Leptoceratops is known from five skulls and parts of skeletons from North America. However, an almost identical limb bone has been found in Australia.

Distribution: Alberta, Canada, to Wyoming, USA.
Classification: Marginocephalia, Ceratopsia, Neoceratopsia.
Meaning of name: Slender horned face.
Named by: Brown, 1914.
Time: Maastrichtian stage of the late Cretaceous.
Size: 3m (10ft).
Lifestyle: Low browser.
Species: *L. gracilis*.

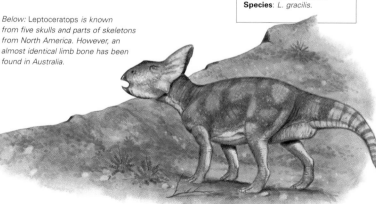

SHORT-FRILLED CERATOPSIDS

The big horned dinosaurs of the late Cretaceous period belonged to the group called the Ceratopsidae. They were only found in North America, with the possible exception of Turanosaurus *from Kazakhstan. They nearly all had massive rhinoceros-like bodies and heavy, armoured heads. They can be divided into two subgroups, the short-frilled Centrosaurinae and the long-frilled Chamosaurinae.*

Avaceratops

Although the ceratopsids are generally big animals, *Avaceratops* is quite small. It is known from an almost complete skeleton missing only the hip bones, much of the tail and, frustratingly, the roof of the skull including the horn cores. The skeleton found is not an adult, since most of the skull came apart before it fossilized, but it was almost fully grown when it died.

Features: This small ceratopsid has a short frill that is quite thick. Like other centrosaurines it has a short, deep snout, a powerful lower jaw with batteries of double-rooted shearing teeth, and a beak like that of a parrot. It is assumed that like other centrosaurines, it has a bigger horn on the nose than above the eyes. It may be a juvenile or subadult of some other genus such as *Monoclonius*.

Distribution: Montana, USA.
Classification: Marginocephalia, Ceratopsia, Ceratopsidae, Centrosaurinae.
Meaning of name: Ava's horned face (from Ava Cole, the wife of the discoverer).
Named by: Dodson, 1986.
Time: Campanian stage of the late Cretaceous.
Size: 2.5m (8ft), but this was a juvenile. The grown animal was probably 4m (13ft).
Lifestyle: Low browser.
Species: *A. lammersi.*

Right: Avaceratops *looked like a diminutive version of its giant contemporaries.*

Centrosaurus

Distribution: Alberta, Canada.
Classification: Marginocephalia, Ceratopsia, Ceratopsidae, Centrosaurinae.
Meaning of name: Pointed lizard.
Named by: Lambe, 1904.
Time: Campanian stage of the late Cretaceous.
Size: 6m (19½ft).
Lifestyle: Low browser.
Species: *C. cutleri,
C. apertus.*

The animal to which the short-frilled group owes its name is known from at least 15 skulls and pieces of bone from animals of all stages of growth. The first part of the skeleton to be found was the back of the neck shield, with its hook-shaped horns which give the animal its name (not the single nose horn).

Features: *Centrosaurus* is noted for the big, single horn on its snout, as well as smaller horns over the eyes and others like hooks on the neck shield. The edge of the shield has bony growths. The big horn curves forward in some individuals, leans back in others, and yet sticks straight up in others, a variation that palaeontologists do not seem to think significant. A pair of openings, or *fenestrae*, on the neck shield keeps the weight down.

SKIN TEXTURE

It is not often that skin impressions are found from dinosaurs. However, there are skin impressions known from ceratopsians. These formed as the dead animal sank into river mud, and the impressions left in the mud solidified as the skin itself rotted away. The skin impressions known come from the area of the hips. These show plates of about 5cm (2in) in diameter, set in irregular rows 5cm (2in) apart, separated by a mass of 1cm (½in) scales. The plates may have been covered in horn, like those on the back of a crocodile. It is possible that the skin had a totally different texture on different parts of the body, especially the underside.

Unfortunately there is no evidence of the kind of skin that covered the face or shield, or even of the shapes of the keratinous parts of the horns, something that would be so valuable to artists of dinosaur restorations.

Below: A ceratopsian skin impression.

Monoclonius

For a long time there has been confusion over the name *Monoclonius*. Many specimens that had been attributed to it have now been reclassified as *Centrosaurus*, the uniting feature being the single nose horn that curved backwards. The spectacular *Centrosaurus* mount in the American Museum of Natural History was labelled *Monoclonius* until 1992, long after the palaeontologists knew better.

Features: *Monoclonius* is an average-size centrosaurine, but the skull is particularly long, more like that of a long-frilled ceratopsine. The single horn on the nose curves backwards, and there are no signs of horns above the eyes or across the shield. The edge of the shield is scalloped. The shield is rather thin compared with others of the group, suggesting it was used for display.

Distribution: Alberta, Canada, to Montana, USA.
Classification: Marginocephalia, Ceratopsia, Ceratopsidae, Centrosaurinae.
Meaning of name: Single stem, referring to a single-rooted tooth found with the original specimen but which turned out to be something else.
Named by: Cope, 1876.
Time: Campanian stage of the late Cretaceous.

Size: 6m (19½ft).
Lifestyle: Low browser.
Species: *M. crassus, M. fissus, M. recurvicornis.*

Styracosaurus

Bonebeds of several thousand *Styracosaurus* individuals are known, but there are few undamaged skulls, the only good one being the one on which Lawrence Lambe based the original description in 1913. Bonebeds incorporating masses of charcoal suggest that herds of these animals were forced into rivers by forest fires, and perished in the water.

Features: The obvious feature of *Styracosaurus* is the array of horns all around the edge of the frill, developed from the bony ornamentation often seen on other centrosaurines. There are six big horns on the shield, and a series of smaller knobs. A massive horn on the nose rounded off this apparition. There seem to have been no horns at all above the eyes, but it is possible that they were present in juveniles and disappeared as the animal reached adulthood.

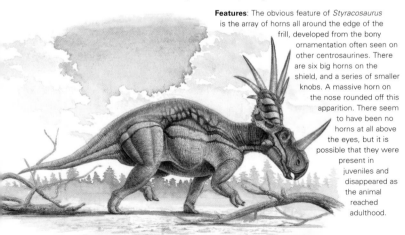

Distribution: Alberta, Canada, to Montana, USA.
Classification: Marginocephalia, Ceratopsia, Ceratopsidae, Centrosaurinae.
Meaning of name: Spike lizard.
Named by: Lambe, 1913.
Time: Campanian to Maastrichtian stages of the late Cretaceous.
Size: 5.5m (18ft).
Lifestyle: Low browser.
Species: *S. albertensis, S. ovatus, S. sphenocerus.*

CENTROSAURINES

Although the Centrosaurinae, a subgroup of the Ceratopsidae, are collectively known as the short-frilled ceratopsians, it does not mean that the frills were small. Most frills were ornamented with spikes and horns that made them look very imposing, and give the impression of belonging to a much larger animal. They tended to have small brow horns, the main horn being a big one on the nose.

Pachyrhinosaurus

The massive lump of bone on the nose of *Pachyrhinosaurus* gave the impression of its being a horned dinosaur with no significant horn. This bony mass may have been used as a battering ram when sparring, or as the base of a massive horn built of keratin that has not fossilized.

Features: The distinctive feature of *Pachyrhinosaurus*, and the one that gives it its name, is the massive shelf of bone on the nose where, in other centrosaurines, there is a horn. It also has a small horn above the eyes and along the centre line of the neck shield, and hook-shaped horns at the shield's top edge. The rest of the body is exactly the same shape as in that of other centrosaurines.

Distribution: Alberta, Canada, to Alaska.
Classification: Marginocephalia, Ceratopsia, Ceratopsidae, Centrosaurinae.
Meaning of name: Thick-nosed lizard.
Named by: C. M. Sternberg, 1950.
Time: Maastrichtian stage of the late Cretaceous.
Size: 7m (23ft).
Lifestyle: Low browser.
Species: *P. canadensis*.

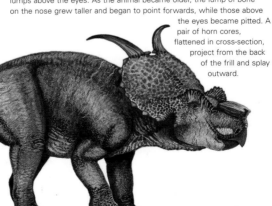

Left: The broad bony lump on the snout of Pachyrhinosaurus *is similar to that of the modern rhinoceros – and that supports a keratinous horn.*

Achelousaurus

A horned dinosaur with bony knobs on its head instead of horns, *Achelousaurus* is known from a huge bonebed in the Two Medicine Formation in Montana, USA. As in *Pachyrhinosaurus*, there is a possibility that these knobs were the bases of keratinous horns.

Features: The distinctive horn arrangement of *Achelousaurus* consists of a bony lump over the nose, smaller than that of *Pachyrhinosaurus* but deeply wrinkled. It also has a pair of smaller lumps above the eyes. As the animal became older, the lump of bone on the nose grew taller and began to point forwards, while those above the eyes became pitted. A pair of horn cores, flattened in cross-section, project from the back of the frill and splay outward.

Distribution: Montana, USA.
Classification: Marginocephalia, Ceratopsia, Ceratopsidae, Centrosaurinae.
Meaning of name: Lizard of Acheloo (a Greek god whose horns were snapped off by Hercules).
Named by: Sampson, 1995.
Time: Campanian to Maastrichtian stages of the late Cretaceous.
Size: 6m (19½ft).
Lifestyle: Low browser.
Species: *A. horneri*.

Right: It is possible that Achelousaurus *is a species of* Pachyrhinosaurus; *the two are definitely closely related.*

Einiosaurus

This dinosaur *Einiosaurus*, together with *Achelousaurus* was established in 1995 when American palaeontologist Scott Sampson revised the Centrosaurinae. He established that the centrosaurines consisted of an evolutionary line with *Pachyrhinosaurus* at the basal end, and *Centrosaurus* and *Styracosaurus* at the other. *Einiosaurus* and *Achelousaurus* occupied a position between the two. The classification of these animals is determined by the ornamentation of the head.

Features: The nasal horn of this centrosaurine is large, compressed from side to side, and curved forwards and downwards. Two blades, rather than horns, protruded above the eyes. At the back of the frill two horns, round in cross-section, projected straight backwards, adding to the length of the skull and making it a spectacular apparition when the animal had its head down and was seen from the front.

Right: Einiosaurus *was known before Sampson's study, but it was thought to have been a species of* Styracosaurus.

Distribution: Montana, USA.
Classification: Marginocephalia, Ceratopsia, Ceratopsidae, Centrosaurinae.
Meaning of name: Bison lizard (in the local Blackfoot language).
Named by: Samson, 1995.
Time: Campanian stage of the late Cretaceous.
Size: 6m (19½ft).
Lifestyle: Low browser.
Species: *E. procurvicornis.*

HEADS FOR SHOW

The only thing that distinguishes one genus of centrosaurine from another is the arrangement of features on the skull, the layout of horns and the shape of the neck shield. The bodies of the animals are, to all intents and purposes, identical.

The group was very restricted in number when it existed at the end of the Cretaceous period in its native North America. The geology of the area shows that each genus evolved very quickly, taking between half a million and one million years. They roamed the plains between the modern Rocky Mountains, USA, and the shallow continental sea in herds, with different ceratopsians keeping to their own group. As among the herds of grass-eaters on the African plains today, each type of animal kept to its own herd, avoiding the others. The distinctive frills and horns of ceratopsians probably evolved as a means of recognition, so that each member of each herd could identify its own kind.

The fact that the horns and shields were tough and heavy suggests that they were also used in combat. This was likely to have been combat within the species, with big males struggling for dominance in the herd, rather than against predators.

Brachyceratops

This dinosaur is known only from six subadult skeletons. Current understanding of the growth and development of the horn and shield arrangement in centrosaurines suggests that they may represent immature specimens of established genera such as *Styracosaurus* or *Monoclonius*.

Features: *Brachyceratops* has a small nose horn. The immature nature of even the largest of the specimens is shown by the fact that the horn is not fully fused to the skull. The shield is thin but broad in relation to the rest of the skull and lacks the holes seen in adults of the group. The tooth row is short, and all the skeletal remains are smaller than would be expected.

Distribution: Montana, USA.
Classification: Marginocephalia, Ceratopsia, Ceratopsidae, Centrosaurinae.
Meaning of name: Short-horned head.
Named by: Gilmore, 1914.
Time: Campanian to Maastrichtian stage of the late Cretaceous.
Size: 1.8m (6ft) as found, but perhaps 4m (13ft) as adult.
Lifestyle: Low browser.
Species: *B. montanaensis.*

Left: With only juvenile remains to work with, it is impossible to determine the exact classification of a *Brachyceratops.*

CHASMOSAURINAE

The chasmosaurinae represent the second of the two groups of advanced ceratopsians. They were also rhinoceros-sized beasts, but with long neck shields. They generally had long snouts, and the brow horns were usually bigger than the nose horn. Like the Centrosaurinae, they were confined to the North American continent where they lived in herds migrating across the open plains.

Chasmosaurus

Distribution: Alberta, Canada, to Texas, USA.
Classification: Marginocephalia, Ceratopsia, Ceratopsidae, Chasmosaurinae.
Meaning of name: Chasm reptile.
Named by: Lambe, 1904.
Time: Maastrichtian stage of the late Cretaceous.
Size: 6m (19½ft).
Lifestyle: Low browser.
Species: *C. belli*, *C. russelli*.

In the 1880s the dinosaur-bearing beds of Canada were being opened up. Many spectacular horned dinosaurs were found, and by the turn of the century about half-a-dozen species of the spectacularly frilled *Chasmosaurus* had been identified. More detailed study by Canadians Stephen Godfrey and Robert Holmes in 1995 whittled these species down to two.

Features: The obvious feature of *Chasmosaurus* was the vast frill, like a huge triangular sail, around the back of its head. The weight of this structure was kept to a minimum by the holes (or chasms, hence the name) that reduced it essentially to a framework of struts. In life this shield would have been covered in skin, and was probably brightly coloured to act as a display organ.

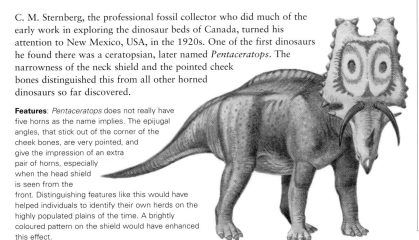

Pentaceratops

C. M. Sternberg, the professional fossil collector who did much of the early work in exploring the dinosaur beds of Canada, turned his attention to New Mexico, USA, in the 1920s. One of the first dinosaurs he found there was a ceratopsian, later named *Pentaceratops*. The narrowness of the neck shield and the pointed cheek bones distinguished this from all other horned dinosaurs so far discovered.

Features: *Pentaceratops* does not really have five horns as the name implies. The epijugal angles, that stick out of the corner of the cheek bones, are very pointed, and give the impression of an extra pair of horns, especially when the head shield is seen from the front. Distinguishing features like this would have helped individuals to identify their own herds on the highly populated plains of the time. A brightly coloured pattern on the shield would have enhanced this effect.

Distribution: New Mexico and Colorado, USA.
Classification: Marginocephalia, Ceratopsia, Ceratopsidae, Chasmosaurinae.
Meaning of name: Five-horned face.
Named by: Osborn, 1923.
Time: Campanian to Maastrichtian stages of the late Cretaceous.
Size: 6m (19½ft).
Lifestyle: Low browser.
Species: *P. sternbergi*.

Anchiceratops

The famous fossil collector Barnum Brown found the first remains of *Anchiceratops* in the Red Deer River Valley, Canada, in 1912. Charles Sternberg found another in 1924. This second one had a longer snout, smaller horns and a much thinner shield. They are thought to have been two species, *A. ornatus* and *A. longisostris*. They are now regarded as a male and female of *A. ornatus*.

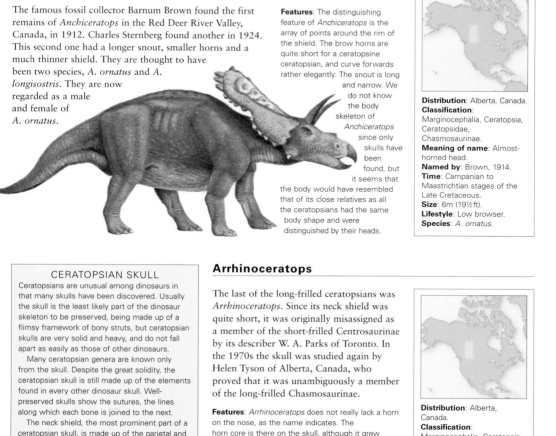

Features: The distinguishing feature of *Anchiceratops* is the array of points around the rim of the shield. The brow horns are quite short for a ceratopsine ceratopsian, and curve forwards rather elegantly. The snout is long and narrow. We do not know the body skeleton of *Anchiceratops* since only skulls have been found, but it seems that the body would have resembled that of its close relatives as all the ceratopsians had the same body shape and were distinguished by their heads.

Distribution: Alberta, Canada.
Classification: Marginocephalia, Ceratopsia, Ceratopsidae, Chasmosaurinae.
Meaning of name: Almost-horned head.
Named by: Brown, 1914.
Time: Campanian to Maastrichtian stages of the Late Cretaceous.
Size: 6m (19½ft).
Lifestyle: Low browser.
Species: *A. ornatus*.

CERATOPSIAN SKULL

Ceratopsians are unusual among dinosaurs in that many skulls have been discovered. Usually the skull is the least likely part of the dinosaur skeleton to be preserved, being made up of a flimsy framework of bony struts, but ceratopsian skulls are very solid and heavy, and do not fall apart as easily as those of other dinosaurs.

Many ceratopsian genera are known only from the skull. Despite the great solidity, the ceratopsian skull is still made up of the elements found in every other dinosaur skull. Well-preserved skulls show the sutures, the lines along which each bone is joined to the next.

The neck shield, the most prominent part of a ceratopsian skull, is made up of the parietal and the squamosal bones. The *fenestrae*, or the holes in the shield, are gaps between these bones. In some ceratopsians, such as *Arrhinoceratops*, the *fenestrae* are quite small (and in *Triceratops* they are missing altogether) but in others, such as *Chasmosaurus*, they are so big that they make up the greater part of the area of the shield. In life the *fenestrae* would have been covered by skin, and may even have been filled with the muscles that powered the jaws.

Below: The bones that make up the skull of a ceratopsian.

Arrhinoceratops

The last of the long-frilled ceratopsians was *Arrhinoceratops*. Since its neck shield was quite short, it was originally misassigned as a member of the short-frilled Centrosaurinae by its describer W. A. Parks of Toronto. In the 1970s the skull was studied again by Helen Tyson of Alberta, Canada, who proved that it was unambiguously a member of the long-frilled Chasmosaurinae.

Features: *Arrhinoceratops* does not really lack a horn on the nose, as the name indicates. The horn core is there on the skull, although it grew from a slightly different position compared with the chasmosaurines, and is considerably smaller than that of most other ceratopsians. The face is rather short. The neck shield is quite thick and carries only small holes. Only a single skull has been found, and the rest of the skeleton is unknown.

Distribution: Alberta, Canada.
Classification: Marginocephalia, Ceratopsia, Ceratopsidae, Chasmosaurinae.
Meaning of name: Lacking a nose horn.
Named by: Parks, 1925.
Time: Maastrichtian stage of the Late Cretaceous.
Size: 6m (19½ft).
Lifestyle: Low browser.
Species: *A. brachyops*.

Left: Only one fossil of Arrhinoceratops has been found. It was probably one of the rarest of the ceratopsians on the North American plains.

MORE CHASMOSAURINAE

The last of the ceratopsians were the long-frilled types. They included the biggest and best-known of them all, Triceratops. Some almost reached the size of modern elephants, and they lived right at the end of the Age of Dinosaurs. They ranged across North America and Canada, from Colorado in the south to Alberta and Saskatchewan in the north.

Triceratops

Although *Triceratops* was the biggest of the long-frilled chasmosaurine ceratopsians (the living animal weighed something like 4.5 tonnes (tons)), its frill wasn't as long as that of its relatives. It was more in the proportion of its short-frilled cousins, the centrosaurines.

When it was discovered, it was only known from a pair of horn cores. However, the whole skulls were so solid that they began to turn up quite regularly as complete fossils. Over the years so many different skulls of *Triceratops* have been unearthed that at one time there were 16 species attributed to the genus. These have now been combined, so that only the two given here are acknowledged, the common *T. horridus* and the bigger, but rarer, *T. prorsus.* Some authorities regard these as male and female *T. horridus.*

Features: This is the biggest and best-known of all ceratopsians. Its three magnificent horns give it its name. The horn cores on the skull are only cores – they are covered in horny sheaths that make them seem much bigger. The neck shield is massive, with no holes in it, and is bordered by little knobs of bone. The teeth are arranged to work as shears, and powered by strong jaw muscles.

Distribution: Wyoming, Montana, South Dakota, and Colorado, USA, and Alberta and Saskatchewan, Canada.
Classification: Marginocephalia, Ceratopsia, Ceratopsidae, Chasmosaurinae.
Meaning of name: Three-horned face.
Named by: Marsh, 1889.
Time: Maastrichtian stage of the late Cretaceous.
Size: 9m (30ft).
Lifestyle: Low browser.
Species: *T. horridus, T. prorsus.*

Diceratops

Diceratops was one of the 16 species of *Triceratops*. It was acknowledged as a separate genus when it was found by R. S. Lull in 1905, but since then it has become *Triceratops hatcheri* with the distinctive skull features put down to disease. Work by Catherine Forster of the University of Pennsylvania on the species in 1990 reinstated it as a genus in its own right.

Features: *Diceratops* is a very large example of a chasmosaurine. As the meaning of the name suggests, it has two horns on its skull, above the eyes. There is only a trace of a nose horn. Another feature is the presence of holes, or *fenestrae*, in the neck shield, showing that it is a different animal from *Triceratops*. The shape and arrangement of the individual bones that make up the neck shield are also different from *Triceratops*, but these would not be visible on the living animal.

Distribution: Wyoming, USA.
Classification: Marginocephalia, Ceratopsia, Ceratopsidae, Chasmosaurinae.
Meaning of name: Two-horned head.
Named by: Lull, 1907.
Time: Maastrichtian stage of the late Cretaceous.
Size: 9m (30ft).
Lifestyle: Low browser.
Species: *D. hatcheri*.

CERATOPSIAN POSTURE

We speak of the ceratopsians as rhinoceros-like, a reference to the horns on the head. However, for a long time the rest of the body was depicted as being very unrhinoceros-like. The standard depiction of *Triceratops*, and of any other big ceratopsian – for they all had almost identical bodies and limbs – has the heavy body supported mostly by straight hind legs. The legs were held, pillar-like, under the body like those of heavy mammals such as elephants and, indeed, rhinoceri. This is the standard modern view of dinosaur leg deployment. The front legs, however, were always shown splayed, with the upper leg more or less horizontal and the elbow bent at right angles. This seemed to make a lot of sense, since such a position would give flexibility, allowing the front part of the body to turn quickly, being pivoted at the hips with strong hind legs, so that the shield and horns could be presented to the enemy whatever its tactics. However, the modern view is that the front legs were held straight like the hind legs, as in all other big dinosaurs. This would give a body that did look like a rhinoceros.

Footprints have been found in the foothills of the Rocky Mountains near Denver, USA, that have been identified as ceratopsian trackways. The front prints fall very slightly outside the track of the back prints. If we reconstruct the body of a ceratopsian as having narrower shoulders than hips, then it does seem as if the front legs were at least slightly splayed.

Torosaurus

This dinosaur had the distinction of having the longest skull of any land-living animal. Much of the length is due to the enormous sweep of the neck shield (although a recent *Pentaceratops* skull may be longer). One specimen shows evidence of cancerous growths forming lesions within the bone of the neck shield.

Features: The skull is the only part of *Torosaurus* that is known. It is very much like that of *Triceratops*, to which it is closely related, having three stout, forward-pointing horns, two long horns over the eyes and another shorter one on the snout. The shield, however, is quite different, having a pair of huge *fenestrae* that kept the weight down. The vast area presented by the frill was undoubtedly brightly coloured and used for display.

Distribution: Wyoming, Montana, South Dakota, Colorado, Utah, New Mexico, and Texas, USA, and Saskatchewan, Canada.
Classification: Marginocephalia, Ceratopsia, Ceratopsidae, Chasmosaurinae.
Meaning of name: Punctured lizard.
Named by: Marsh, 1891.
Time: Maastrichtian stage of the late Cretaceous.
Size: 6m (19½ft).
Lifestyle: Low browser.
Species: *T. latus*.

Above: Torosaurus had the longest skull of any dinosaur.

NODOSAURIDS

The nodosaurids are (generally) the ankylosaurs without the tail clubs. They were more primitive than the ankylosaurids and came slightly earlier in the Cretaceous period. They tended to have armour that consisted of spines and spikes, and narrow mouths indicating that they had a more specialized diet and were more selective about their food than their club-tailed cousins.

Edmontonia

One of the best-known of the nodosaurids is *Edmontonia*. The original specimen was found in 1924 in Alberta, Canada, by collector George Paterson, but others, some almost complete, have since been found all across North America. It was once regarded as the same animal as *Panoplosaurus*. One species, *E. schlessmani*, is often given its own genus, *Denversaurus*, and it may be the same as *E. rugosidens*.

Features: This is a classic nodosaurid, with a broad back covered in armour and a wicked array of huge spikes sticking outwards, forwards and slightly downwards from each shoulder. The largest spike is split, giving it two points. The tail is very long. The skull is long and narrow, and angled downwards, adapted for grazing low vegetation and being selective about what it ate. Acoustical studies suggest it had the ability to make honking sounds.

Distribution: Alberta, Canada; Montana, South Dakota, Texas and Alaska, USA.
Classification: Thyreophora, Ankylosauria, Nodosauridae.
Meaning of name: From Edmonton.
Named by: C. M. Sternberg, 1928.
Time: Campanian to Maastrichtian stage of the late Cretaceous.
Size: 7m (23ft).
Lifestyle: Low browser.
Species: *E. longiceps*, *E. australis*, *E. rugosidens*, *E. schlessmani*.

Right: The two tines on the end of the main shoulder spike may have interlocked with those of rivals in mating contests, resulting in a trial of strength between big males.

Niobrarasaurus

The skeleton of *Niobrarasaurus* was found in 1930 by Virgil Cole, prospecting for oil-bearing rocks in Kansas, USA. He thought it was a plesiosaur, but had it shipped to his old college at the University of Missouri. There it was identified as a dinosaur and named *Hierosaurus* by Dr. M. G. Mehl. It was renamed as *Niobrarasaurus* by Ken Carpenter and his team in 1995.

Features: *Niobrarasaurus* is a typical nodosaurid, with a broad, armoured back and a slim tail. The armour consists of broad plates along the back and short spines along the sides. The foot bones are very short. As the original description was not very scientific, it was studied again in the 1990s and renamed. In 2003 a piece of the leg left behind by Cole 70 years earlier was discovered at the same Kansas site.

Distribution: Alberta, Canada, to Texas, USA.
Classification: Thyreophora, Ankylosauria, Nodosauridae.
Meaning of name: Lizard from the Niobrara chalk.
Named by: Carpenter, Delkes and Weishampel, 1995 (but originally as *Hierosaurus* by Mehl, 1935).
Time: Campanian stage of the late Cretaceous.
Size: 5m (16½ft).
Lifestyle: Low browser.
Species: *N. coleii*.

Animantarx

This dinosaur has the distinction of being the first to be found by radiometric survey. Ramal Jones, a technician at the University of Utah, USA, knowing fossil bones to be slightly radioactive, surveyed a likely fossil site in Utah and persuaded the university to excavate the spot where low-level radiation seemed strongest.

Features: *Animantarx* is known from remains that consist of a partial skull with its jawbone, and a partial skeleton consisting of backbones, ribs, shoulder structure and parts of front and rear legs. It is a medium-size nodosaurid that resembles *Pawpawsaurus* with armour plates like upturned boats. The skull has a very high cranium and two pairs of short horns, one pair behind the eyes and another on the cheeks.

Distribution: Utah, USA.
Classification: Thyreophora, Ankylosauria, Nodosauridae.
Meaning of name: Animated living fortress.
Named by: Carpenter, Kirkland, Burge and Bird, 1999.
Time: Cenomanian to Turonian stages of the late Cretaceous.
Size: 3m (10ft).
Lifestyle: Low browser.
Species: *A. ramaljonesi.*

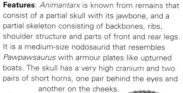

Top: Pawpawsaurus.

Left: "A 12ft-long dinosaur, looking like an armadillo but bigger than a cow," was how Don Burge, one of the team that studied Animantarx, described it.

BROAD AND NARROW MOUTHS

The main difference between the head of a nodosaurid and that of an ankylosaurid is the breadth of the mouth. A nodosaurid tended to have a pear-shaped skull that narrowed towards the jaws, unlike the hourglass-shaped jaws of an ankylosaurid with its broad beak. At the time these dinosaurs thrived, flowering plants had evolved, and there would have been plenty of leafy and seed-bearing undergrowth to be grazed. The nodosaurids, with their narrower mouths, must have been selective in their food choices, unlike the broad-mouthed ankylosaurids.

Common to both groups is the presence of a palate in the roof of the mouth (typical in mammals but rare in dinosaurs). The palate is a shelf that separates the airways of the nostrils from the foodways of the mouth, allowing that animal to eat and breathe at the same time. This would have speeded up the eating process.

The teeth of both groups were quite small, shaped like little hands and designed for chopping. Primitive types had pointed teeth at the front as well.

Below: The difference in the shape of the head between a nodosaurid (left) and an ankylosaurid (right) is obvious in top view.

Struthiosaurus

This is the smallest known of the nodosaurids. Its remains have been found all across Europe, in areas that are known to have been parts of an island chain during late Cretaceous times. Its small stature is taken to be proof that animals on islands tend to develop dwarf forms to make the best use of limited resources.

Features: *Struthiosaurus* resembles its larger relatives, but is more lightly built. The armour consists of three pairs of sideways-projecting spikes on the neck, at least one pair of tall spines over the shoulders and a double row of triangular plates sticking up along the tail. The back is covered in keeled scutes separated by a groundmass of ossicles, and there seems to be a well-marked boundary between the armoured back and the skin of the underside.

Distribution: Austria, France and Hungary.
Classification: Thyreophora, Ankylosauria, Nodosauridae.
Meaning of name: Ostrich lizard.
Named by: Bunzel, 1871.
Time: Campanian stage of the late Cretaceous.
Size: 2m (6½ft).
Lifestyle: Low browser.
Species: *S. austriacus, S. ludgunensis, S. transylvanicus.*

MORE NODOSAURIDS

The nodosaurids spread across the Northern Hemisphere. Their remains have been found in Europe, Asia and North America. They do not seem to have penetrated to Gondwana, however, as hardly any ankylosaurs have been found in the southern continents. The Australian primitive ankylosaur, Minmi, and the scattered remains in Argentina are the only exceptions.

Anoplosaurus

When *Anoplosaurus* was found, early in the history of dinosaur discovery, it was thought to be a relative of *Iguanodon*, one of the few dinosaurs known at that time. It is now known to be a nodosaurid, closely related to the American genera *Silvisaurus* and *Texasestes*. It was once put into the wastebasket taxon *Acanthophoplis*.

Features: All that is really known of the first species of this animal, *A. curtonotus*, is a few vertebrae from the neck, but some of the remains are mixed with the bones of an iguanodont, hence the confusion. It seems to be a primitive nodosaurid. *A. major* is better known, with parts of the jawbone, vertebrae of the neck and back, pieces of rib and bones of the legs and toes found.

Distribution: Cambridgeshire, England.
Classification: Thyreophora, Ankylosauria, Nodosauridae.
Meaning of name: Lizard without a weapon.
Named by: Seeley, 1878.
Time: Cenomanian stage of the late Cretaceous.
Size: 5m (16ft).
Lifestyle: Low browser.
Species: *A. curtonotus*, *A. major*, *A. tanyspondulus*.

Panoplosaurus

Panoplosaurus was one of the last of the nodosaurids before they died out and the armoured dinosaur niche was filled by the club-tailed ankylosaurids. It is one of the best-known, with two partial skeletons and three skulls discovered. Until recently it was thought to have been a species of *Edmontonia*, but now it is regarded as a genus in its own right.

Features: *Panoplosaurus* has a wide, pear-shaped skull, toothless at the front and with broad nostrils. It is similar to *Edmontonia*, but differs in having no long spines projecting from the shoulders. Instead, the armour is restricted to a series of thick, bony plates, each with a prominent keel along the centre. The plates are broadest over the shoulder and neck. Short spikes run along the sides of the body and tail.

Distribution: Alberta, Canada, to Montana, USA.
Classification: Thyreophora, Ankylosauria, Nodosauridae.
Meaning of name: Totally armoured reptile.
Named by: Lambe, 1919.
Time: Campanian stage of the late Cretaceous.
Size: 7m (23ft).

Left: As with most ankylosaurs, the armour plates of Panoplosaurus are fused to the skull, rather than being merely embedded in the skin.

Lifestyle: Low browser.
Species: *P. mirus*.

Sauropelta

The best-known of the nodosaurids is *Sauropelta*, and it is known from several almost complete skeletons. It appeared quite early in the geological succession, and had a number of primitive and unspecialized features. It is also one of the largest of the group, with a long tail that accounted for much of its length.

Features: *Sauropelta* is a nodosaurid with a very long tail, which is more highly armoured than was originally thought. The notable feature about its armour is the array of four pairs of spines projecting up from the neck. The armour is arranged in transverse rows of big, bony studs interspersed with smaller, pebbly armour. Its primitive features consist of the lack of a palate in the mouth and unfused neck bones, unlike its relatives.

Distribution: Wyoming, Montana and Utah, USA.
Classification: Thyreophora, Ankylosauria, Nodosauridae.
Meaning of name: Lizard shield.
Named by: Ostrom, 1970.
Time: Aptian to Cenomanian stages of the early and late Cretaceous.
Size: 8m (26ft).
Lifestyle: Low browser.
Species: *S. edwardsorum*.

Left: Sauropelta lived in the same area and at the same time as many of the sickle-clawed dromaeosaurids. It would have needed its armour to guard against them.

WHO WAS PALAEOSCINCUS?

In old dinosaur books, up to the late 1970s, an ankylosaur referred to as *Palaeoscincus* may appear. This dinosaur genus was based on a single tooth found in 1855 in Montana, USA, by fossil collector F. V. Hayden and named by the pioneer American palaeontologist Joseph Leidy in Philadephia, USA, the following year. The name means "ancient skink" and refers to the similarity between this tooth and that of the modern skink lizard. Apart from that there were no clues as to what this animal was, since dinosaur studies were in their infancy at that time.

As the fossil discoveries of the late nineteenth century gained momentum, armoured dinosaurs of various kinds started coming to light, all with this distinctive kind of tooth. Soon *Palaeoscincus* began to be depicted as a generalized ankylosaur, complete with nodosaurid side spines and an ankylosaurid tail club. It was invariably shown in a sprawling posture, with the legs held out to the side like those of a lizard. This chimerical animal slipped into the public consciousness and continued to turn up in books for over a century until the demand for more accurate and scientific popular literature put it to rest relatively recently.

Silvisaurus

We know *Silvisaurus* from the front end of the skeleton including the skull. It was found in a stream bed where it had lain exposed and damaged by being trampled underfoot by drinking cattle. It was partly protected by being embedded in an iron nodule, but this also meant a great deal of effort in extracting and preparing it.

Features: *Silvisaurus* has a short beak at the front of the mouth with small, pointed teeth. It has large cheekbones and quite a long neck. The air-passages through the skull are cavernous, and suggest the ability to make loud honking noises for signalling. The armour consists of rows of thick, rounded plates over the back and sharp spines on its shoulders. It is possible that there were also spines down each side of its tail.

Distribution: Kansas, USA.
Classification: Thyreophora, Ankylosauria, Nodosauridae.
Meaning of name: Forest lizard.
Named by: Eaton, 1960.
Time: Aptian to Cenomanian stages of the early and late Cretaceous.
Size: 4m (13ft).
Lifestyle: Low browser.
Species: *S. condrayi*.

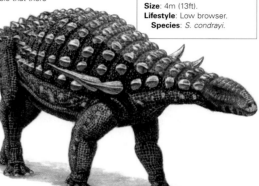

Right: In reality we know nothing about the appearance of Silvisaurus behind the shoulder region. It is reasonable to surmise that it would be similar to other nodosaurids.

BIG ANKYLOSAURIDS

The proliferation of ankylosaurids across both Asia and North America suggests that these animals lived in an era when other big meat-eaters lived and against which the ankylosaurids needed to be protected. Indeed this was the time and the stalking ground of the great, formidable tyrannosaurids. Slow-moving beasts such as the ankylosaurids would need heavy armour to protect themselves against such hunters.

Shanxia

Shanxia is known from a single partial skeleton found in 1993 by the Hebei Geological Survey. It consists of a partial skull, some of the backbone and limb bones and, sadly, only one piece of armour. It was found in river deposits with few other fossils, and has been impossible to date accurately.

Features: What distinguishes *Shanxia* from the other ankylosaurids is the shape of the horns on the rear of the skull. There are two pairs of horns and they are flattened, pointed and swept sideways and backwards rather than pointing straight out at the side as in other ankylosaurids. However, the way the skull articulates to the neck is more like that of a nodosaurid, but palaeontologists do not think that significant in the classification.

Distribution: China.
Classification: Thyreophora, Ankylosauria, Ankylosauridae.
Meaning of name: From Shanxi Province.
Named by: Barrett, You, Upchurch and Burton, 1998.
Time: Late Cretaceous.
Size: 3.5m (11½ ft).
Lifestyle: Low browser.
Species: *S. tianzhenensis*.

Left: Shanxia *was probably similar in appearance to the other club-tailed ankylosaurs of the time.*

Tsagantegia

Tsagantegia is known from a single skull indicating that it was a medium-size ankylosaurid that must have been very similar to *Shamosaurus* and *Talarurus*. The skull is distinctive enough to show that it is a separate genus, but the lack of any fossils from the rest of the skeleton is very frustrating.

Features: The skull, the only known part of this animal, is long and flat for an ankylosaur. There is a prominent ring of bone around the eye socket. The eye sockets are situated just behind the midpoint of the skull. There is armour on the roof of the skull, formed from masses of small, bony knobs, but this is quite low and not very prominent. The snout is wider than that of *Shamosaurus* or *Cedarpelta*. *Tsagantegia* was found in the Gobi Desert, close to the border between Mongolia and China.

Distribution: Mongolia.
Classification: Thyreophora,
Ankylosauria, Ankylosauridae.
Meaning of name: From Tsagan Teg.
Named by: Tumanova, 1993.
Time: Cenomanian stage of the late Cretaceous.
Size: 6m (19½ ft).
Lifestyle: Low browser.
Species: *T. longicranialis*.

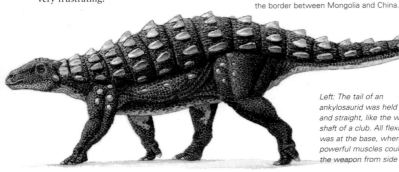

Left: The tail of an ankylosaurid was held stiff and straight, like the wooden shaft of a club. All flexibility was at the base, where powerful muscles could swing the weapon from side to side.

ANKYLOSAURID BRAIN

The brain cavity of an ankylosaur is quite well-known, since the armour-covered heads were well protected and often fossilized. The part of the brain that controlled movement and general activity is quite small compared with that of other dinosaurs, such as the ornithopods. This indicates that they were quite slow in moving about. The shapes of the legs, with a long thigh bone compared to the lower leg, is that of a slow-moving animal too.

The most highly developed part of the brain is that which deals with the sense of smell. This, combined with the complex maze of nasal passages that we find in many ankylosaurid skulls, suggests that this may have been the primary sense on which they relied. The nasal passages were probably also lined with membranes to moisten and warm the air as it passed down to the lungs. Most ankylosaurids, certainly the Asian forms, lived in very dry environments and would benefit from this. It is also possible that the nasal passages were for generating sounds for communication. The nodosaurids, on the other hand, did not have such complex nasal passages, merely paired tubes that passed from the nostrils straight back to the throat.

Ankylosaurus

The ankylosaur dynasty, both nodosaurid and ankylosaurid, reached its climax in *Ankylosaurus* itself. This is the most familiar of the ankylosaurids, but is only known from a skull, some vertebrae and pieces of armour and a few teeth. Until the ankylosaurs were re-examined in the 1970s, *Ankylosaurus* was portrayed as a mixture of ankylosaurid types (with the tail club), and the nodosaurids (with the side spikes). It was the largest and last of the ankylosaurs.

Features: The armour consists of ovals of embedded bone, each one supporting a horny covering. They are quite smooth compared with those of other ankylosaurids. The tail, stiffened by having the vertebrae lashed together by fused, bony tendons, carries a bony club at the end. The skull is broad, with two pairs of sideways-pointing horns at the rear corners. There are no teeth at the front of the mouth, just the broad beak.

Distribution: Alberta, Canada, and Wyoming, Texas, USA.
Classification: Thyreophora, Ankylosauria, Ankylosauridae.
Meaning of name: Fused lizard.
Named by: Brown, 1908.
Time: Maastrichtian stage of the late Cretaceous.
Size: 11m (36ft).
Lifestyle: Low browser.
Species: *A. magniventris*.

Below: Large plates on the neck and shoulders, and smaller plates in rows along the sides, characterize the armour Ankylosaurus *and the other ankylosaurids.*

THE LAST ANKYLOSAURIDS

Some members of the ankylosaurid group, as well as being the last dinosaurs to evolve, and taking over from their cousins the nodosaurids, existed right up to the end of the Age of Dinosaurs. They could be regarded as the climax of the dinosaur dynasty that lasted 160 million years. Whatever the disaster that wiped dinosaurs out, it was observed by the ankylosaurs, tyrannosaurs, ceratopsians and hadrosaurids.

Maleevus

All that is known of *Maleevus* is the jawbone and part of the skull, found by E. A. Maleev in 1952. These are so similar to *Talarurus* that many palaeontologists regard it as a species of this genus. Some even think that there is so little information about the fragments that it cannot be identified as anything with any certainty.

Features: *Maleevus* is regarded as being almost identical to its relative *Talarurus*, except for details of the hind portion of the skull. Like *Talarurus* it is a large ankylosaurid. It has armour in transverse bands, and a club formed by three masses of fused bones in two lobes at the end of a long tail which is stiffened by solidified tendons. Its broad mouth has a cutting beak adapted for biting off large mouthfuls of food indiscriminately.

Distribution: Mongolia.
Classification: Thyreophora, Ankylosauria, Ankylosauridae.
Meaning of name: From E. A. Maleev (a Russian palaeontologist).
Named by: Turmanova, 1987.
Time: Cenomanian to Turonian stages of the late Cretaceous.
Size: 6m (19½ft).
Lifestyle: Low browser.
Species: M. disparoserratus.

Left: The genus name honours the Russian E. A. Maleev who did much to open Mongolia to palaeontologists in the 1950s.

Talarurus

Talarurus is known from the remains of at least five individuals. At the time Maleev, who discovered it, also found the skull that is now regarded as *Maleevus disparoserratus*, naming it *Syrmosaurus disparoserratus*. The name *Talarurus* means "wicker tail" and refers to the interwoven tendons that kept the tail straight and stiff, forming a rigid shaft for the club.

Features: *Talarurus* has a smaller tail club than *Euoplocephalus*, to which it is closely related. The completeness of the remains shows that the hind foot has four toes. This is possibly the condition that existed among the primitive ankylosaurids, while the more advanced forms, such as *Euoplocephalus*, had three. The armour is arranged in transverse bands, with no sign of the upward-pointing spines of *Euoplocephalus*. As in other ankylosaurids, half of the tail consisted of fused vertebrae.

Distribution: Mongolia.
Classification: Thyreophora, Ankylosauria, Ankylosauridae.
Meaning of name: Wicker tail.
Named by: Maleev, 1952.
Time: Cenomanian to Turonian stages of the late Cretaceous.
Size: 6m (19½ft).
Lifestyle: Low browser.
Species: T. plicatospineus.

Right: The remains of Talarurus are so similar to those of Maleevus, that some scientists think that the two are the same genus. If this is so, then Talarurus must take priority, having been named first.

Euoplocephalus

This is without doubt the ankylosaurid that is best known to science. There are more than 40 specimens known, including 15 skulls, suggesting that this was the most common ankylosaurid in North America at the time. They have never been found in groups, and were probably solitary foragers. The forelegs were quite supple, suggesting that this animal could dig for roots and buried stems.

Below: Different species appear to have differently shaped clubs. E. tutus has a broad heavy club, while E. acutosquameus has a smaller pointed club.

Features: The eyelids are armoured with movable slabs of bone, the first time this is seen in an ankylosaur. The skull is quite light with tortuous air passages, probably to warm or moisten the air before it reached the lungs. The back is armoured with heavy, bony nodules set into leathery skin. Spines, up to 15cm (6in) long, stick up from the neck and shoulders.

Distribution: Alberta, Canada, to Montana, USA.
Classification: Thyreophora, Ankylosauria, Ankylosauridae.
Meaning of name: Completely well-armoured head.
Named by: Lambe, 1910.
Time: Campanian to Maastrichtian stages of the late Cretaceous.
Size: 6m (19½ft).
Lifestyle: Low browser.
Species: E. tutus, E. acutosquameus.

GEOGRAPHICAL LOCATION OF ANKYLOSAURS

As we have noted, the ankylosaurs were confined almost entirely to the northern continents. However, there have been isolated bones found in late Cretaceous rocks of Argentina that seem to belong to ankylosaurs – the muscle scars and the shape of the articulations show this. There are also isolated armour plates.

The remains are too scrappy for any identification, but they seem to come from nodosaurids rather than ankylosaurids. They occur in the same rock sequences as duckbills, another rare group for South America. It appears that about this time there was some land connection with the continent of North America and, for a brief period, there was an exchange of animals between the two. However, this invasion does not seem to have been much of a success, since ankylosaur remains appear to be very limited in the Southern Hemisphere. As the world passed out of the Age of Dinosaurs, the continents were beginning to separate and diverge, and were developing their distinctive suites of animal types.

Pinacosaurus

More than 15 specimens are known of this genus, a good reflection of how well the armoured backs fossilize. It is one of the few ankylosaurs for which the juveniles are known, after two finds of several of them huddled together, killed by a sandstorm. Such finds give us a good idea of the relation of the armour to the skeleton, something that is difficult to understand from an adult specimen.

Features: The skeleton of this ankylosaurid is relatively light, with more slender limb bones and smaller feet than others of the group. The front foot has five toes while the rear foot has four. The shoulder blades are much more slender than in any other ankylosaurid, but we cannot determine the significance of this. The skull has more openings than is usual in an ankylosaurid. The armour consists of deeply keeled plates. One species, P. mephistocephalus, has a pair of devil-like horns, hence the species name.

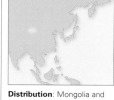

Distribution: Mongolia and China.
Classification: Thyreophora, Ankylosauria, Ankylosauridae.
Meaning of name: Plank lizard.
Named by: Gilmore, 1933.
Time: Santonian to Campanian stages of the late Cretaceous.
Size: 5.5m (18ft).
Lifestyle: Low browser.
Species: P. grangeri, P. mephistocephalus.

Left: As in other ankylosaurids the unarmoured belly would have been this animal's weak point. However its weight would have made it difficult to turn over and so it would not often have been vulnerable.

GLOSSARY

Abelisaurid A group of theropods of the late Cretaceous, mostly from the Southern Hemisphere.

Alvarezsaurid Long-legged running dinosaurs with diminutive forelimbs, often classed as primitive birds.

Amphibious A creature able to survive on land or in the water.

Ankylosaur Quadrupedal herbivorous ornithischian dinosaurs from the late Cretaceous, making up the suborder Ankylosauria.

Ankylosaurid A member of the Ankylosauridae, a family of the suborder Ankylosauria.

Antorbital fossa A hole in the skull between the snout and the eye socket.

Archosaur A member of the diapsid group of reptiles that includes the crocodiles, the pterosaurs and the dinosaurs – the so-called "ruling reptiles".

Arthropod A member of the invertebrate group with chitinous shells and jointed legs, including the crustaceans, insects, arachnids and centipedes.

Atrophy Wasting away of an organ that is no longer important, as a result of evolutionary development.

Belemnite A common Mesozoic cephalopod resembling a squid but having a bullet-shaped internal shell.

Bipedal Two-footed animal.

Caenagnathid A group of theropods related to the oviraptorids.

Carapace A thick, hard shell or shield that covers the body of some animals.

Carnosaur In old terminology, any big theropod, but in more modern terms a theropod belonging to the group that contains Allosaurus and its relatives.

Cartilaginous Referring to a skeleton composed entirely of cartilage, a tough, elastic tissue.

Cassowary A large flightless bird with a horny head crest and black plumage, from northern Australia.

Cenomanian A stage of the late Cretaceous period lasting from about 97 to 90 million years ago.

Cephalopod A mollusc with the limbs very close to the head, such as a squid or an octopus.

Ceratopsian Horned dinosaur.

Clade A group with common ancestry.

Cladogram A diagram illustrating the development of a clade.

Cololite A trace fossil consisting of the fossilized remains of the contents of an animal's digestive system.

Convergent evolution The evolutionary development of similar features on different animals that share the same environment.

Coprolite A trace fossil consisting of the fossilized remains of an animal's droppings.

Crepuscular Active at twilight or dawn.

Crest A tuft of fur, feathers or skin or a ridge of bone along the top of the head.

Cretaceous The last period of time in the Mesozoic era, which lasted 81 million years.

Cycad A tropical or subtropical plant with unbranched stalk and fern-like leaves crowded together at the top.

Denticle A small tooth or tooth-like part.

Dermal Relating to the skin.

Diapsid A member of a major group of the reptiles, classed by the presence of two holes in the skull behind the eye socket, and comprising the majority of modern reptiles including the lizards, snakes and crocodiles.

Digitigrade Walking so that only the toes touch the ground.

Diplodocid A herbivorous quadrupedal dinosaur from the late Jurassic or early Cretaceous periods, with a long neck and tail.

Dorsal Relating to the back or spine.

Fibula The outer thin bone from the knee to the ankle.

Fluke A blade-like projection at the end of the tail used for swimming, as in whales.

Gastralia A set of extra ribs covering the stomach area, as seen in some dinosaurs.

Gastrolith A stone in the stomach, deliberately swallowed to aid in digestion or buoyancy.

Gavial A type of fish-eating crocodile from South-east Asia.

Genus (genera pl.) A taxonomic group into which a family is divided and containing one or more species, all with a common characteristic.

Gizzard The thick-walled part of the stomach in which food is broken up by muscles and possibly gastroliths.

Gondwana (sometimes called Gondwanaland) The southern of two ancient continents, comprising modern-day Africa, Australia, South America, Antarctica and the Indian subcontinent. It was formed from the break-up of the supercontinent Pangaea 200 million years ago.

Groundmass A matrix of rock in which larger crystals are found.

Homeotherimic "Warm blooded", having the ability to keep the body at an almost constant temperature despite changes in the environment, as in mammals and birds, and probably some of the dinosaurs.

Humanoid As an adjective, human-like in appearance, or as a noun, a member of the group to which humans belong.

Humerus The bone from the shoulder to the elbow.

Ichnogenus A genus based only on fossil footprints.

Ichnology The study of fossil footprints.

Ichnospecies A species based only on fossil footprints.

Ichthyologist A scientist who studies fish.

Ischium A section of the hip bone which, in reptiles, sweeps backwards.

Isotope A form of the atom of a chemical element in which the atomic number is different from that of other atoms of the same element.

Jurassic The second period of the Mesozoic era and lasting for approximately 45 million years.

Keeled In a scale or an armour plate or a bone, having a ridge running along its length.

Keratinous Made up of keratin – a horny substance similar to fingernails.

Laurasia One of the two supercontinents formed by the break up of Pangaea 200 million years ago. It comprises modern North America, Greenland, Europe and Asia.

Lias The lowest series of rocks in the Jurassic system.

Maastrichtian The last age of the Cretaceous period, from 74 to 65 million years ago.

Megalosaur Large Jurassic or Cretaceous carnivorous bipedal dinosaur.

Mesozoic The era of geological time lasting from 245 to 65 million years ago and consisting of the Triassic, Jurassic and Cretaceous periods.

Mosasaur Cretaceous giant marine lizards with paddle-like limbs.

Nodule A small knot or lump – as a piece of armour bone embedded in the skin of an animal or as a mineral occurrence embedded in rock.

Nomen dubium Literally – "dubious name" – a name given to an animal that is not fully supported by scientific study.

Olecranon Bony projection behind the elbow joint.

Olfactory Relating to the sense of smell.

Olfactory bulb The point from which the nerves concerned with the sense of smell originate.

Ornithischian An order of dinosaurs that includes the ornithopods, stegosaurs, ankylosaurs and marginocephalians – characterized by the hip bones, which are arranged like those of a bird.

Ornithomimid Bird-like, ostrich-mimic.

Ornithopod A herbivorous bipedal ornithischian dinosaur.

Ossicle A small bone.

Pachystasis A thickening of the bones, particularly the ribs, in some aquatic animals, such as the modern walrus, that helps the buoyancy of the animal in water.

Palaeogeography The study of what the geography was like in the past – the arrangement of the continents, the distribution of land and sea, and the climatic zones.

Palaeozoic The era of geological time that began 600 million years ago and lasted for 375 million years.

Paleontologist A scientist who studies fossils and the life of the past.

Pangaea The ancient supercontinent comprising all the present continents before they broke up 200 million years ago.

Pangolin A mammal from tropical Africa, southern Asia and Indonesia, with a long snout and a body covered in overlapping horny scales.

Patagium A web of skin between the neck, limbs or tail in gliding animals, that assists flight in place of a wing.

Petrifaction A process of forming fossils, particularly the process in which the organic matter of each cell of the creature is replaced by mineral.

Phalange A bone in the finger or toe.

Plantigrade Walking with the entire sole of the foot in contact with the ground.

Plastron Bony plate forming the underpart of the shell of a turtle or similar animal.

Plate A thin sheet that forms an overlapping layer of protection.

Plesiosaur A marine reptile with a long neck, short tail and paddle like limbs form Jurassic and Cretaceous times.

Polydactyly With more than the usual number of digits.

Polyphylangy With more than the usual number of bones in the digit.

Prehensile Adapted for grasping.

Premaxilla The front bone of the upper jaw of dinosaurs.

Pterodactyl The popular name for a member of the pterosaur group.

Pterosaur A flying reptile from Jurassic and Cretaceous times.

Pygidium The tail shield of an arthropod.

Pygostyle The bony structure formed from fused tail bones that is used as the base for the tail feathers of a bird.

Quadrupedal An animal that walks on all four limbs.

Refugium A geographical region that has remained unaltered by climate change.

Rostral Beak- or snout-like.

Saurischian An order of dinosaurs that includes the theropods, therizinosaurs, prosauropods and sauropods – characterized by the arrangement of hip bones, similar to those of a lizard.

Sauropod A herbivorous quadrupedal saurischian dinosaur, including *Apatosaurus, Diplodocus* and *Brachiosaurus*. Smallheads and long necks and tails characterize the group.

Sclerotic A ring of bone inside the eye, used for adjusting focus or for adjusting pressure in a swimming animal.

Scute A horny plate that makes up part of an armour.

Seismic Relating to earthquakes or earth tremors.

Silica The oxide of the element silicon, which is a major constituent of the minerals of the Earth's crust.

Species A taxonomic group into which a genus is divided.

Stegosaurid A quadrupedal herbivorous ornithischian dinosaur, with bony plates and armour.

Stomach stone A gastrolith.

Swallow hole A depression in limestone terrain, usually into which a river or a stream disappears underground.

Synapsid A member of a major group of the reptiles, classed by the presence of a single combined hole in the skull behind the eye socket, and comprising the mammal-like reptiles from which the mammals evolved.

Symphysis A growing together of parts joined by an intermediary layer, particularly the join at the front of the lower jaw.

Taphonomy The study of what happens to a dead organism before it becomes buried and fossilized.

Taxonomy A system of classification of organisms.

Thecodont A reptile of Triassic times with teeth set in sockets. They gave rise to dinosaurs, crocodiles, pterodactyls and birds.

Theropod A bipedal carnivorous saurischian dinosaur.

Titanosaur A herbivorous quadrupedal dinosaur.

Triassic The first period of the Mesozoic era, which lasted 37 million years.

Turbinal A folded bone inside the nose of some animals, supporting a membrane used to adjust the temperature or humidity of breathed air.

Ulna The inner and longer bone of the forearm.

Viviparous Giving birth to live offspring.

Wastebasket genus A genus to which many dubious fossils are attributed.

INDEX

PICTURE ACKNOWLEDGEMENTS

The Publisher would like to thank the following picture agencies for granting permission to use their photographs in this book:

Alamy 28t, 32, 43, 53, 55, 113, 129.
Ardea 10, 22, 28bl, 28br, 30t, 30b, 37tl, 37tr, 58, 62t, 62b, 63b, 131.
Corbis 45, 59br, 37br.
The Natural History Museum 11, 33, 37bl, 63tr.
David Varrichio 34.

All illustration credits as follows:
Andrey Atuchin 24–5 main, 24bl, 25t, 27b, 61t, 144–5, 186–95, 222–3, 236–7.
Peter Barrett 2, 8–9, 10, 11, 12–17, 21tr, 21bl, 23b, 25c, 26br, 27bc, 30, 31, 32, 33, 34, 35, 36, 38, 39b, 40–1, 42–3, 44, 45, 46–7, 48–9, 50–1, 52–3, 54–5, 58–9, 60b, 61bl, 61br, 64, 66–7, 69–91, 93–131, 134–5, 138–41, 146–7, 149–53, 162–3, 176–81, 198–9, 206–7, 210–15, 218–21, 224–5, 230–3, 238–41, 250–1, 252, 253, 254t, 255t, 256c.

Stuart Carter 6b, 24bl, 24bc, 24br, 25bl, 25br, 25bc, 26bl, 26t, 56–7, 154–61, 164–75.
Julius T. Csotonyi www.csotonyi.com julius@ualberta.net, 1, 3, 6tr, 7t, 7b, 20–1 main, 21br, 21tl, 22–3 main, 23t, 27t, 27tl, 27br, 29, 39t, 182–5, 196–7, 208–9, 216–7, 226–9, 234–5, 242–9, 255b, 256t.
Anthony Duke all maps and timelines.
Alain Beneteau 60t, 132–3, 136–7, 142–3, 254b.

Key: l = left, r = right, t = top, c = centre, b = bottom